Infinite Tropics

An Alfred Russel Wallace Anthology

Edited by Andrew Berry
With a preface by Stephen Jay Gould

VERSO

London • New York

First published by Verso 2002
© Verso 2002
Editor's copy © Andrew Berry 2002
Preface © Stephen Jay Gould 2002

Paperback edition first published by Verso 2003
© Verso 2003
Editor's copy © Andrew Berry 2003
Preface © Stephen Jay Gould 2003

1 3 5 7 9 10 8 6 4 2

Verso
UK: 6 Meard Street, London W1F 0EG
USA: 180 Varick Street, New York, NY 10014–4606
www.versobooks.com

Verso is the imprint of New Left Books

ISBN 1–85984–478–2

British Library Cataloguing in Publication Data
Wallace, Alfred Russel, 1823–1913
   Infinite tropics : an Alfred Russel Wallace anthology
   1. Evolution (Biology) 2. Natural selection
   I. Title II. Berry, Andrew
   576.8

ISBN 1859844782

Library of Congress Cataloging-in-Publication Data
A catalog record for this book is available from the Library of Congress

Typeset in ITC New Baskerville by SetSystems Ltd, Saffron Walden, Essex
Printed and bound in Great Britain by Biddles Ltd
www.biddles.co.uk

Contents

List of Illustrations

Preface

In analysing the skills and achievements of the people who shape our history, our worst failures stem from our inability to grapple with the cardinal factor of luck itself, for how can we understand what we can neither predict nor control? In one truly vital sense, Alfred Russel Wallace (1823–1913) was a lucky man indeed, for he lived to be ninety years old and remained productive until the end. (And we must remember that we underestimate the role of luck in history largely because so many true geniuses died young – and we know them not at all, even though they might have so altered the course of history.)

But in most other palpable ways, Wallace was a most unlucky man indeed. (I define "luck" here in the moral sense of things imposed beyond one's control, and not in the technical or philosophical sense of things random or things uncaused.) His father enjoyed good social and intellectual standing in an age when such factors, albeit always crucial, mattered far more than in our own times. But his father's finances and status failed and Wallace grew up in a poverty (and with limited formal education) more characteristic of truly working-class people of his generation. Then, in his first major venture as a naturalist, Wallace lost virtually all his collections, representing scores of new species and years of assiduous work in remote rainforests of South America, when his ship sank in the Atlantic (nearly drowning him as well) on his return.

But, in history's judgment, Wallace's worst bit of luck also secured him the permanent footnote of truly lasting fame that also ranked him, so unfairly, as an appendage and Johnny come lately – thus converting one of the most brilliant biologists and interesting men of nineteenth-century science into a second rater. (For who would not prefer the greater anonymity of professionally acknowledged brilliance as one's own person to a one-liner of popular recognition as a frill to someone else's accomplishment?)

Perhaps all cultures do not judge in this unfair manner but, in our system, winning or being first takes all the kudos and wins all renown, whereas even the most honorable second-place finish spells oblivion or, even worse, a grudging memory as an also-ran, even when your ranking did not reflect a true beating by the "winner," but only recorded the happenstance of age or logistics. In my favourite American example, everyone knows Jackie Robinson as the first African-American player in Major League baseball (for the Brooklyn Dodgers). But who even recognizes the name of Larry Doby – a splendid ballplayer and human being – who entered the game just a few months later as the first black to play in the other division of the totality, the American League?

Wallace, as we all know – but we should know so much more about him! – devised the theory of natural selection, independently of Darwin, in 1858, writing out his ideas feverishly (literally in the midst of a malarial attack) in a short paper, while doing fieldwork in Indonesia. He sent the manuscript to Darwin, knowing about his senior colleague's interest in evolution, but having no inkling that Darwin had devised effectively the identical theory long before in 1838, when Wallace was still a teenager. Darwin had then refined his ideas and collected data in privacy for twenty years (revealing the content only to a handful of most trusted friends), and had already written several hundred pages of a projected long book of several volumes on the subject (that Darwin would have called *Natural Selection*, had not Wallace's prod spurred a decision to quicken the pace and produce the single-volumed, and still substantial, *Origin of Species*, published in 1859).

Wallace therefore became the Larry Doby of biology, known and admired to all professionals, but effectively invisible in the public eye, except as a factoid or footnote. (One positive recent biography of Wallace, for example, even perpetuated the tradition of unconscious denigration in the title itself, stripping Wallace of his name and calling the book *Darwin's Moon*.)

But to these unlucky circumstances of Wallace's ironic fate, one must also, in fairness, mention an equally substantial set of factors that we attribute to poor choices rather than pure bad fortune (in the same moral rather than technical sense, because we view them as active and alterable decisions rather than as happenstances beyond a subject's control – although the distinction breaks down when we recognize that much of a person's basic temperament also derives from inheritance, and not only from hard-won learning about gumption and willpower). Wallace was a shy, awkward, and self-effacing man, who had no instinct for what our modern age would call self-promotion. (Neither did Darwin, for that matter, but Darwin enjoyed benefits of private wealth, dignity of bearing, and the simple luxury of being first, largely because he enjoyed a nearly fifteen-year advantage of earlier birth.)

Wallace wrote some wonderfully expert technical books (with his two-volume *Geographic Distribution of Animals*, published in 1876, as his masterpiece and the foundational document of evolutionary biogeography, one of the most important subdisciplines of modern organismal biology). But, unlike Darwin, who worked with the tenacity of a terrier and the patience of Job, Wallace's restless intellect allowed him no similar harbour to ponder and build. Moreover, Wallace's charming and almost childish naïveté, closely allied to a trusting nature that precluded the hard and relentless skepticism so essential to residence at the highest levels of achievement in science, led him to spread his efforts so thin, and so promiscuously, that his voluminous outpouring in support, for example, of spiritualism, the nationalization of land, and phrenology, and against the possibility of life on Mars (or anywhere else in the universe besides Earth), and compulsory vaccination for smallpox, made him seem more cranky than learned to many of his contemporaries.

How then shall we rescue the reputation, and reveal the sheer delight and fascination, of this passionate scholar who had the misfortune to fall under Darwin's shadow? Two pathways suggest themselves. One might follow the all-too-human tendency to search for nonexistent conspiracies, and argue that, somehow, Darwin and his pals cheated Wallace of just deserts and true priority. Recent books by Brooks and Brackman have explored this non-starter of a tactic (and, if I may state my one criticism of this fine book, I think that Andrew Berry has granted this option too much weight and credence). For no one can doubt that Darwin had been developing natural selection for twenty years when Wallace brilliantly derived the same basic concept independently. Moreover, one cannot find a word in any of Wallace's utterances, including the most private and uncensored comments, to indicate that he felt anything other than delight and gratitude for Darwin's generosity of sharing and citation (while Darwin properly pointed out his undeniable priority).

The second and proper tactic – so beautifully and skillfully accomplished by Berry in this book – is both so simple, and so obviously right: provide an intelligent and comprehensive sampling of Wallace's unique and passionate writing. Show us the remarkable man as he was, and as he spoke, in all his multifariousness, warts and all. No one will be able to read this fine selection of Alfred Russel Wallace and not feel the power, pleasure and instruction in the force of his words and in the example of his sometimes painful, often flawed, but utterly fascinating existence.

Stephen Jay Gould
Museum of Comparative Zoology
Harvard University
August 2001

Editor's Note

Wallace was extraordinarily prolific. The latest bibliography of his writings[1] lists 764 publications, including 21 books, four of which stretch to two volumes, each of at least 400 pages. Wallace was much more than just a biologist; he published on everything from the nationalization of the railways to the real identity of Shakespeare. However, despite his importance in the history of science (as co-discoverer, with Charles Darwin, of natural selection), virtually none of his work is currently in print. This anthology is an attempt to set that to rights, with a selection of Wallace's writing culled from his books, articles, and published letters.

In condensing this enormous body of work, I have aimed at a collection that is as representative as possible of all of Wallace's wide-ranging interests. My purpose is not academic or archival, but rather to introduce contemporary readers to an important and unjustly neglected thinker. This means that the treatment of any single topic is inevitably somewhat superficial, but in each case I have tried to provide Wallace's most succinct summary of his views. Because there is so much material to choose from, however, Wallace-philes should be warned that they will doubtless find that I have failed to include their favourite passages. Also, space limititations have compelled me to pick and choose among the seeming infinity of topics Wallace addressed. Fans of particular aspects of his work – his views, say, on the role of natural selection in promoting infertility between incipient species, or on the gold standard – may therefore be disappointed.

I have tried to keep the excerpts short and the commentary minimal. However, I have included the full text of what are generally agreed, from a historical standpoint, to be Wallace's three most important articles: his "Sarawak Law" (1855), in which he first explicitly addressed evolutionary issues; the "Ternate Paper" (1858), in which he unveiled natural selection; and the "Origin of Human Races" paper (1864), in which, for the first time, he applied natural selection to humans. In addition, the "science" section of the anthology includes a number of long and occasionally technical excerpts from other particularly influential publications. I encourage readers only casually interested in the issues addressed in these articles to skim them, and to move on to more accessible topics dealt with later in the volume.

I have divided Wallace's work up thematically. This in itself is a problematic exercise: Wallace was such an inter-disciplinary thinker that categorizing pieces of his work is often a difficult or even arbitrary exercise. For example, one of his most important contributions was to bring an explicitly evolutionary approach to biogeography, the study of the geographical distribution of living things, and yet I have retained separate sections for "Evolution" and "Biogeography". This problem is in fact more systemic than might at first sight appear. Wallace came to view all aspects of his work – science, society, and spiritualism – merely as different components of a single integrated whole. Wallace's vision and inter-disciplinary ideals notwithstanding, however, a thematic organization does at least bring some order to the plethora of material.

Each section has a loose biographical organization so that the development of Wallace's thinking on each topic can be tracked. I nevertheless start with a sketch of Wallace's life based on excerpts from his autobiography, *My Life* (1905). Because his intellectual development is treated in each of the main sections of the anthology, the biographical sketch offers a lopsided account that concentrates on those aspects of Wallace's life – like education and career – not covered elsewhere in the volume. Only by reading the entire book will the reader have a comprehensive perspective on Wallace's life.

I have used acronyms for Wallace's major works in the text and endnotes. Please refer to the bibliography for a key to the abbreviations. To avoid overburdening the text with references, I have given only a date at the end of each excerpt, with full references given in the endnotes. The dates generally refer to the date of the publication from which the quotation is taken, but may occasionally differ, such as when the passage is taken from a letter quoted in the source. When short articles have been reprinted (for example, in Wallace's two-volume collection of essays, *Studies Scientific and Social*), I have given the full original reference *and* the page numbers of the extract in the reprint, because readers interested in the context of an excerpt will in general find reprints easier to track down than originals.

Wallace, like many Victorians, was inconsistent in his use of punctuation, especially commas. I have preserved his eccentricities except in the few instances when they cause ambiguity or confusion. Similarly, wherever possible, I have preserved the general features of his presentation, such as his use of italics and sub-headings.

Although the vast bulk of Wallace's work is no longer in print, many of his most important writings are available on the web at http://www.wku.edu/~smithch/home.htm. Charles Smith, editor of an invaluable anthology of Wallace's shorter writings,[2] single-handedly maintains the site – which should serve as a model for all web-based archival projects. Smith's site is an essential port of call for anyone interested in Wallace, and is a special boon to scanner-weary anthologizers.

For their help and enthusiasm, I thank Christine Bastoni, Richard Bondi, Mike Davis, Justin Dyer, Martin Fichman, Steve Gould, Jane Hindle, Jim Mallet, Naomi Pierce and Charles Smith.

Andrew Berry
Cambridge, MA
August 2001

A Biographical Sketch

Alfred Russel Wallace (1823–1913) is best known for writing the letter that panicked Charles Darwin into rushing out the *Origin of Species*. Wallace had independently discovered natural selection, and Darwin, who had been quietly incubating his ideas on the subject for some twenty years, had to scramble to assert his precedence. However, Wallace was much more than just Darwin's goad. He was an important scientist in his own right, and is justly considered the father of evolutionary biogeography – the study of the geographical distribution of animals and plants – to which his most famous contribution was the identification of what came to be called "Wallace's Line", the boundary between the faunas of the Australasian and Oriental biogeographic regions. Having worked for twelve years in the field in the Amazon and South-east Asia, he was the pre-eminent tropical biologist of his day, and contributed significantly to what we now know as "ecology" but was then a discipline without a name. His studies of adaptation (especially the colours of insects) and speciation remain influential today. Wallace's scientific interests ranged across disciplinary boundaries, and, in addition to being a biologist, he may legitimately be regarded as an anthropologist and geologist.

Unlike the majority of his Victorian scientific colleagues, Wallace grew up poor and was largely an autodidact. However, his real introduction to the economic and social injustices of the age came in the early 1840s when he was working as a surveyor in rural South Wales, which was then in a state of virtual open rebellion as agricultural prices plummeted and government cash demands on farmers climbed. His travels overseas additionally exposed him to (and gave him great respect for) people from non-European cultures, or "savages", to use the era's term of choice, and to the systems of colonial exploitation these people were subject to. It is not surprising that Wallace's lifelong identification with the underdog resulted ultimately in his becoming a socialist. He wrote extensively on social issues, remaining an anti-establishment firebrand right up to his death shortly before his ninety-first birthday. Wallace's excellence as a scientist, his embrace of egalitarian and humanitarian ideals in a Britain at height of empire, and his willingness to defend those ideals at any cost combined to make him truly one of the most remarkable Victorians.

"A very dull, ignorant, and uneducated person"[1]

Alfred Russel Wallace was born into genteel poverty; he was not, as one of the canonical tellings of the Wallace story has it, working class. Prior to marrying, his father Thomas Vere Wallace, a non-practising lawyer of independent means, opted to live "quite idly, so far as being without any systematic occupation, often going to Bath in the season, where he used to tell us he had met the celebrated Beau Brummell".[2] The good life, however, came to an end with Thomas's marriage in 1807 to Mary Ann Greenell: "By 1810 he had two children and the prospects of a large family, [and] he appears to have felt the necessity of increasing his income."[3] He made the mistake of launching a magazine devoted to literature and the arts. Things did not go well, and Wallace's father "had to bear almost the whole loss, and this considerably reduced his already too scanty income".[4]

The expanding family moved to "a place where living was as cheap as possible",[5] Usk, Gwent, where Wallace was born on January 8, 1823, eighth in a family ultimately of nine children. Only two of Wallace's five sisters, Fanny (born 1812) and Eliza (1810), survived into adulthood, and Eliza died of consumption at twenty-two; all three brothers, William (born 1809), John (1818), and Herbert (1828) survived childhood. In 1828 the family moved to Hertford, where Mr Wallace's finances took a further nosedive after he entrusted his remaining money to "a solicitor and friend"[6] but who duly made a series of catastrophic 7investments. Wallace's education was accordingly abruptly curtailed around Christmas 1836, when he was thirteen.

This is how Wallace describes himself as a small child in Usk: "I was exceedingly fair and my long hair was of a very light flaxen tint, so that I was generally spoken of among the Welsh-speaking country people as the 'little Saxon.'"[7] His boyhood, albeit impecunious and occasionally interrupted by bouts of illness, was a happy one. Scarlet fever, the first of three life-threatening illnesses, yielded the consolation of a luxurious convalescence: "For some weeks after this I lived a very enjoyable life in bed, having tea and toast, puddings, grapes, and other luxuries till I was well again."[8]

Wallace's cricket bat, a status symbol in the eyes of his peers,[9] was a favourite in an era "when the practice of overhand bowling was just beginning, and there was much controversy as to whether or not it should be allowed".[10] Explosives, however, were more fun: "In connection with fireworks, we were fond of making miniature cannon out of keys. For this purpose we begged of our friends any discarded box or other keys with rather large barrels, and by filing a touch-hole, filing off the handle, and mounting them on block carriages, we were able to fire off salutes or startle our sister or the servant to our great satisfaction."[11]

He must have excelled at Hertford Grammar School because, to save on school fees during his final year there, he taught younger boys reading, dictation, arithmetic and writing.[12] Nevertheless, Wallace's education was perfunctory, consisting of the Victorian staples of Latin, some geography and history, and some mathematics. Wallace gained little from his exposure to the classics – "so we blundered through our forty or fifty lines" – but did gain a lifelong appreciation of poetry from a translation of *The Aeneid*, whose verse he considered "clear and melodious".[13] Nevertheless he had access to books both at home and at the town library, where his father was a librarian.

❧ During the time I lived in Hertford I was subject to influences which did more for my real education than the mere verbal training I received at school. My father belonged to a book club, through which we had a constant stream of interesting books, many of which he used to read aloud in the evening. Among these I remember Mungo Park's[14] travels and those of Denham and Clapperton[15] in West Africa. We also had *Hood's Comic Annual*[16] for successive years, and I well remember my delight with "The Pugsley Papers" and "A Tale of the Great Plague," while as we lived first at a No. 1, I associated Hood's "Number One" with our house, and learnt the verses by heart when I was about seven years old. Ever since those early experiences I have been an admirer of Hood in all his various moods, from his inimitable mixture of pun and pathos in his "Sea Spell," to the exquisite poetry of "The Haunted House," "The Elm Tree," and "The Bridge of Sighs."

We also had some good old standard works in the house, "Fairy Tales," "Gulliver's Travels," "Robinson Crusoe," and the "Pilgrim's Progress," all of which I read over again and again with constant pleasure. We also had "The Lady of the Lake," "The Vicar of Wakefield," and some others; and among the books from the club I well remember my father reading to us Defoe's wonderful "History of the Great Plague." We also had a few highly educational toys, among which were large dissected maps of England and of Europe, which we only had out as a special treat now and then, and which besides having the constant charm of a puzzle, gave us a better knowledge of topographical geography than all our school teaching, and also gave me that love of good maps which has continued with me throughout life.[17] ଛ

Early in 1837, Wallace joined his brother John, a builder, in London, where he heard, and was heavily influenced by, the teachings of Robert Owen, proto-socialist and social reformer. Later in the same year he moved to Bedfordshire, where his brother William had established a land surveying business. Wallace acquired the rudiments of surveying and map-making[18] and, apart from a brief detour in 1839 into the watchmaking trade – which taught him that he should "never have succeeded as a man of business, for which I am not fitted by nature"[19] – during a lull in surveying opportunities, would remain a surveyor until 1843. He and William moved around, following the available work.

As Wallace wrote to his boyhood friend George Silk, he enjoyed the outdoor life.

ଛ I think you would like land-surveying, about half indoors and half outdoors work. It is delightful on a fine Summer's day to be (literally) cutting all over the country, following the chain [surveyor's measure] and admiring the beauties of nature, breathing the fresh and pure air on the hills, or in the noontide heat enjoying our luncheon of bread-and-cheese in a pleasant valley by the side of a rippling brook. Sometimes, indeed, it is not quite so pleasant on a cold winter's day to find yourself on the top of a bare hill, not a house within a mile, and the wind and sleet chilling you to the bone. But it is all made up for in the evening; and those who are in the house all day can have no idea of the pleasure there is in sitting down to a good dinner and feeling hungry enough to eat plates, dishes, and all.[20] &

Surveying also stimulated Wallace to compensate for the short-comings of his education.

& These wonderfully accurate measurements and calculations impressed me greatly, and with my practical work at surveying and learning the use of that beautiful little instrument, the pocket-sextant, opened my mind to the uses and practical applications of mathematics, of which at school I had been taught nothing whatever, although I had learnt some Euclid and algebra. This glimmer of light made me want to know more, and I obtained some of the cheap elementary books published by the Society for the Diffusion of Useful Knowledge.[21] &

Wallace's peripatetic life exposed him to a cross-section of rustic British society. He duly debated the pressing issues of the day, like an "objection to the success of the railway": "the hold of the engine on the rails would not be sufficient to draw heavy trucks or carriages – . . . in fact, the wheels would whizz round instead of going on."[22] Agricultural topics, however, were the staple.

& A young farmer was complaining of the poor crop of wheat he had got from one of his best fields, and he said be could not make it out. One of the large farmers, who was looked up to as an authority, asked, "What did you do to the field?" "Well," said the young man, "I ploughed it" (a pause) "I ploughed it twice." "Ah!" said the expert, "that's where you lost your crop." The rest looked approval. Some said, "That's it;" others said, "Ah!" The young man said nothing, but looked gloomy. Evidently the oracle had spoken, and nothing more was to be said; but I have often wondered since if that really was the cause of the bad crop of wheat.[23] &

Most significantly, though, surveying supplied Wallace with opportunities to nurture his nascent interest in natural history.

& It was here, too, that during my solitary rambles I first began to feel the influence of nature and to wish to know more of the various flowers, shrubs, and trees I daily met with, but of which for the most part I did not even know the English names. At that time I hardly

realized that there was such a science as systematic botany, that every
flower and every meanest and most insignificant weed had been
accurately described and classified, and that there was any kind of
system or order in the endless variety of plants and animals which I
knew existed.[24]

. . .

It must be remembered that my ignorance of plants at this time was
extreme. I knew the wild rose, bramble, hawthorn, buttercup, poppy,
daisy, and foxglove, and a very few others equally common and
popular, and this was all. I knew nothing whatever as to genera and
species, nor of the large numbers of distinct forms related to each
other and grouped into natural orders. My delight, therefore, was
great when I was now able to identify the charming little eyebright,
the strange-looking cow-wheat and louse-wort, the handsome mullein
and the pretty creeping toad-flax, and to find that all of them, as well
as the lordly foxglove, formed parts of one great natural order, and
that under all their superficial diversity of form there was a similarity
of structure which, when once clearly understood, enabled me to
locate each fresh species with greater ease.[25] ❧

In 1843, the year of his father's death, Wallace's work with
William came to an end: "As he had no work in prospect it was
necessary that I should leave him and look out for myself."[26]

❧ I therefore determined to try for some post in a school to teach
English, surveying, elementary drawing, etc. Through some school
agency I heard of two vacancies that might possibly suit. The first
required, in addition to English, junior Latin and algebra. Though I
had not looked at a Latin book since I left school, I thought I might
possibly manage; and as to algebra, I could do simple equations, and
had once been able to do quadratics, and felt sure I could keep ahead
of beginners. So with some trepidation I went to interview the master,
a rather grave but kindly clergyman. I told him my position, and what
I had been doing since I left school. He asked me if I could translate
Virgil, at which I hesitated, but told him I had been through most of
it at school. So he brought out the book and gave me a passage to
translate, which, of course, I was quite unable to do properly. Then
he set me a simple equation, which I worked easily. Then a quadratic,
at which I stuck. So he politely remarked that I required a few months'
hard work to be fitted for his school, and wished me good-morning.

My next attempt was more hopeful, as drawing, surveying, and mapping were required. Here, again, I met a clergyman, but a younger man, and more easy and friendly in his manner. I had taken with me a small coloured map I had made at Neath to serve as a specimen, and also one or two pencil sketches. These seemed to satisfy him, and as I was only wanted to take the junior classes in English reading, writing, and arithmetic, teach a very few boys surveying, and beginners in drawing, he agreed to engage me. I was to live in the house, preside over the evening preparation of the boarders (about twenty in number) and to have, I think, thirty or forty pounds a year, with which I was quite satisfied. I was to begin work in about a fortnight. My employer was the Rev. Abraham Hill, headmaster of the Collegiate School at Leicester.[27]   ৯৯

Wallace spent only one year in Leicester, but it had a "determining influence"[28] on his future. His first encounters with mesmerism and phrenology provoked an interest in metaphysics that culminated, later in life, in Wallace's embrace of spiritualism. He also met his future travelling companion.

৯৯   How I was introduced to Henry Walter Bates[29] [in 1844] I do not exactly remember, but I rather think I heard him mentioned as an enthusiastic entomologist, and met him at the library. I found that his specialty was beetle collecting, though he also had a good set of British butterflies. Of the former I had scarcely heard, but as I already knew the fascinations of plant life I was quite prepared to take an interest in any other department of nature. He asked me to see his collection, and I was amazed to find the great number and variety of beetles, their many strange forms and often beautiful markings or colouring, and was even more surprised when I found that almost all I saw had been collected around Leicester, and that there were still many more to be discovered. If I had been asked before how many different kinds of beetles were to be found in any small district near a town, I should probably have guessed fifty or at the outside a hundred, and thought that a very liberal allowance. But I now learnt that many hundreds could easily be collected, and that there were probably a thousand different kinds within ten miles of the town. He also showed me a thick volume containing descriptions of more than three thousand species inhabiting the British Isles. I also learnt from him in what an infinite variety of places beetles may be found, while some may be collected all the year round, so I at once determined to begin collecting, as I did not find a great many new plants about Leicester.

I therefore obtained a collecting bottle, pins, and a store-box; and in order to learn their names and classification I obtained, at wholesale price through Mr. Hill's bookseller, Stephen's "Manual of British Coleoptera [1839]," which henceforth for some years gave me almost as much pleasure as Lindley's Botany [1835], with my MS. descriptions [notes], had already done.[30] &

Wallace had to leave Leicester precipitously in early 1846 on the death of his brother William, who "speedily succumbed", having "caught a severe cold by being chilled in a wretched third class carriage, succeeded by a damp bed at Bristol".[31] Wallace's experience in Neath, Wales, where he had to sort out William's affairs, convinced him that he was not cut out for a life in business, and was one of the reasons for his eventually going abroad.[32]

& One gentleman, whose account was a few pounds, declared he had paid it, and asked us to call on him. We did so, and, instead of producing the receipt as we expected, he was jocose about it, asked us what kind of business men we were to want him to pay twice; and when we explained that it was not shown so in my brother's books, and asked to look at the receipt, he coolly replied, "Oh, I never keep receipts; never kept a receipt in my life, and never was asked to pay a bill twice till now!" In vain we urged that we were bound as trustees for the rest of the family to collect all debts shown by my brother's books to be due to him, and if he did not pay it, we should have to lose the amount ourselves. He still maintained that he had paid it, that he remembered it distinctly, and that he was not going to pay it twice. At last we were obliged to tell him that if he did not pay it we *must* put it in the hands of a lawyer to take what steps he thought necessary; then he gave way, and said, "Oh, if you are going to law about such a trifle, I suppose I must pay it again!" and, counting out the money, added, "There it is; but I paid it before, so give me a receipt *this* time," apparently considering himself a very injured man.[33] &

Now in partnership with his brother John, Wallace returned to surveying, and undertook minor architectural projects – including the design of the Neath Mechanics' Institution.[34] They also spent a great deal of time "wandering about this beautiful district",[35] adding to their natural history collections. One of the reasons they had so much time on the hands was that surveying work in the region had been interrupted by civil unrest: the "Rebecca Disturbances" swept through South Wales during the early 1840s. Nominally the "Rebeccaites" were protesting tolls on public roads, but the attacks were symptomatic of widespread hardship (and even starvation) caused by increases in tithe charges, and the Poor Law Amendment Act of 1834.[36]

Among the expeditions that Wallace and his brother undertook was a trip to the local mountains, the Brecon Beacons, that involved sleeping rough in a cave, where the "harsh contact of [his] bones with the rock or pebbles" led Wallace to remark that, "while in health I have never passed a more uncomfortable night".[37] Contrarily, however, this seems to have increased his desire to experience life away from the comforts of civilization. The major impetus to travel, though, came from his reading:

In a letter [to Bates] dated April 11, 1846, there occur the following remarks on two books about which there has been little difference of opinion, and whose authors I had at that time no expectation of ever calling my friends. "I was much pleased to find that you so well appreciated Lyell.[38] I first read Darwin's 'Journal'[39] three or four years ago, and have lately re-read it. As the journal of a scientific traveller, it is second only to Humboldt's 'Personal Narrative'[40] – as a work of general interest, perhaps superior to it. He is an ardent admirer and most able supporter of Mr. Lyell's views. His style of writing I very much admire, so free from all labour, affectation, or egotism, and yet so full of interest and original thought . . . I quite envy you, who have friends near attached to the same pursuits. I know not a single person in this little town who studies any one branch of natural history, so that I am quite alone in this respect." My references to Darwin's "Journal" and to Humboldt's "Personal Narrative" indicate, I believe, the two works to whose inspiration I owe my determination to visit the tropics as a collector.[41]

Journeys

What decided our going to Pará [Belém] and the Amazon rather than to any other part of the tropics was the publication in 1847, in Murray's Home and Colonial Library, of "A Voyage up the Amazon," by Mr. W. H. Edwards.[42] This little book was so clearly and brightly written, described so well the beauty and the grandeur of tropical vegetation, and gave such a pleasing account of the people, their

Alfred Russel Wallace aged
around twenty-five, a portrait
taken at about the time he went
to the Amazon (from Wallace's
My Life)

kindness and hospitality to strangers, and especially of the English
and American merchants in Para, while expenses of living and of
travelling were both very moderate, that Bates and myself at once
agreed that this was the very place for us to go to if there was any
chance of paying our expenses by the sale of our duplicate collections.
I think we read the book in the latter part of the year (or very early in
1848), and we immediately communicated with Mr. Edward Doubleday[43] who had charge of the butterflies at the British Museum, for his
advice upon the matter. He assured us that the whole of northern
Brazil was very little known, that some small collections they had
recently had from Para and Pernambuco contained many rarities and
some new species, and that if we collected all orders of insects, as well
as landshells, birds, and mammals, there was no doubt we could easily
pay our expenses. Thus encouraged, we determined to go to Para,
and began to make all the necessary arrangements.[44]   ⇌

Before departing, Wallace and Bates "were fortunate in finding
an excellent and trustworthy agent in Mr. Samuel Stevens, an
enthusiastic collector of British Coleoptera and Lepidoptera".[45]
Stevens was crucial to the venture. Wallace and Bates were going
as commercial collectors: their expenses would be paid by the

sale, by Stevens, of their specimens. Happily he proved an ideal choice: "During the whole period of our business relations, extending over more than fifteen years, I cannot remember that we ever had the least disagreement about any matter whatever."[46]

They arrived in Brazil at the end of May 1848. Wallace returned to England in late 1852; Bates stayed in the Amazon until 1859. At first Wallace and Bates travelled and collected together, but after only about nine months in Brazil they decided to go their separate ways.[47] The reasons for the split are unclear: possibly they simply had different objectives – certainly, there is no hint of personal animosity in their extensive subsequent correspondence. Whereas Bates headed up the Amazon proper from Manaus, close to its confluence with the Rio Negro, Wallace chose to concentrate on the Rio Negro and its tributaries. Wallace's younger brother Herbert came out to join him to see if "he had sufficient taste for natural history to become a good collector."[48] He did not: "he took little interest in birds or insects, and without enthusiasm in the pursuit [of biological collecting] he would not have been likely to succeed."[49] Wallace made arrangements for him to return, and "had little doubt that he would get home without difficulty. But I never saw him again."[50] Herbert died of yellow fever in Belém. Wallace, miles away

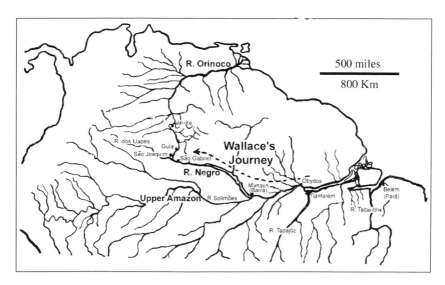

Map of Wallace's travels in the Amazon, 1848–52 (modified from Wilma George, *Biologist Philosopher*)

upstream, heard the news piecemeal: first that his brother was sick, and then, many months later, that he was dead.

Wallace published a full account of his Amazon travels in *A Narrative of Travels on the Amazon and Rio Negro* (1853); he provides the following synopsis in his autobiography.

≈ᵇ The remainder of my South American travels consisted of two voyages up the Rio Negro. On the first I went beyond the boundaries of Brazil, and crossed by a road in the forest to one of the tributaries of the Orinoko. Returning thence I visited a village up a small branch of the Rio Negro, where there is an isolated rocky mountain, the haunt of the beautiful Cock of the Rock; afterwards going up the Uaupés as far as the second cataract at Juaurité. I then returned with my collections to Barra [Manaus], having determined to go much farther up the Uaupés, in order to obtain, if possible, the white umbrella bird which I had been positively assured was found there; and also in the hopes of finding some new and better collecting ground near the Andes. These journeys were made, but the second was cut short by delays and the wet season. My health also had suffered so much by a succession of fevers and dysentery that I did not consider it prudent to stay longer in the country.

Although during the last two journeys in the Rio Negro and Orinoko districts I had made large miscellaneous collections, and especially of articles of native workmanship, I never found any locality at all comparable with Para as a collecting ground. The numerous places I visited along more than a thousand miles of river, all alike had that poverty of insect and bird-life which characterized Barra itself, a poverty which is not altogether explicable. The enormous difficulties and delays of travel made it impossible to be at the right place at the right season; while the excessive wetness of the climate rendered the loss of the only month or two of fine weather irreparable for the whole year. The comparative scantiness of native population at all the towns of the Rio Negro, the small amount of cultivation, the scarcity of roads through the forest, and the want of any guide from the experience of previous collectors, combined to render my numerous journeys in this almost totally unknown region comparatively unproductive in birds and insects.[51] ᵇ≈

Wallace had completed a remarkable journey into regions previously virtually unvisited by Europeans. However, it turned out that the most harrowing leg of the trip was the journey home in 1852: the *Helen* caught fire and sank in the middle of the Atlantic, and, with the crew, Wallace spent ten days in an open boat before being picked up. The rescue ship then itself nearly sank in a storm.

Some of the most alarming incidents, to a landsman, are not mentioned either in this letter or in my published "Narrative." The captain had given the only berths in the cabin to Captain Turner[52] and myself, he sleeping on a sofa in fine weather, and on a mattress on the floor of the cabin when rough. On the worst night of the storm I saw him, to my surprise, bring down an axe and lay it beside him, and on asking what it was for, he replied, "To cut away the masts in case we capsize in the night." In the middle of the night a great sea smashed our skylight and poured in a deluge of water, soaking the poor captain, and then slushing from side to side with every roll of the ship. Now, I thought, our time is come; and I expected to see the captain rush up on deck with his axe. But he only swore a good deal, sought out a dry coat and blanket, and then lay down on the sofa as if nothing had happened. So I was a little reassured.

Not less alarming was the circumstance of the crew coming aft in a body to say that the forecastle was uninhabitable, as it was constantly wet, and several of them brought handfuls of wet rotten wood which they could pull out in many places. This happened soon after the first gale began; so the two captains and I went to look, and we saw sprays and squirts of water coming in at the joints in numerous places, soaking almost all the men's berths, while here and there we could see the places where they had pulled out rotten wood with their fingers. The captain then had the sail-room amidships cleared out for the men to sleep in for the rest of the voyage.

One day in the height of the storm, when we were being flooded with spray and enormous waves were coming up behind us, Captain Turner and I were sitting on the poop in the driest place we could find, and, as a bigger wave than usual rolled under us and dashed over our sides, he said quietly to me, "If we are pooped by one of those waves we shall go to the bottom," then added, "We were not very safe in our two small boats, but I had rather be back in them where we were picked up than in this rotten old tub." It is, therefore, I think, quite evident that we *did* have a very narrow escape.[53]

The situation was enormously exacerbated for Wallace by a customs mix-up in Manaus which had resulted in an earlier consignment of his specimens being held up there until his final departure. The fire thus claimed a large proportion of all of Wallace's labours over those four years:

What I had hitherto sent home had little more than paid my expenses, and what I had with me in the *Helen* I estimated would have realized about £500. But even all this might have gone with little regret had not by far the richest part of my own private collection gone also. All my private collection of insects and birds since I left Para [in August 1849] was with me, and comprised hundreds of new and beautiful species, which would have rendered (I had fondly hoped) my cabinet, as far as regards American species, one of the finest in Europe.[54]

Wallace realized he had to do it all over again, and opted this time for South-east Asia. Meanwhile, despite his setback, he was now a member – albeit a minor one – of the scientific community: "As my collections had now made my name well known to the authorities of the Zoological and Entomological Societies, I received a ticket from the former giving me admission to their gardens while I remained in England, and I was a welcome visitor at the scientific meetings of both societies."[55] It was during this period that Wallace first met the biological notables of his day, including Huxley, and – fleetingly, in the backroom insect collections at the Natural History Museum – Darwin. In addition, Wallace sorted his remaining collections, took a brief trip to Switzerland, wrote up his Amazon experiences, and planned his South-east Asian venture. Through the offices of Sir Roderick Murchison, President of the Royal Geographical Society, he received a free ticket to Singapore; he arrived on April 20 1854, a mere eighteen months after his return from Brazil.

In the course of the next eight years, Wallace logged an extraordinary "fourteen thousand miles within the Archipelago, and made some sixty or seventy separate journeys", and collected an astonishing 125,660 specimens.[56] Somewhat apologetically, in true Wallace fashion, he mentions that because all the travelling involved "some preparation and loss of time", he thought that fewer "than six years were really occupied in collecting".[57] The journey, immortalized in *The Malay Archipelago*, yielded Wallace's two most important biological insights: natural selection and the boundary of the biogeographic discontinuity between the Asian

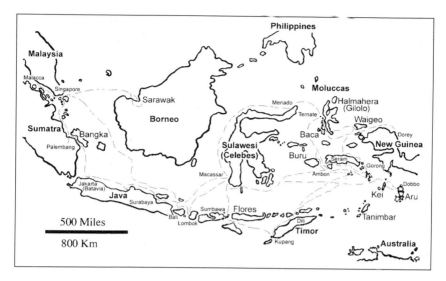

Map of Wallace's journeys in the Malay Archipelago, 1854–62 (modified from Wilma George, *Biologist Philosopher*)

and Australasian faunas that has become known as "Wallace's Line".

After collecting in Peninsular Malaysia, Wallace spent more than a year in Sarawak, Borneo, where, through his studies of the orang-utan, he became one of the few biologists of his time to have observed one of the Great Apes in its natural habitat. He repaid his host, the "White Rajah" of Sarawak, Sir James Brooke, by naming "perhaps the most elegant butterfly in the world",[58] *Ornithoptera Brookeana*,[59] after him. From Borneo he headed, via Lombok, to Sulawesi and on to the Aru Islands off the south-east coast of New Guinea, where he found the collecting opportunities plentiful – "This journey was the most successful of any that I undertook."[60] After visits to Sulawesi, Timor, Banda (south of Seram), and Ambon, Wallace reached Ternate in the Moluccas (Spice Islands) in January 1858, and had written his paper on natural selection by the end of the following month.[61] His much anticipated trip to mainland New Guinea was a disappointment.

I obtained a passage to Dorey Harbour on the north coast of New Guinea [in the western half, modern Irian Jaya], in a trading schooner, which left me there, and called for me three or four months later to bring me back to Ternate. I was the first European who had lived alone on this great island; but partly owing to an accident which confined me to the house for a month, and partly because the locality was not a good one, I did not get the rare species of birds of paradise I had expected. I obtained, however, a number of new and rare birds and a fine collection of insects, though not so many of the larger and finer kinds as I expected. The weather had been unusually wet, and the place was unhealthy. I had four Malay servants with me, three of whom had fever as well as myself, and one of my hunters died, and though I should have liked to have stayed longer, we were all weak or unwell, and were very glad when the schooner arrived and took us back to Ternate [in August 1858].[62] ප

Wallace spent the next three years in a frenzy of island-hopping throughout the southern portion of what today is Indonesia, often risking his life in small open boats. The following voyage, from Manowolko (Manawoka, Kepualan Gorong) to the Mata-bello Islands (Kepualuan Watubela), took place in April 1860.

ප ... a miserable boat and five men were found, and with some difficulty I stowed away in it such baggage as it was absolutely necessary for me to take, leaving scarcely sitting or sleeping room. The sailing qualities of the boat were highly vaunted, and I was assured that at this season a small one was much more likely to succeed in making the journey. ... I did not much like the look of the heavy sky and rather rough sea, and my men were very unwilling to make the attempt; but as we could scarcely hope for a better chance, I insisted upon trying. The pitching and jerking of our little boat soon reduced me to a state of miserable helplessness, and I lay down, resigned to whatever might happen.[63] ප

By July 1861, Wallace was back in the relative civilization of Java and Sumatra, and in April 1862 he was in England, where he could at last hand over to the zoo the two birds of paradise he had carefully tended throughout the journey home.

Intellectual Life: "Satisfaction, Retrospection and Work"[64]

Wallace's South-east Asian collections and publications assured him on his return of a place in Britain's scientific elite. The famous recluse Darwin extended to him the rare honour of an invitation to spend the night at his home, Down House, soon after he arrived back in England,[65] and Wallace eventually came to look upon such exalted figures as Huxley, Lyell, Hooker[66] and Spencer[67] as his friends.

As Wallace worked through his collections, he wrote a number of articles, many of them destined to become landmarks in their respective fields. In particular, in 1864 he brought natural selection to bear on the evolution of humans[68] for the first time – Darwin had studiously avoided the issue in the *Origin* – and, one year later, he published an extraordinarily ahead-of-its-time paper on what today would be termed the ecology, population genetics and evolution of South-east Asian swallow-tailed butterflies.[69]

Wallace attended his first seance in 1865, and was publishing on spiritualism within a year. It is likely that this new interest influenced his biology, and by 1869 he was willing to state publicly that he believed that more than natural selection was involved in the evolution of humans. This divergence of opinion – indeed, defection – duly alarmed Darwin. Wallace's biological and metaphysical views subsequently evolved in concert, culminating in his final major work, *The World of Life* (1910), in which he outlines a thoroughly teleological view of the world, yet with natural selection still mechanistically preeminent.

Strangely it was a book about far-off lands, *The Malay Archipelago* (1869), that, by attracting the interest of the economist and philosopher J. S. Mill, initiated Wallace's involvement in domestic social issues.[70] In 1881 he became the founding president of the Land Nationalization Society, and campaigned passionately for the rest of his life for government ownership of land – the first step, he was convinced, towards a just society. His longstanding left-wards leaning was upgraded in 1889 to socialism. Wallace wrote extensively on issues ranging from the reform of the Church of England, public health (he was a prominent campaigner, on epidemiological grounds, against mandatory small-pox vaccination), imperialism, and the inequities in the distribution of wealth.[71] In *My Life* he wrote "To allow one child to be born a millionaire and another a pauper is a crime against humanity, and, for those who believe in a deity, a crime against God."[72]

Many of these views were published in his somewhat ironically titled *Wonderful Century* (1898) and in *Studies Scientific and Social* (1900), a two-volume collection of essays and articles.

In 1876, Wallace published his *meisterwerk*, the monumental *Geographical Distribution of Animals*, which essentially established the science of zoogeography and brought evolutionary factors fully into the analysis of the geographical distribution of animals.[73] He followed this in 1880 with *Island Life*, which, despite its title, did much to consolidate his reputation in geography and geology through its chapters on the ice ages and their causes.

For most of a year, 1886–87, Wallace travelled though North America, lecturing on biological, spiritualist, and occasionally political topics. The biological lectures later formed the basis of *Darwinism* (1889), his summary of the state of late-nineteenth-century evolutionary biology.

The political talks Wallace gave on his American tour were not always well received.

✑ [The address] on "Social Economy versus Political Economy" . . . was given . . . to a large audience of gentlemen and ladies. It was an attempt to show how and why the old "political economy" was effete and useless, in view of modern civilization and modern accumulations of individual wealth. Its one end, aim, and the measure of its success, was the accumulation of wealth, without considering who got the wealth, or how many of the producers of the wealth starved. What we required now was a science of "social economy," whose success should be measured by the good of all. Under this system, not only should no worker ever be in want, but labour must be so organized that every worker, without exception, must receive as the product of his labour all the essentials of a healthy and happy life; must have ample relaxation, adequate change of occupation, the means of enjoying the beauty and the solace of nature on the one hand, and of literature and art on the other. This must be a first charge on the labour of the community; till this is produced there must be no labour expended on luxury, no private accumulations of wealth in order that unborn generations may live lives of idleness and pleasure.

This paper was altogether too revolutionary for many of my hearers, and the general feeling was perhaps expressed in the following passage from the *Washington Post*: "It is astounding that a man who really possesses the power of induction and ratiocination, and who, in physical synthesis has been a leader of his generation, should express notions of political economy, which belong only or mainly to savage tribes." At that time, however, there was hardly a professed socialist in America.[74] ✑

Wallace was introduced to a virtual Who's Who of America: he dined with Oliver Wendell Holmes; toured the redwood forests of northern California with John Muir, the naturalist and conservationist; and met the president, Grover Cleveland. The latter was not a sparkling occasion: "I had nothing special to say to him, and he had nothing special to say to me, the result being that we were both rather bored, and glad to get it over with as soon as we could."[75] In San Francisco, he caught up with his brother John, who had emigrated in 1849 while Wallace was in Brazil.

Wallace's radical positions inevitably made him unpopular with the establishment, but he was nevertheless elected a Fellow of the Royal Society in 1893, and, in 1908, awarded the Order of Merit:[76] ". . . and as to the Order of Merit – to be given to a red-hot Radical, Land Nationaliser, Socialist, Anti-Militarist, etc., etc., etc., is quite astounding and unintelligible!"[77] In 1915, two years after his death, a medallion in his honour was unveiled in Westminster Abbey, and Wallace found himself for one final time once again in Darwin's shadow. Darwin was actually buried in the Abbey – Wallace had been a pall-bearer[78] – not merely commemorated.

Domestic Life: "Gardening and rural walks"[79]

Wallace returned from South-east Asia with both his reputation and fortune made. Unfortunately, the latter was not to last.

A large proportion of my insects and birds were either wholly new or of extreme rarity in England; and as many of them were of large size and of great beauty, they brought very high prices. My agent had invested the proceeds from time to time in Indian guaranteed railway stock, and a year after my return I found myself in possession of about £300 a year. Besides this, I still possessed the whole series of private collections, including large numbers of new or very rare species, which, after I had made what use of them was needed for my work, produced an amount which in the same securities would have produced about £200 a year more.

But I never reached that comfortable position. Owing to my never before having had more than enough to supply my immediate wants, I was wholly ignorant of the numerous snares and pitfalls that beset the ignorant investor, and I unfortunately came under the influence of two or three men who, quite unintentionally, led me into trouble. Soon after I came home I made the acquaintance of Mr. R.,[80] who held a good appointment under Government, and had, besides, the expectation of a moderate fortune on the death of an uncle. I soon became intimate with him, and we were for some years joint investigators of spiritualistic phenomena. He was, like myself at that time, an agnostic, well educated, and of a more positive character than myself. He had for some years saved part of his income, and invested it in various foreign securities at low prices, selling out when they rose in value, and in this way he assured me he had in a few years doubled the amount he had saved. He studied price-lists and foreign news, and assured me that it was quite easy, with a little care and judgment, to increase your capital in this way. He quite laughed at the idea of allowing several thousand pounds to lay idle, as he termed it, in Indian securities, and so imbued me with an idea of his great knowledge of the money market, that I was persuaded to sell out some of my bonds and debentures and buy others that he recommended, which brought in a higher interest, and which he believed would soon rise considerably in value. This change went on slowly with various success for several years, till at last I had investments in various English, American, and foreign railways, whose fluctuations in value I was quite unable to comprehend, and I began to find, when too late, that almost all my changes of investment brought me loss instead of profit, and later on, when the great depression of trade of 1875–85 occurred, the loss was so great as to be almost ruin.[81]

Wallace lost money on just about his every venture. There were the Welsh slate quarries which "if properly developed would be profitable". They were not: "I was the loser of about a thousand pounds."[82] There were the lead mines in Shropshire and Montgomeryshire where Wallace "went down the shafts by endless perpendicular ladders, and examined the veins and workings"; and where he "gradually had a large proportion of . . . capital invested in them". Competition from mines in Nevada caused the price to drop like lead: "By 1880 a large part of the money I had earned at the risk of health and life was irrecoverably lost."[83] There was the dishonest builder who failed to complete work on Wallace's house and then, years later, sued him on the grounds that he had not been allowed to finish the job![84]

The most galling of Wallace's miserable financial experiences, however, was also the strangest: an encounter with a lunatic flat-earther that cost him "fifteen years of continued worry, litigation, and persecution, with the final loss of several hundred pounds".[85] In 1870 a Mr John Hampden "challenged scientific men to prove the convexity of the surface of any inland water, offering to stake £500 on the result".[86] Drawing on his surveying skills, Wallace duly provided an excellent proof on the Old Bedford canal in Norfolk, but discovered quickly that proponents of the flat earth theory do not subscribe to normal scientific notions of "proof". The £500 soon became the subject of litigation, and Mr Hampden went on the offensive, referring to Wallace rather splendidly as "that degraded blackleg",[87] and a swindler who has "dared to make a fraudulent attempt to support the globular theory".[88] He even involved Wallace's wife.

Mrs. Wallace,
Madame – If your infernal thief of a husband is brought home some day on a hurdle,[89] with every bone of his head smashed to pulp, you will know the reason. Do you tell him from me he is a lying infernal thief, and as sure as his name is Wallace he never dies in his bed.
You must be a miserable wretch to be obliged to live with a convicted felon. Do not let him think I have done with him.
John Hampden[90]

Wallace's attempts to solve his financial problems with a job were as ill-fated as his investments. He applied in 1864 to be assistant secretary of the Royal Geographic Society, but the position went instead to Bates, who, Wallace insisted, was "much better qualified" than himself.[91] In 1869 Wallace had hopes that he would be appointed director of a new branch of the British Museum to be established in Bethnal Green, east London, but in the end no independent director was appointed. In 1878 he applied to be superintendent of Epping Forest, an area of publicly owned land to the north-east of London, publishing his own management plan,[92] but again it was not to be. Apparently the overseeing committee preferred "to encourage excursions and school treats, allowing swings, round-abouts, and other such amusements more suited to a beer-garden or village fair than to a tract of land secured at enormous cost and much hardship to individuals in order to preserve an example of the wild natural woodland wastes of our country for the enjoyment and instruc-

tion of successive generations of nature-lovers".[93] Darwin con-
soled Wallace: "I suppose it was too much to hope that such a
body of men [a committee of the Corporation of London, which
owned the land] should make a good selection."[94] After this
latest disappointment, it was only the intervention of a relative –
"Miss Roberts, of Epsom, a cousin of my mother's, with whose
family I had been intimate from my boyhood" – that prevented
Wallace from being in "absolute want".[95] The only regular job
Wallace held was as an examiner in "Physical Geography under
the Science and Art Department" – about as menial a task as
academic life had to offer. He did it every year from 1871 to
1897.

Wallace, however, received a financial lifeline in 1881.

From this ever-increasing [financial] anxiety I was relieved
through the grant of a Civil Service pension of £200, which came
upon me as a very joyful surprise. My most intimate and confidential
friend at this time was Mrs. Fisher (then Miss Buckley),[96] and to her
alone I mentioned my great losses, and my anxiety as to any sure
source of income. Shortly afterwards she was visiting Darwin, and
mentioned it to him, and he thought that a pension might be granted
me in recognition of my scientific work. Huxley most kindly assisted
in drawing up the necessary memorial to the Prime Minister, Mr.
Gladstone, to whom Darwin wrote personally. He promptly assented,
and the next year, 1881, the first payment was made. Other of my
scientific friends, I believe, signed the memorial, but it is especially to
the three named that I owe this very great relief from anxiety for the
remainder of my life.[97]

Darwin wrote to Wallace with the news, telling him that "I will
keep this note [from Gladstone, apprising Darwin of the award
to Wallace] carefully, as, if the present Government were to go
out, I do not doubt that it would be binding on the next
Government."[98] Wallace, needless to say, was pleased: "To-day is
my fifty-eighth birthday, and it is a happy omen that your letter
should have arrived this morning."[99]

Wallace's luck in love was only marginally better than his luck
in financial matters. In a letter written in Singapore at the end of
his South-east Asian journey, he revealed that years of the solitary
travelling life had diminished his conjugal ambitions.

On the question of marriage we probably differ much. I believe a good wife to be the greatest blessing a man can enjoy, and the only road to happiness, but the qualifications I should look for are probably not such as would satisfy you. My opinions have changed much on this point: I now look at intellectual companionship as quite a secondary matter, and should my good stars ever send me an affectionate, good-tempered and domestic wife, I shall care not one iota for accomplishments or even for education.[100]

Despite a promising start, Wallace's first romance, at age thirty-nine, was disastrous.

Soon after my return home in the spring of 1862 my oldest friend and schoolfellow, Mr. George Silk, introduced me to a small circle of his friends, who had formed a private chess club, and thereafter, while I lived in the vicinity of Kensington, I was invited to attend the meetings of the club. One of these friends was a Mr. L-,[101] a widower with two daughters, and a son who was at Cambridge University. I sometimes went there with Silk on Sunday afternoons, and after a few months was asked to call on them whenever I liked in the evening to play a game with Mr. L-. On these occasions the young ladies were present, and we had tea or supper together, and soon became very friendly. The eldest Miss L- was, I think, about seven or eight and twenty, very agreeable though quiet, pleasant looking, well educated and fond of art and literature, and I soon began to feel an affection for her, and to hope that she would become my wife. In about a year after my first visit there, thinking I was then sufficiently known, and being too shy to make a verbal offer, I wrote to her, describing my feelings and asking if she could in any way respond to my affection. Her reply was a negative, but not a very decided one. Evidently my undemonstrative manner had given her no intimation of my intentions. She concluded her letter, which was a very kind one, by begging that I would not allow her refusal to break off my visits to her father. At first I was inclined not to go again, but on showing the letter to my sister and mother, they thought the young lady was favourably disposed, and that I had better go on as before, and make another offer later on. Another year passed, and thinking I saw signs of a change in her feelings towards me, but fearing another refusal, I wrote to her father stating the whole circumstances, and asking him to ascertain his daughter's wishes, and if she was now favourable, to grant me a private interview. In reply I was asked to call on Mr. L-, who inquired as to my means, etc., told me that his daughter had a small income of her own, and asked that I should settle an equal amount on her. This was satisfactorily arranged, and at a subsequent meeting we were engaged. Everything went on smoothly for some months. We met two or three times a week, and after delays, owing to Miss L-'s ill-health and other causes, the wedding day was fixed and all details arranged. I had brought her to visit my mother and sister, and I was quite unaware of any cause of doubt or uncertainty when one day, on making my usual call, I was informed by the servant that Miss L- was not at home, that she had gone away that morning, and would write. I came home completely staggered, and the next morning had a letter from Mr. L-, saying that his daughter wished to break off the engagement and would write to me shortly. The blow was very severe, and I have never in my life experienced such intensely painful emotion.

When the letter came I was hardly more enlightened. The alleged cause was that I was silent as to myself and family, that I seemed to have something to conceal, and that I had told her nothing about a widow lady, a friend of my mother's, that I had almost been engaged to. All this was to me the wildest delusion. The lady was the widow of an Indian officer, very pleasant and good-natured, and very gossipy, but as utterly remote in my mind from all ideas of marriage as would have been an aunt or a grandmother. As to concealment, it was the furthest thing possible from my thoughts, but it never occurs to me at any time to talk about myself, even my own children say that they know nothing about my early life; but if anyone asks me and wishes to know, I am willing to tell all that I know or remember. I was dreadfully hurt.[102]   ॐ

Fortunately natural history intervened. "Mr. Mitten was an enthusiastic botanist and gardener, and knew every wild plant in the very rich district which surrounds the village [Hurstpierpoint, Sussex], and all his family were lovers of wild flowers. . . . This similarity of taste led to a close intimacy, and in the spring of the following year [1866] I was married to Mr. Mitten's eldest daughter [Annie], then about eighteen years old."[103] They had three children, Herbert Spencer (sic.), who was born in 1867 and died aged four, Violet, born 1868, and William, 1871. Perhaps Annie had no idea when she signed on with Wallace that his return to England had not fully exorcized her husband's wanderlust. The

Alfred Russel Wallace in later life
(from Wallace's *Revolt of
Democracy*)

family moved at least seven times, always within south-eastern
England, perhaps because each shift gave Wallace a new opportunity to engage "in the occupation I most enjoy - making a
garden".[104]

Despite the privations he endured in the Tropics, Wallace's
health was generally good. However, in 1887, on the way back
from Canada, he "was very unwell during the whole voyage, with
chest oppression, and asthma".[105] By 1896, he had "no expectation of living much longer"; his asthma had become chronic,
"together with violent palpitations on the least sudden exertion,
and frequent colds almost invariably followed by bronchitis".[106]
However, in 1899, he "obtained relief . . . in an altogether accidental way, if there are any 'accidents' in our lives".[107] A serendipitous encounter with a Dr Salisbury resulted in Wallace adopting
a low-starch diet consisting mainly of beef. It apparently worked,
but it must been something of an indignity for Wallace, a champion of vegetarianism.

＊ I am not a vegetarian, but I believe in it as certain to be adopted
in the future, and as essential to a higher social and moral state of
society. My reasons are:

(1) That far less land is needed to supply vegetable than to supply animal food.

(2) That the business of a butcher is, and would be, repulsive to all refined natures.

(3) That with proper arrangements for variety and good cookery, vegetable food is better for health of body and mind.[108] &

Wallace remained astonishingly productive throughout his life, slowing down only in his final year, 1913, when he was ninety. In a letter dated August 26 1913, he wrote, "The papers are wrong about me. I am not writing anything now; perhaps shall write no more. Too many letters and home business. Too much bothered with many slight ailments, which altogether keep me busy attending to them."[109] He died in his sleep at his home, "Old Orchard", in Wimborne, Dorset, on November 7 1913.

For Wallace, a fervent spiritualist to the end, death was merely a gateway to another form of existence.

& With all my belief in, and knowledge of, Spiritualism, I have, however, occasional qualms of doubt, the remnants of my original deeply ingrained scepticism; but my reason goes to support the psychical and spiritualistic phenomena in telling me that there *must* be a hereafter for us all.[110] &

ANTHOLOGY

Science

Evolution

Beginnings

Wallace traced his interest in evolution back to 1847. That summer, Wallace, then twenty-four and living in Neath, South Wales, was visited for a week by Henry Walter Bates. Their time was spent "chiefly in beetle collecting and in discussing various matters, and it must have been at this time that we talked over a proposed collecting trip to the tropics, but had not then decided where to go".[1] Over the following months, their plans for their trip to Brazil crystallized while they read about, and became interested in, contemporary ideas on evolution, or "the species question".

≈§ In a letter written [to Bates] November 9, I finish by asking: "Have you read 'Vestiges of the Natural History of Creation,'[2] or is it out of your line?" And in my next letter (December 28), having had Bates' reply to the question, I say: "I have rather a more favourable opinion of the 'Vestiges' than you appear to have. I do not consider it a hasty generalization, but rather as an ingenious hypothesis strongly supported by some striking facts and analogies, but which remains to be proved by more facts and the additional light which more research may throw upon the problem. It furnishes a subject for every observer of nature to attend to; every fact he observes will make either for or against it, and it thus serves both as an incitement to the collection of facts, and an object to which they can be applied when collected. Many eminent writers support the theory of the progressive development of animals and plants."[3]

. . .

These extracts from my early letters to Bates suffice to show that the great problem of the origin of species was already distinctly formulated in my mind; that I was not satisfied with the more or less vague solutions at that time offered; that I believed the conception of evolution through natural law so clearly formulated in the "Vestiges" to be, so far as it went, a true one; and that I firmly believed that a full and careful study of the facts of nature would ultimately lead to a solution of the mystery.[4] [1905] ❧

The Leicester town library had in 1844 supplied Wallace with another critical piece of background reading: "But perhaps the most important book I read was Malthus's 'Principles of Population,'[5] which I greatly admired for its masterly summary of facts and logical induction to conclusions. It was the first work I had yet read treating any of the problems of philosophical biology, and its main principles remained with me as a permanent possession, and twenty years later gave me the long-sought clue to the effective agent in the evolution of organic species."[6]

With his departure for the Tropics, Wallace substituted rainforests, rivers, and savannahs for books in his biological education. His four years collecting in the Amazon were followed in short order by an eight-year stint in South-east Asia. Considering the hardships he had to endure – disease and the logistical difficulties of biological collecting in the Tropics – and that he was all the time busy collecting specimens for a living,[7] it is remarkable that Wallace was able to think so creatively about the major biological conundrum of the day.

The "Sarawak Law"

Wallace's first major contribution to evolution came in 1855 with the so-called "Sarawak Law" paper. Having been introduced to Sir James Brooke, the independent ruler of the northern Borneo province of Sarawak, when he arrived in Singapore in 1854, Wallace frequently benefited from the self-styled Rajah's hospitality. It seems to have been a mutually appreciated arrangement as Brooke enjoyed having Wallace around. Spencer St John, Brooke's secretary and biographer, records Wallace's visits: "We had at this time in Sarawak the famous naturalist, traveller and philosopher, Mr. Alfred Wallace, who was then elaborating in his mind the theory which was simultaneously worked out by Darwin – the theory of the origin of species; if he could not convince us that our ugly neighbours, the orang-utans, were our ancestors, he pleased, instructed and delighted us by his clever and inexhaustible flow of talk – really good talk. The Rajah was pleased to have so clever a man with him and it excited his mind and brought out his brilliant ideas. . . . Our discussions were always either philosophical or religious. Fast and furious would flow the argument. . . ."[8]

Among Brooke's properties was a small cottage at the base of a fine forest-clad hill, Santubong, not far from his capital (now called Kuching). It was in this cottage, during the wet season, that Wallace wrote his "Sarawak Law" paper:

   I was quite alone, with one Malay boy as cook, and during the evenings and wet days I had nothing to do but to look over my books and ponder over the problem which was rarely absent from my thoughts. Having always been interested in the geographical distribution of animals and plants, having studied Swainson[9] and Humboldt, and having now myself a vivid impression of the fundamental differences between the Eastern and Western tropics; and having also read through such books as Bonaparte's "Conspectus,"[10] . . . and several catalogues of insects and reptiles in the British Museum (which I almost knew by heart), giving a mass of facts as to the distribution of animals over the whole world, it occurred to me that these facts had never been properly utilized as indications of the way in which species had come into existence. The great work of Lyell had furnished me with the main features of the succession of species in time, and by combining the two I thought that some valuable conclusions might be reached. I accordingly put my facts and ideas on paper, and the result seeming me to be of some importance, I sent it to *The Annals and Magazine of Natural History*, in which it appeared in the following September (1855). Its title was "On the Law which has regulated the Introduction of New Species," which law was briefly stated (at the end) as follows: "*Every species has come into existence coincident both in space and time with a pre-existing closely-allied species.*" This clearly pointed to some kind of evolution. It suggested the *when* and the *where* of its occurrence, and that it could only be through natural generation, as was also suggested in the "Vestiges"; but the *how* was still a secret only to be penetrated some years later.[11] [1905]

The "Sarawak Law" is an astonishing paper. Wallace's publications hitherto had been strictly empirical; they had not gone beyond mere descriptions of his observations. Even his book chronicling his entire Amazon experience seldom departs from the facts staring the author in the face. Suddenly, fully fledged, we see here for the first time Wallace the theorizer, the synthe-

Forest trees in Borneo (from
Wallace's *The Malay Archipelago*)

sizer. The basic premise of the paper is that the close relatives of
a given species tend to be found nearby both in space and in
time. Thus all the species within a genus – a group of closely
related species – are typically distributed in the same geographical
region, rather than being scattered all around the globe. Thus
too the closest relative of a fossil species is typically found in strata
from a nearby time horizon rather than from a vastly more
ancient or more modern one. The rationale is simple: an evolutionary process generates diversity, so we expect to find related
species close both geographically and chronologically. It is this
interplay of space and time that would mark much of Wallace's
biological thinking, and it is here, in his first full articulation of
it, that the idea is most clearly expressed.

Cut off from the scientific world, Wallace must have wondered
what impact his evolutionary bombshell was having in the scientific salons of Europe and the US. He was destined to be disappointed. The paper created little excitement, even among those,
like Darwin, who were best placed to appreciate its significance.
In fact, the first feedback Wallace received was from his agent
in London, Samuel Stevens, who was responsible for selling
the material Wallace sent back. "Soon after this article appeared,
Mr. Stevens wrote me that he had heard several naturalists

express regret that I was 'theorizing,' when what we had to do was to collect more facts."[12] One can hardly blame Stevens. He could sell specimens but not "theorizing".

Bates, writing from Brazil, was more appreciative: "I received about six months ago a copy of your paper in the *Annals* on 'The Laws which have Governed the Introduction of New Species' [*sic*]. I was startled at first to see you already ripe for the enunciation of the theory. You can imagine with what interest I read and studied it, and I must say that I think it is perfectly well done. The idea is like truth itself, so simple and obvious that those who read and understand it will be struck by its simplicity; and yet it is perfectly original. The reasoning is close and clear, and although so brief an essay, it is quite complete, embraces the whole difficulty, and anticipates and annihilates all objections. Few men will be in a condition to comprehend and appreciate the paper, but it will infallibly create for you a high and sound reputation."[13]

Before responding to Bates, Wallace heard from Darwin: "By your letter and even still more by your paper in the *Annals*, a year or more ago, I can plainly see that we have thought much alike and to a certain extent have come to similar conclusions. In regard to the paper in the *Annals*, I agree to the truth of almost every word of your paper; and I daresay that you agree with me that it is very rare to find oneself agreeing pretty closely with any theoretical paper; for it is lamentable how each man draws his own conclusions from the very same fact."[14]

A later letter from Darwin addressed Wallace's concerns about the lack of attention garnered by the "Sarawak Law" paper: "You say that you have been somewhat surprised at no notice having been taken of your paper in the *Annals*. I cannot say that I am; for so very few naturalists care for anything beyond the mere description of species. But you must not suppose that your paper has not been attended to: two very good men, Sir C. Lyell and Mr. E. Blyth[15] at Calcutta, specially called my attention to it."[16]

Wallace wrote to Bates from Amboyna (Ambon, Indonesia):

To persons who have not thought much on the subject I fear my paper on the succession of species will not appear so clear as it does to you. That paper is, of course, only the announcement of the theory, not its development. I have prepared the plan and written portions of an extensive work embracing the subject in all its bearings and endeavouring to prove what, in the paper, I have only indicated. . . .

I have been much gratified by a letter from Darwin, in which he says that he agrees with "almost every word" of my paper. He is now preparing for publication his great work on species and varieties, for which he has been collecting information twenty years. He may save me the trouble of writing the second part of my hypothesis by proving that there is no difference in nature between the origin of species and varieties, or he may give me trouble by arriving at another conclusion, but at all events his facts will be given for me to work upon. Your collections and my own will furnish most valuable material to illustrate and prove the universal applicability of the hypothesis.[17] [1858]   ❧

❧ ON THE LAW WHICH HAS REGULATED THE INTRODUCTION OF NEW SPECIES

Every naturalist who has directed his attention to the subject of the geographical distribution of animals and plants, must have been interested in the singular facts which it presents. Many of these facts are quite different from what would have been anticipated, and have hitherto been considered as highly curious, but quite inexplicable. None of the explanations attempted from the time of Linnæus[18] are now considered at all satisfactory; none of them have given a cause sufficient to account for the facts known at the time, or comprehensive enough to include all the new facts which have since been, and are daily being added. Of late years, however, a great light has been thrown upon the subject by geological investigations, which have shown that the present state of the earth, and the organisms now inhabiting it, are but the last stage of a long and uninterrupted series of changes which it has undergone, and consequently, that to endeavour to explain and account for its present condition without any reference to those changes (as has frequently been done) must lead to very imperfect and erroneous conclusions.

The facts proved by geology are briefly these:– That during an immense, but unknown period, the surface of the earth has undergone successive changes; land has sunk beneath the ocean, while fresh land has risen up from it; mountain chains have been elevated; islands have been formed into continents, and continents submerged till they have become islands; and these changes have taken place, not once merely, but perhaps hundreds, perhaps thousands of times:– That all these operations have been more or less continuous, but unequal in their progress, and during the whole series the organic life of the earth has undergone a corresponding alteration. This alteration also has been gradual, but complete; after a certain interval not a single species existing which had lived at the commencement of the period. This complete renewal of the forms of life also appears to have occurred several times:– That from the last of the Geological epochs to the present or Historical epoch, the change of organic life has been gradual: the first appearance of animals now existing can in many cases be traced, their numbers gradually increasing in the more recent formations, while other species continually die out and disappear, so that the present condition of the organic world is clearly derived by a natural process of gradual extinction and creation of species from that of the latest geological periods. We may therefore safely infer a like gradation and natural sequence from one geological epoch to another.

Now, taking this as a fair statement of the results of geological inquiry, we see that the present geographical distribution of life upon the earth must be the result of all the previous changes, both of the surface of the earth itself and of its inhabitants. Many causes no doubt have operated of which we must ever remain in ignorance, and we may therefore expect to find many details very difficult of explanation, and in attempting to give one, must allow ourselves to call into our service geological changes which it is highly probable may have occurred, though we have no direct evidence of their individual operation.

The great increase of our knowledge within the last twenty years, both of the present and past history of the organic world, has accumulated a body of facts which should afford a sufficient foundation for a comprehensive law embracing and explaining them all, and giving a direction to new researches. It is about ten years since the idea of such a law suggested itself to the writer of this paper, and he has since taken every opportunity of testing it by all the newly ascertained facts with which he has become acquainted, or has been able to observe himself. These have all served to convince him of the correctness of his hypothesis. Fully to enter into such a subject would occupy much space, and it is only in consequence of some views having been lately promulgated, he believes in a wrong direction, that he now ventures to present his ideas to the public, with only such obvious illustrations of the arguments and results as occur to him in a place far removed from all means of reference and exact information.

The following propositions in Organic Geography and Geology give the main facts on which the hypothesis is founded.

GEOGRAPHY

1. Large groups, such as classes and orders, are generally spread over the whole earth, while smaller ones, such as families and genera, are frequently confined to one portion, often to a very limited district.

2. In widely distributed families the genera are often limited in range; in widely distributed genera, well-marked groups of species are peculiar to each geographical district.

3. When a group is confined to one district, and is rich in species, it is almost invariably the case that the most closely allied species are found in the same locality or in closely adjoining localities, and that therefore the natural sequence of the species by affinity is also geographical.

4. In countries of a similar climate, but separated by a wide sea or lofty mountains, the families, genera and species of the one are often represented by closely allied families, genera and species peculiar to the other.

GEOLOGY

5. The distribution of the organic world in time is very similar to its present distribution in space.

6. Most of the larger and some small groups extend through several geological periods.

7. In each period, however, there are peculiar groups, found nowhere else, and extending through one or several formations.

8. Species of one genus, or genera of one family occurring in the same geological time are more closely allied than those separated in time.

9. As generally in geography no species or genus occurs in two very distant localities without being also found in intermediate places, so in geology the life of a species or genus has not been interrupted. In other words, no group or species has come into existence twice.

10. The following law may be deduced from these facts:– *Every species has come into existence coincident both in space and time with a pre-existing closely allied species.*

This law agrees with, explains and illustrates all the facts connected with the following branches of the subject:– 1st. The system of natural affinities. 2nd. The distribution of animals and plants in space. 3rd. The same in time, including all the phenomena of representative groups, and those which Professor Forbes[19] supposed to manifest polarity. 4th. The phenomena of rudimentary organs. We will briefly endeavour to show its bearing upon each of these.

If the law above enunciated be true, it follows that the natural series of affinities will also represent the order in which the several species came into existence, each one having had for its immediate antitype[20] a closely allied species existing at the time of its origin. It is evidently possible that two or three distinct species may have had a common antitype, and that each of these may again have become the antitypes from which other closely allied species were created. The effect of this would be, that so long as each species has had but one new species formed on its model, the line of affinities will be simple, and may be represented by placing the several species in direct succession in a straight line. But if two or more species have been independently formed on the plan of a common antitype, then the series of affinities will be compound, and can only be represented by a forked or many-branched line. Now, all attempts at a Natural classification and arrangement of organic beings show that both these plans have obtained in creation. Sometimes the series of affinities can be well represented for a space by a direct progression from species to species or from group to group, but it is generally found impossible so to continue. There constantly occur two or more modifications of an organ or modifications of two distinct organs, leading us on to two distinct series of species, which at length differ so much from each other as to form distinct genera or families. These are the parallel series or representative groups of naturalists, and they often occur in different countries, or are found fossil in different formations. They are said to have an analogy to each other when they are so far removed from their common antitype as to differ in many important points of structure, while they still preserve a family resemblance. We thus see how difficult it is to determine in every case whether a given relation is an analogy or an affinity, for it is evident that as we go back along the parallel or divergent series, towards the common antitype, the analogy which existed between the two groups becomes an affinity. We are also made aware of the difficulty of arriving at a true classification, even in a small and perfect group – in the actual state of nature it is almost impossible, the species being so numerous and the modifications of form and structure so varied, arising probably from the immense number of species which have served as antitypes for the existing species, and thus produced a complicated branching of the lines of affinity, as intricate as the twigs of a gnarled oak or the vascular system of the human body. Again, if we consider that we have only fragments of this vast system, the stem and main branches being represented by extinct species of which we have no knowledge, while a vast mass of limbs and boughs and minute twigs and scattered leaves is what we have to place in order, and determine the true position each originally occupied with regard to the others, the whole difficulty of the true Natural System of classification becomes apparent to us.

We shall thus find ourselves obliged to reject all those systems of classification which arrange species or groups in circles, as well as those which fix a definite number for the divisions of each group. The latter class have been very generally rejected by naturalists, as contrary to nature, notwithstanding the ability with which they have been advocated; but the circular system of affinities seems to have obtained a deeper hold, many eminent naturalists having to some extent adopted it. We have, however, never been able to find a case in which the circle has been closed by a direct and close affinity. In most cases a palpable analogy has been substituted, in others the affinity is very obscure or altogether doubtful. The complicated branching of the lines of affinities in extensive groups must also afford great facilities for giving a show of probability to any such purely artificial arrangements. Their death-blow was given by the admirable paper of the lamented Mr. Strickland,[21] published in the "Annals of Natural History," in which he so clearly showed the true synthetical method of discovering the Natural System.

If we now consider the geographical distribution of animals and plants upon the earth, we shall find all the facts beautifully in accordance with, and readily explained by, the present hypothesis. A country having species, genera, and whole families peculiar to it, will be the necessary result of its having been isolated for a long period, sufficient for many series of species to have been created on the type of pre-existing ones, which, as well as many of the earlier-formed species, have become extinct, and thus made the groups appear isolated. If in any case the antitype had an extensive range, two or more groups of species might have been formed, each varying from it in a different manner, and thus producing several representative or analogous groups. The *Sylviadæ* of Europe and the *Sylvicolidæ* of North America [birds], the *Heliconidæ* of South America and the *Euplœas* of the East [butterflies], the group of *Trogons* [birds] inhabiting Asia, and that peculiar to South America, are examples that may be accounted for in this manner.

Such phenomena as are exhibited by the Galapagos Islands, which contain little groups of plants and animals peculiar to themselves, but most nearly allied to those of South America, have not hitherto received any, even a conjectural explanation. The Galapagos are a volcanic group of high antiquity, and have probably never been more closely connected with the continent than they are at present. They must have been first peopled, like other newly-formed islands, by the action of winds and currents, and at a period sufficiently remote to have had the original species die out, and the modified prototypes only remain. In the same way we can account for the separate islands having each their peculiar species, either on the supposition that the same original emigration peopled the whole of the islands with the same species from which differently modified prototypes were created, or that the islands were successively peopled from each other, but that new species have been created in each on the plan of the pre-existing ones. St. Helena is a similar case of a very ancient island having obtained an entirely peculiar, though limited, flora. On the other hand, no example is known of an island which can be proved geologically to be of very recent origin (late in the Tertiary, for instance), and yet possesses generic or family groups, or even many species peculiar to itself.

When a range of mountains has attained a great elevation, and has so remained during a long geological period, the species of the two sides at and near their bases will be often very different, representative species of some genera occurring, and even whole genera being peculiar to one side only, as is remarkably seen in the case of the Andes and Rocky Mountains. A similar phænomenon occurs when an island has been separated from a continent at a very early period. The shallow sea between the Peninsula of Malacca [Peninsular Malaysia], Java, Sumatra and Borneo was probably a continent or large island at an early epoch, and may have become submerged as the volcanic ranges of Java and Sumatra were elevated. The organic results we see in the very considerable number of species of animals common to some or all of these countries, while at the same time a number of closely allied representative species exist peculiar to each, showing that a considerable period has elapsed since their separation. The facts of geographical distribution and of geology may thus mutually explain each other in doubtful cases, should the principles here advocated be clearly established.

In all those cases in which an island has been separated from a continent, or raised by volcanic or coralline action from the sea, or in which a mountain-chain has been elevated, in a recent geological epoch, the phenomena of peculiar groups or even of single representative species will not exist. Our own island is an example of this, its separation from the continent being geologically very recent, and we have consequently scarcely a species which is peculiar to it; while the Alpine range, one of the most recent mountain elevations, separates faunas and floras which scarcely differ more than may be due to climate and latitude alone.

The series of facts alluded to in Proposition 3, of closely allied species in rich groups being found geographically near each other, is most striking and important. Mr. Lovell Reeve[22] has well exemplified it in his able and interesting paper on the Distribution of the *Bulimi* [snails]. It is also seen in the Humming-birds and Toucans, little groups of two or three closely allied species being often found in the same or closely adjoining districts, as we have had the good fortune of personally verifying. Fishes give evidence of a similar kind: each great river has its peculiar genera, and in more extensive genera its groups of closely allied species. But it is the same throughout Nature; every class and order of animals will contribute similar facts. Hitherto no attempt has been made to explain these singular phenomena, or to show how they have arisen. Why are the genera of Palms and of Orchids in almost every case confined to one hemisphere? Why are the closely allied species of brown-backed Trogons all found in the East, and the green-backed in the West? Why are the Macaws and the Cockatoos similarly restricted? Insects furnish a countless number of analogous examples – the *Goliathi* [beetles] of Africa, the *Ornithopteræ* [butterflies] of the Indian islands, the *Heliconidæ* [butterflies] of South America, the *Danaidæ* [butterflies] of the East, and in all, the most closely allied species found in geographical proximity. The question forces itself upon every thinking mind, – why are these things so? They could not be as they are, had no law regulated their creation and dispersion. The law here enunciated not merely explains, but necessitates the facts we see to exist, while the vast and long-continued geological changes of the earth readily account for the exceptions and apparent discrepancies that here and there occur. The writer's object in putting forward his views in the present imperfect manner is to submit them to the test of other minds, and to be made aware of all the facts supposed to be inconsistent with them. As his hypothesis is one which claims acceptance solely as explaining and connecting facts which exist in nature, he expects facts alone to be brought to disprove it; not *à priori* arguments against its probability.

The phenomena of geological distribution are exactly analogous to those of geography. Closely allied species are found associated in the same beds, and the change from species to species appears to have been as gradual in time as in space. Geology, however, furnishes us with positive proof of the extinction and production of species, though it does not inform us how either has taken place. The extinction of species, however, offers but little difficulty, and the *modus operandi* has been well illustrated by Sir C. Lyell in his admirable "Principles."[23] Geological changes, however gradual, must occasionally have modified external conditions to such an extent as to have rendered the existence of certain species impossible. The extinction would in most cases be effected by a gradual dying-out, but in some instances there might have been a sudden destruction of a species of limited range. To discover how the extinct species have from time to time been replaced by new ones down to the very latest geological period, is the most difficult, and at the same time the most interesting problem in the natural history of the earth. The present inquiry, which seeks to eliminate from known facts a law which has determined, to a certain degree, what species could and did appear at a given epoch, may, it is hoped, be considered as one step in the right direction towards a complete solution of it.

Much discussion has of late years taken place on the question, whether the succession of life upon the globe has been from a lower to a higher degree of organization? The admitted facts seem to show that there has been a general, but not a detailed progression. Mollusca and Radiata [simple radially symmetrical animals like corals] existed before Vertebrata, and the progression from Fishes to Reptiles and Mammalia, and also from the lower mammals to the higher, is indisputable. On the other hand, it is said that the Mollusca and Radiata of the very earliest periods were more highly organized than the great mass of those now existing, and that the very first fishes that have been discovered are by no means the lowest organized of the class. Now it is believed the present hypothesis will harmonize with all these facts, and in a great measure serve to explain them; for though it may appear to some readers essentially a theory of progression, it is in reality only one of gradual change. It is, however, by no means difficult to show that a real progression in the scale of organization is perfectly consistent with all the appearances, and even with apparent retrogression, should such occur.

Returning to the analogy of a branching tree, as the best mode of representing the natural arrangement of species and their successive creation, let us suppose that at an early geological epoch any group (say a class of the Mollusca) has attained to a great richness of species and a high organization. Now let this great branch of allied species, by geological mutations, be completely or partially destroyed. Subsequently a new branch springs from the same trunk, that is to say, new species are successively created, having for their antitypes the same lower organized species which had served as the antitypes for the former group, but which have survived the modified conditions which destroyed it. This new group being subject to these altered conditions, has modifications of structure and organization given to it, and becomes the representative group of the former one in another geological formation. It may, however, happen, that though later in time, the new series of species may never attain to so high a degree of organization as those preceding it, but in its turn become extinct, and give place to yet another modification from the same root, which may be of higher or lower organization, more or less numerous in species, and more or less varied in form and structure than either of those which preceded it. Again, each of these groups may not have become totally extinct, but may have left a few species, the modified prototypes of which have existed in each succeeding period, a faint memorial of their former grandeur and luxuriance. Thus every case of apparent retrogression may be in reality a progress, though an interrupted one: when some monarch of the forest loses a limb, it may be replaced by a feeble and sickly substitute. The foregoing remarks appear to apply to the case of the Mollusca, which, at a very early period, had reached a high organization and a great development of forms and species in the Testaceous Cephalopoda [ammonites]. In each succeeding age modified species and genera replaced the former ones which had become extinct, and as we approach the present æra but few and small representatives of the group remain, while the Gasteropods and Bivalves have acquired an immense preponderance. In the long series of changes the earth has undergone, the process of peopling it with organic beings has been continually going on, and whenever any of the higher groups have become nearly or quite extinct, the lower forms which have better resisted the modified physical conditions have served as the antitypes on which to found the new races. In this manner alone, it is believed, can the representative groups at successive periods, and the risings and fallings in the scale of organization, be in every case explained.

The hypothesis of polarity, recently put forward by Professor Edward Forbes to account for the abundance of generic forms at a very early period and at present, while in the intermediate epochs there is a gradual diminution and impoverishment, till the minimum occurred at the confines of the Palæozoic and Secondary epochs, appears to us quite unnecessary, as the facts may be readily accounted for on the principles already laid down. Between the Palæozoic and Neozoic periods of Professor Forbes, there is scarcely a species in common, and the greater part of the genera and families also disappear to be replaced by new ones. It is almost universally admitted that such a change in the organic world must have occupied a vast period of time. Of this interval we have no record; probably because the whole area of the early formations now exposed to our researches was elevated at the end of the Palæozoic period, and remained so through the interval required for the organic changes which resulted in the fauna and flora of the Secondary period. The records of this interval are buried beneath the ocean which covers three-fourths of the globe. Now it appears highly probable that a long period of quiescence or stability in the physical conditions of a district would be most favourable to the existence of organic life in the greatest abundance, both as regards individuals and also as to variety of species and generic groups, just as we now find that the places best adapted to the rapid growth and increase of individuals also contain the greatest profusion of species and the greatest variety of forms, the tropics in comparison with the temperate and arctic regions. On the other hand, it seems no less probable that a change in the physical conditions of a district, even small in amount if rapid, or even gradual if to a great amount, would be highly unfavourable to the existence of individuals, might cause the extinction of many species, and would probably be equally unfavourable to the creation of new ones. In this too we may find an analogy with the present state of our earth, for it has been shown to be the violent extremes and rapid changes of physical conditions, rather than the actual mean state in the temperate and frigid zones, which renders them less prolific than the tropical regions, as exemplified by the great distance beyond the tropics to which tropical forms penetrate when the climate is equable, and also by the richness in species and forms of tropical mountain regions which principally differ from the temperate zone in the uniformity of their climate. However this may be, it seems a fair assumption that during a period of geological repose the new species which we know to have been created would have appeared, that the creations would then exceed in number the extinctions, and therefore the number of species would increase. In a period of geological activity, on the other hand, it seems probable that the extinctions might exceed the creations, and the number of species consequently diminish. That such effects did take place in connexion with the causes to which we have imputed them, is shown in the case of the Coal formation, the faults and contortions of which show a period of great activity and violent convulsions, and it is in the formation immediately succeeding this that the poverty of forms of life is most apparent. We have then only to suppose a long period of somewhat similar action during the vast unknown interval at the termination of the Palæozoic period, and then a decreasing violence or rapidity through the Secondary period, to allow for the gradual repopulation of the earth with varied forms, and the whole of the facts are explained. We thus have a clue to the increase of the forms of life during certain periods, and their decrease during others, without recourse to any causes but those we know to have existed, and to effects fairly deducible from them. The precise manner in which the geological changes of the early formations were effected is so extremely obscure, that when we can explain important facts by a retardation at one time and an acceleration at another of a process which we know from its nature and from observation to have been unequal, – a cause so simple may surely be preferred to one so obscure and hypothetical as polarity.

I would also venture to suggest some reasons against the very nature of the theory of Professor Forbes. Our knowledge of the organic world during any geological epoch is necessarily very imperfect. Looking at the vast numbers of species and groups that have been discovered by geologists, this may be doubted; but we should compare their numbers not merely with those that now exist upon the earth, but with a far larger amount. We have no reason for believing that the number of species on the earth at any former period was much less than at present; at all events the aquatic portion, with which geologists have most acquaintance, was probably often as great or greater. Now we know that there have been many complete changes of species; new sets of organisms have many times been introduced in place of old ones which have become extinct, so that the total amount which have existed on the earth from the earliest geological period must have borne about the same proportion to those now living, as the whole human race who have lived and died upon the earth, to the population at the present time. Again, at each epoch, the whole earth was no doubt, as now, more or less the theatre of life, and as the successive generations of each species died, their exuviæ and preservable parts would be deposited over every portion of the then existing seas and oceans, which we have reason for supposing to have been more, rather than less, extensive than at present. In order then to understand our possible knowledge of the early world and its inhabitants, we must compare, not the area of the whole field of our geological researches with the earth's surface, but the area of the examined portion of each formation separately with the whole earth. For example, during the Silurian period all the earth was Silurian, and animals were living and dying, and depositing their remains more or less over the whole area of the globe, and they were probably (the species at least) nearly as varied in different latitudes and longitudes as at present. What proportion do the Silurian districts bear to the whole surface of the globe, land and sea (for far more extensive Silurian districts probably exist beneath the ocean than above it), and what portion of the known Silurian districts has been actually examined for fossils? Would the area of rock actually laid open to the eye be the thousandth or the ten-thousandth part of the earth's surface? Ask the same question with regard to the Oolite [limestone] or the Chalk, or even to particular beds of these when they differ considerably in their fossils, and you may then get some notion of how small a portion of the whole we know.

But yet more important is the probability, nay, almost the certainty, that whole formations containing the records of vast geological periods are entirely buried beneath the ocean, and for ever beyond our reach. Most of the gaps in the geological series may thus be filled up, and vast numbers of unknown and unimaginable animals, which might help to elucidate the affinities of the numerous isolated groups which are a perpetual puzzle to the zoologist, may there be buried, till future revolutions may raise them in their turn above the waters, to afford materials for the study of whatever race of intelligent beings may then have succeeded us. These considerations must lead us to the conclusion, that our knowledge of the whole series of the former inhabitants of the earth is necessarily most imperfect and fragmentary, – as much so as our knowledge of the present organic world would be, were we forced to make our collections and observations only in spots equally limited in area and in number with those actually laid open for the collection of fossils. Now, the hypothesis of Professor Forbes is essentially one that assumes to a great extent the *completeness* of our knowledge of the *whole series* of organic beings which have existed on the earth. This appears to be a fatal objection to it, independently of all other considerations. It may be said that the same objections exist against every theory on such a subject, but this is not necessarily the case. The hypothesis put forward in this paper depends in no degree upon the completeness of our knowledge of the former condition of the organic world, but takes what facts we have as fragments of a vast whole, and deduces from them something of the nature and proportions of that whole which we can never know in detail. It is founded upon isolated groups of facts, recognizes their isolation, and endeavours to deduce from them the nature of the intervening portions.

Another important series of facts, quite in accordance with, and even necessary deductions from, the law now developed, are those of *rudimentary organs*. That these really do exist, and in most cases have no special function in the animal œconomy, is admitted by the first authorities in comparative anatomy. The minute limbs hidden beneath the skin in many of the snake-like lizards, the anal hooks of the boa constrictor, the complete series of jointed finger-bones in the paddle of the Manatus [Manatee] and whale, are a few of the most familiar instances. In botany a similar class of facts has long been recognized. Abortive stamens, rudimentary floral envelopes and undeveloped carpels [reproductive parts of flowers], are of the most frequent occurrence. To every thoughtful naturalist the question must arise, What are these for? What have they to do with the great laws of creation? Do they not teach us something of the system of Nature? If each species has been created independently, and without any necessary relations with pre-existing species, what do these rudiments, these apparent imperfections mean? There must be a cause for them; they must be the necessary results of some great natural law. Now, if, as it has been endeavoured to be shown, the great law which has regulated the peopling of the earth with animal and vegetable life is, that every change shall be gradual; that no new creature shall be formed widely differing from anything before existing; that in this, as in everything else in Nature, there shall be gradation and harmony, – then these rudimentary organs are necessary, and are an essential part of the system of Nature. Ere the higher Vertebrata were formed, for instance, many steps were required, and many organs had to undergo modifications from the rudimental condition in which only they had as yet existed. We still see remaining an antitypal sketch of a wing adapted for flight in the scaly flapper of the penguin, and limbs first concealed beneath the skin, and then weakly protruding from it, were the necessary gradations before others should be formed fully adapted for locomotion. Many more of these modifications should we behold, and more complete series of them, had we a view of all the forms which have ceased to live. The great gaps that exist between fishes, reptiles, birds and mammals would then, no doubt, be softened down by intermediate groups, and the whole organic world would be seen to be an unbroken and harmonious system.

It has now been shown, though most briefly and imperfectly, how the law that *"Every species has come into existence coincident both in time and space with a pre-existing closely allied species,"* connects together and renders intelligible a vast number of independent and hitherto unexplained facts. The natural system of arrangement of organic beings, their geographical distribution, their geological sequence, the phenomena of representative and substituted groups in all their modifications, and the most singular peculiarities of anatomical structure, are all explained and illustrated by it, in perfect accordance with the vast mass of facts which the researches of modern naturalists have brought together, and, it is believed, not materially opposed to any of them. It also claims a superiority over previous hypotheses, on the ground that it not merely explains, but necessitates what exists. Granted the law, and many of the most important facts in Nature could not have been otherwise, but are almost as necessary deductions from it, as are the elliptic orbits of the planets from the law of gravitation.[24] [1855] &

Following up his comments on "rudimentary organs", Wallace observed, in an article on the natural history of the orang-utan, that the secret to understanding evolution lay in determining the significance of both adaptive and non-adaptive characters ("anomalies"). Some characteristics serve little function, and persist solely because they have been inherited from ancestors for which, presumably, they did serve a purpose. Wallace notably abandoned this sophisticated viewpoint in favour of a more simple-minded view – that indeed "every modification exists solely for some such use" – in his later arguments on human evolution.

& The separate species of which the organic world consists being parts of a whole, we must suppose some dependence of each upon all; some general design which has determined the details, quite independently of individual necessities. We look upon the anomalies, the eccentricities, the exaggerated or diminished development of certain parts, as indications of a general system of nature, by a careful study of which we may learn much that is at present hidden from us; and we believe that the constant practice of imputing, right or wrong, some use to the individual, of every part of its structure, and even of inculcating the doctrine that every modification exists solely for some such use, is an error fatal to our complete appreciation of all the variety, the beauty, and the harmony of the organic world.[25] [1856] ৯১

The "Ternate Paper"

Wallace's most famous contribution to science, his discovery of natural selection, came in February 1858 while he was collecting in the Moluccan Islands. Because the article is signed "Ternate, February, 1858", it has traditionally been known as the "Ternate Paper". Close analysis of his field notes and other writings[26] has, however, suggested that the Ternate paper was in fact written not on Ternate, but on the neighbouring island of Gilolo (modern Helmahera). He probably referred to Ternate simply because his main base was there at the time. Wallace wrote several accounts of what happened, but all of them a good many years after the events. This one comes from *My Life*:

১৯ It was while waiting at Ternate in order to get ready for my next journey, and to decide where I should go, that the idea already referred to [i.e. natural selection] occurred to me. It has been shown how, for the preceding eight or nine years, the great problem of the origin of species had been continually pondered over, and how my varied observations and study had been made use of to lay the foundation for its full discussion and elucidation. My paper written at Sarawak rendered it certain to my mind that the change had taken place by natural succession and descent – one species becoming changed either slowly or rapidly into another. But the exact process of the change and the causes which led to it were absolutely unknown and appeared almost inconceivable. The great difficulty was to understand how, if one species was gradually changed into another, there continued to be so many quite distinct species, so many which differed from their nearest allies by slight yet perfectly definite and constant characters. One would expect that if it was a law of nature that species were continually changing so as to become in time new and distinct species, the world would be full of an inextricable mixture of various slightly different forms, so that the well-defined and constant species we see would not exist. Again, not only are species, as a rule, separated from each other by distinct external characters, but they almost always differ also to some degree in their food, in the places they frequent, in their habits and instincts, and all these characters are quite as definite and constant as are the external characters. The problem then was, not only how and why do species change, but how and why do they change into new and well-defined species, distinguished from each other in so many ways; why and how do they become so exactly adapted to distinct modes of life; and why do all the intermediate grades die out (as geology shows they have died out) and leave only clearly defined and well-marked species, genera, and higher groups of animals.

Now, the new idea or principle which Darwin had arrived at twenty years before, and which occurred to me at this time, answers all these questions and solves all these difficulties, and it is because it does so, and also because it is in itself self-evident and absolutely certain, that it has been accepted by the whole scientific world as affording a true solution of the great problem of the origin of species.

At the time in question I was suffering from a sharp attack of intermittent fever [probably malaria], and every day during the cold and succeeding hot fits had to lie down for several hours, during which time I had nothing to do but to think over any subjects then particularly interesting me. One day something brought to my recollection Malthus's "Principles of Population," which I had read about twelve years before. I thought of his clear exposition of "the positive checks to increase" disease, accidents, war, and famine – which keep down the population of savage races to so much lower an average than that of more civilized peoples. It then occurred to me that these causes or their equivalents are continually acting in the case of animals also, and as animals usually breed much more rapidly than does mankind, the destruction every year from these causes must be enormous in order to keep down the numbers of each species, since they evidently do not increase regularly from year to year, as otherwise the world would long ago have been densely crowded with those that breed most quickly. Vaguely thinking over the enormous and constant destruction which this implied, it occurred to me to ask the question, Why do some die and some live? And the answer was clearly, that on the whole the best fitted live. From the effects of disease the most healthy escaped; from enemies, the strongest, the swiftest, or the most cunning; from famine, the best hunters or those with the best digestion; and so on. Then it suddenly flashed upon me that this self-acting process would necessarily *improve the race*, because in every generation the inferior would inevitably be killed off and the superior would remain – that is, *the fittest would survive*. Then at once I seemed to see the whole effect of this, that when changes of land and sea, or of climate, or of food supply, or of enemies occurred – and we know that such changes have always been taking place – and considering the amount of individual variation that my experience as a collector had shown me to exist, then it followed that all the changes necessary for the adaptation of the species to the changing conditions would be brought about; and as great changes in the environment are always slow, there would be ample time for the change to be effected by the survival of the best fitted in every generation. In this way every part of an animal's organization could be modified exactly as required, and in the very process of this modification the unmodified would die out, and thus the *definite* characters and the clear *isolation* of each new species would be explained. The more I thought over it the more I became convinced that I had at length found the long-sought-for law of nature that solved the problem of the origin of species. For the next hour I thought over the deficiencies in the theories of Lamarck[27] and of the author of the "Vestiges," and I saw that my new theory supplemented these views and obviated every important difficulty. I waited anxiously for the termination of my fit so that I might at once make notes for a paper on the subject. The same evening I did this pretty fully, and on the two succeeding evenings wrote it out carefully in order to send it to Darwin by the next post, which would leave in a day or two.

I wrote a letter to him in which I said that I hoped the idea would be as new to him as it was to me, and that it would supply the missing factor to explain the origin of species. I asked him if he thought it sufficiently important to show to Sir Charles Lyell, who had thought so highly of my former paper.[28] [1905]

ON THE TENDENCY OF VARIETIES TO DEPART INDEFINITELY FROM THE ORIGINAL TYPE

One of the strongest arguments which have been adduced to prove the original and permanent distinctness of species is, that *varieties* produced in a state of domesticity are more or less unstable, and often have a tendency, if left to themselves, to return to the normal form of the parent species; and this instability is considered to be a distinctive peculiarity of all varieties, even of those occurring among wild animals in a state of nature, and to constitute a provision for preserving unchanged the originally created distinct species.

In the absence or scarcity of facts and observations as to *varieties* occurring among wild animals, this argument has had great weight with naturalists, and has led to a very general and somewhat prejudiced belief in the stability of species. Equally general, however, is the belief in what are called "permanent or true varieties," – races of animals which continually propagate their like, but which differ so slightly (although constantly) from some other race, that the one is considered to be a *variety* of the other. Which is the *variety* and which the original *species*, there is generally no means of determining, except in those rare cases in which one race has been known to produce an offspring unlike itself and resembling the other. This, however, would seem quite incompatible with the "permanent invariability of species," but the difficulty is overcome by assuming that such varieties have strict limits, and can never again vary further from the original type, although they may return to it, which, from the analogy of the domesticated animals, is considered to be highly probable, if not certainly proved.

It will be observed that this argument rests entirely on the assumption, that *varieties* occurring in a state of nature are in all respects analogous to or even identical with those of domestic animals, and are governed by the same laws as regards their permanence of further variation. But this is the object of the present paper to show that this assumption is altogether false, that there is a general principle in nature which will cause many *varieties* to survive the parent species, and to give rise to successive variations departing further and further from the original type, and which also produces, in domesticated animals, the tendency of varieties to return to the parent form.

The life of wild animals is a struggle for existence. The full exertion of all their faculties and all their energies is required to preserve their own existence and provide for that of their infant offspring. The possibility of procuring food during the least favourable seasons, and of escaping the attacks of their most dangerous enemies, are the primary conditions which determine the existence both of individuals and of entire species. These conditions will also determine the population of a species; and by a careful consideration of all the circumstances we may be enabled to comprehend, and in some degree to explain, what at first sight appears so inexplicable – the excessive abundance of some species, while others closely allied to them are very rare.

The general proportion that must obtain between certain groups of animals is readily seen. Large animals cannot be so abundant as small ones; the carnivora must be less numerous than the herbivora; eagles and lions can never be so plentiful as pigeons and antelopes; the wild asses of the Tartarian deserts cannot equal in numbers the horses of the more luxuriant prairies and pampas of America. The greater or less fecundity of an animal is often considered to be one of the chief causes of its abundance or scarcity; but a consideration of these facts will show us that it really has little or nothing to do with the matter. Even the least prolific of animals would increase rapidly if unchecked, whereas it is evident that the animal population of the globe must be stationary, or perhaps, through the influence of man, decreasing. Fluctuations there may be; but permanent increase, except in restricted localities, is almost impossible. For example, our own observation must convince us that birds do not go on increasing every year in a geometrical ratio, as they would do, were there not some powerful check to their natural increase. Very few birds produce less than two young ones each year, while many have six, eight, or ten; four will certainly be below the average; and if we suppose that each pair produce young only four times in their life, that will also be below the average, supposing them not to die either by violence or want of food. Yet at this rate how tremendous would be the increase in a few years from a single pair! A simple calculation will show that in fifteen years each pair of birds would have increased to nearly ten millions! Whereas we have no reason to believe that the number of birds of any country increases at all in fifteen or in one hundred and fifty years. With such powers of increase the population must have reached its limits, and have become stationary, in a very few years after the origin of each species. It is evident, therefore, that each year an immense number of birds must perish – as many in fact as are born; and as in the lowest calculation the progeny are each year twice as numerous as their parents, it follows that, whatever be the average number of individuals existing in any given country, *twice that number must perish annually*, – a striking result, but one which seems at least highly probable, and is perhaps under rather than over the truth. It would therefore appear that, as far as the continuance of the species and the keeping up the average number of individuals are concerned, large broods are superfluous. On the average all above *one* become food for hawks and kites, wild cats and weasels, or perish of cold and hunger as winter comes on. This is strikingly proved by the case of particular species; for we find that their abundance in individuals bears no relation whatever to their fertility in producing offspring. Perhaps the most remarkable instance of an immense bird population is that of the passenger pigeon of the United States, which lays only one, or at most two eggs, and is said to rear generally but one young one. Why is this bird so extraordinarily abundant,[29] while others producing two or three times as many young are much less plentiful? The explanation is not difficult. The food most congenial to this species, and on which it thrives best, is abundantly distributed over a very extensive region, offering such differences of soil and climate, that in one part or another of the area the supply never fails. The bird is capable of a very rapid and long-continued flight, so that it can pass without fatigue over the whole of the district it inhabits, and as soon as the supply of food begins to fail in one place is able to discover a fresh feeding-ground. This example strikingly shows us that the procuring a constant supply of wholesome food is almost the sole condition requisite for ensuring the rapid increase of a given species, since neither the limited fecundity, nor the unrestricted attacks of birds of prey and of man are here sufficient to check it. In no other birds are these peculiar circumstances so strikingly combined. Either their food is more liable to failure, or they have not sufficient power of wing to search for it over an extensive area, or during some season of the year it becomes very scarce, and less wholesome substitutes have to be found; and thus, though more fertile in offspring, they can never increase beyond the supply of food in the least favourable seasons. Many birds can only exist by migrating, when their food becomes scarce, to regions possessing a milder, or at least a different climate, though, as these migrating birds are seldom excessively abundant, it is evident that the countries they visit are still deficient in a constant and abundant supply of wholesome food. Those whose organization does not permit them to migrate when their food becomes periodically scarce, can never attain a large population. This is probably the reason why woodpeckers are scarce with us, while in the tropics they are among the most abundant of solitary birds. Thus the house sparrow is more abundant than the redbreast, because its food is more constant and plentiful, – seeds of grasses being preserved during the winter, and our farm-yards and stubble-fields furnishing an almost inexhaustible supply. Why, as a general rule, are aquatic, and especially sea birds, very numerous in individuals? Not because they are more prolific than the others, generally the contrary; but because their food never fails, the sea-shores and river-banks daily swarming with a fresh supply of small mollusca and crustacea. Exactly the same law applies to mammals. Wild cats are prolific and have few enemies; why then are they never as abundant as rabbits? The only intelligible answer is, that their supply of food is more precarious. It appears evident, therefore, that so long as a country remains physically unchanged, the numbers of its animal population cannot materially increase. If one species does so, some others requiring the same kind of food must diminish in proportion. The numbers that die annually must be immense; and as the individual existence of each animal depends upon itself, those that must die must be the weakest – the very young, the aged, and the diseased, – while those that prolong their existence can only be the most perfect in health and vigour – those who are best able to obtain food regularly, and avoid their numerous enemies. It is, as we commenced by remarking, "a struggle for existence," in which the weakest and least perfectly organized must always succumb.

Now it is clear that what takes place among the individuals of a species must also occur among the several allied species of a group, – viz. that those which are best adapted to obtain a regular supply of food, and to defend themselves against the attacks of their enemies and the vicissitudes of the seasons must necessarily obtain and preserve a superiority in population; while those species which from some defect of power or organization are the least capable of counteracting the vicissitudes of food supply, &c., must diminish in numbers, and, in extreme cases, become altogether extinct. Between these extremes the species will present various degrees of capacity for ensuring the means of preserving life; and it is thus we account for the abundance or rarity of species. Our ignorance will generally prevent us from accurately tracing the effects to their causes; but could we become perfectly acquainted with the organization and habits of the various species of animals, and could we measure the capacity of each for performing the different acts necessary to its safety and existence under all the varying circumstances by which it is surrounded, we might be able even to calculate the proportionate abundance of individuals which is the necessary result.

If now we have succeeded in establishing these two points – 1st, *that the animal population of a country is generally stationary, being kept down by a periodical deficiency of food, and other checks*; and, 2nd, *that the comparative abundance or scarcity of the individuals of the several species is entirely due to their organization and resulting habits, which, rendering it more difficult to procure a regular supply of food and to provide for their personal safety in some cases than in others, can only be balanced by a difference in the population which have to exist in a given area* – we shall be in a condition to proceed to the consideration of varieties, to which the preceding remarks have a direct and very important application.

Most or perhaps all the variations from the typical form of a species must have some definable effect, however slight, on the habits or capacities of the individuals. Even a change of colour might, by rendering them more or less distinguishable, affect their safety; a greater or less development of hair might modify their habits. More important changes, such as an increase in the power or dimensions of the limbs or any of the external organs, would more or less affect their mode of procuring food or the range of country which they inhabit. It is also evident that most changes would affect, either favourably or adversely, the powers of prolonging existence. An antelope with shorter or weaker legs must necessarily suffer more from the attacks of the feline carnivora; the passenger pigeon with less powerful wings would sooner or later be affected in its powers of procuring a regular supply of food; and in both cases the result must necessarily be a diminution of the population of the modified species. If, on the other hand, any species should produce a variety having slightly increased powers of preserving existence, that variety must inevitably in time acquire a superiority in numbers. These results must follow as surely as old age, intemperance, or scarcity of food produces an increased mortality. In both cases there may be many individual exceptions; but on the average the rule will invariably be found to hold good. All varieties will therefore fall into two classes – those which under the same conditions would never reach the population of the parent species, and those which would in time obtain and keep a numerical superiority. Now let some alteration of physical conditions occur in the district – a long period of drought, a destruction of vegetation by locusts, the irruption of some new carnivorous animal seeking "pastures new" – any change in fact tending to render existence more difficult to the species in question, and taking its utmost powers to avoid complete extermination; it is evident that, of all the individuals composing the species, those forming the least numerous and most feebly organized variety would suffer first, and, were the pressure severe, must soon become extinct. The same causes continuing in action, the parent species would next suffer, would gradually diminish in numbers, and with a recurrence of similar unfavourable conditions might also become extinct. The superior variety would then alone remain, and on a return to favourable circumstances would rapidly increase in numbers and occupy the place of the extinct species and variety.

The *variety* would now have replaced the *species*, of which it would be a more perfectly developed and more highly organized form. It would be in all respects better adapted to secure its safety, and to prolong its individual existence and that of the race. Such a variety *could not* return to the original form; for that form is an inferior one, and could never compete with it for existence. Granted, therefore, a "tendency" to reproduce the original type of species, still the variety must ever remain preponderant in numbers, and under adverse physical conditions *again alone survive*. But this new, improved, and populous race might itself, in course of time, give rise to new varieties, exhibiting several diverging modifications of form, any of which, tending to increase the facilities for preserving existence, must, by the same general law, in their turn become predominant. Here, then, we have *progression and continued divergence* deduced from the general laws which regulate the existence of animals in a state of nature, and from the undisputed fact that varieties do frequently occur. It is not, however, contended that this result would be invariable; a change of physical conditions in the district might at times materially modify it, rendering the race which had been the most capable of supporting existence under the former conditions now the least so, and even causing the extinction of the newer and, for a time, superior race, while the old or parent species and its first inferior varieties continued to flourish. Variations in unimportant parts might also occur, having no perceptible effect of the life-preserving powers; and the varieties so furnished might run a course parallel with the parent species, either giving rise to further variations or returning to the former type. All we argue for is, that certain varieties have a tendency to maintain their existence longer than the original species, and this tendency must make itself felt; for though the doctrine of chances or averages can never be trusted to on a limited scale, yet, if applied to high numbers, the results come nearer to what theory demands, and, as we approach to an infinity of examples, becomes strictly accurate. Now the scale on which nature works is so vast – the numbers of individuals and periods of time with which she deals approach so near to infinity – that any cause, however slight, and however liable to be veiled and counter-acted by accidental circumstances, must in the end produce its full legitimate results.

Let us now turn to domesticated animals, and inquire how varieties produced among them are affected by the principles here enunciated. The essential difference in the condition of wild and domestic animals is this – that among the former, their well-being and very existence depend upon the full exercise and healthy condition of all their senses and physical powers, whereas, among the latter, these are only partially exercised, and in some cases are absolutely unused. A wild animal has to search, and often to labour, for every mouthful of food – to exercise sight, hearing, and smell in seeking it, and in avoiding dangers, in procuring shelter from the inclemency of the seasons, and in providing for the subsistence and safety of its offspring. There is no muscle of its body that is not called into daily and hourly activity; there is no sense or faculty that is not strengthened by continual exercise. The domestic animal, on the other hand, has food provided for it, is sheltered, and often confined, to guard against the vicissitudes of the seasons, is carefully secured from the attacks of its natural enemies, and seldom even rears its young without human assistance. Half of its senses and faculties are quite useless; and the other half are but occasionally called into feeble exercise, while even its muscular system is only irregularly called into action.

Now when a variety of such an animal occurs, having increased power or capacity in any organ or sense, such increase is totally useless, is never called into action, and may even exist without the animal ever becoming aware of it. In the wild animal, on the contrary, all its faculties and powers being brought into full action for the necessities of existence, any increase becomes immediately available, is strengthened by exercise, and must even slightly modify the food, the habits, and the whole economy of the race. It creates as it were a new animal, one of superior powers, and which will necessarily increase in numbers and outlive those inferior to it.

Again, in the domesticated animal all variations have an equal chance of continuance; and those which would decidedly render a wild animal unable to compete with its fellows and continue its existence are no disadvantage whatever in a state of domesticity. Our quickly fattening pigs, short-legged sheep, pouter pigeons, and poodle dogs could never have come into existence in a state of nature, because the very first step towards such inferior forms would have led to rapid extinction of the race; still less could they now exist in competition with their wild allies. The great speed but slight endurance of the race horse, the unwieldy strength of the ploughman's team, would both be useless in a state of nature. If turned wild on the pampas, such animals would probably soon become extinct, or under favourable circumstances might each lose those extreme qualities which would never be called into action, and in a few generations would revert to a common type, which must be that in which the various powers and faculties are so proportioned to each other as to be best adapted to procure food and secure safety, – that in which by the full exercise of every part of his organization the animal can alone continue to live. Domestic varieties, when turned wild, *must* return to something near the type of the original wild stock, *or become altogether extinct.*

We see, then, that no inferences as to varieties in a state of nature can be deduced from the observation of those occurring among domestic animals. The two are so much opposed to each other in every circumstance of their existence, that what applies to the one is almost sure not to apply to the other.

Domestic animals are abnormal, irregular, artificial; they are subject to varieties which never occur and never can occur in a state of nature; their very existence depends altogether on human care; so far are many of them removed from that just proportion of faculties, that true balance of organization, by means of which alone an animal left to its own resources can preserve its existence and continue its race.

The hypothesis of Lamarck – that progressive changes in species have been produced by the attempts of animals to increase the development of their own organs, and thus modify their structure and habits – has been repeatedly and easily refuted by all writers on the subject of varieties and species, and it seems to have been considered that when this was done the whole question has been finally settled; but the view here developed renders such an hypothesis quite unnecessary, by showing that similar results must be produced by the action of principles constantly at work in nature. The powerful retractile talons of the falcon- and the cat-tribes have not been produced or increased by the volition of those animals; but among different varieties which occurred in the earlier and less highly organized forms of these groups, *those always survived longest which had the greatest facilities for seizing their prey.* Neither did the giraffe acquire its long neck by desiring to reach the foliage of the more lofty shrubs, and constantly stretching its neck for the purpose, but because any varieties which occurred among its antitypes with a longer neck than usual *at once secured a fresh range of pasture over the same ground as their shorter-necked companions, and on the first scarcity of food were thereby enabled to outlive them.* Even the peculiar colours of many animals, especially insects, so closely resembling the soil or the leaves or the trunks on which they habitually reside, are explained on the same principle; for though in the course of ages varieties of many tints may have occurred, *yet those races having colours best adapted to conceal-ment from their enemies would inevitably survive the longest.* We have also here an acting cause to account for that balance so often observed in nature, – a deficiency in one set of organs always being compen-sated by an increased development of some others – powerful wings accompanying weak feet, or great velocity making up for the absence of defensive weapons; for it has been shown that all varieties in which an unbalanced deficiency occurred could not long continue their existence. The action of this principle is exactly like that of the centrifugal governor of the steam engine, which checks and corrects any irregularities almost before they become evident; and in like manner no unbalanced deficiency in the animal kingdom can ever reach any conspicuous magnitude, because it would make itself felt at the very first step, by rendering existence difficult and extinction almost sure soon to follow. An origin such as is here advocated will also agree with the peculiar character of the modifications of form and structure which obtain in organized beings – the many lines of divergence from a central type, the increasing efficiency and power of a particular organ through a succession of allied species, and the remarkable persistence of unimportant parts such as colour, texture of plumage and hair, form of horns or crests, through a series of species differing considerably in more essential characters. It also furnishes us with a reason for that "more specialized structure" which Professor Owen[30] states to be a characteristic of recent compared with extinct forms, and which would evidently be the result of the progressive modification of any organ applied to a special purpose in the animal economy.

We believe we have now shown that there is a tendency in nature to the continued progression of certain classes of *varieties* further and further from the original type – a progression to which there appears no reason to assign any definite limits – and that the same principle which produces this result in a state of nature will also explain why domestic varieties have a tendency to revert to the original type. This progression, by minute steps, in various directions, but always checked and balanced by the necessary conditions, subject to which alone existence can be preserved, may, it is believed, be followed out so as to agree with all the phenomena presented by organized beings, their extinction and succession in past ages, and all the extraordinary modifications of form, instinct, and habits which they exhibit.

Ternate, February, 1858.[31]   ❧

"Thunderbolt from a cloudless sky":
Darwin and Natural Selection

What followed is one of the most celebrated episodes in the history of science. Darwin received Wallace's manuscript on June 18 1858. Even a brief glance must have revealed that Wallace had independently stumbled on the very idea to which Darwin had dedicated the previous twenty years of his life. He had first started pondering the question of "Transmutation of Species" shortly after his return to England from the *Beagle* voyage in 1836, and was at the time working actively on his "big species book" on the subject. Darwin was experiencing the ultimate scientific nightmare: loss of priority. He wrote to his trusted friend and confidant Charles Lyell that "all my originality, whatever it may amount to, will be smashed".[32] All, however, was not lost. Lyell and Joseph Hooker, another of Darwin's lieutenants, brokered a "delicate arrangement" to rescue Darwin's precedence. Their solution was a joint presentation to the Linnean Society, the pre-eminent biological forum of the day, consisting of previously unpublished manuscripts of Darwin's along with Wallace's paper. Lyell and Hooker made the presentation on July 1 1858 on behalf of the two "indefatigable naturalists", neither of whom was present: one, half a world away in South-east Asia, had no idea what was happening, and the other, Darwin, was burying his baby son, Charles Waring, who had contracted scarlet fever on the night of June 23. It had been a bad month for Darwin.

Wallace scholarship has called into question this standard account of events: John Langdon Brooks[33] and Arnold Brackman[34] have offered revised accounts in which Darwin is willing to bend, or even ignore, the rules of scientific conduct to protect his claims to precedence. Scrutiny of mailing schedules between the Far East and England has suggested that Darwin *may* have received Wallace's letter as much as a month earlier than June 18, the date he publicly acknowledged having received it. He was perhaps strategizing during those missing weeks. Given the general completeness of Darwin's correspondence – he was a keeper, not a discarder, of letters – there exists a remarkable gap in the documentary record for this period. The original Wallace letter and manuscript are, for example, missing; our only record of Wallace's essay comes from the *Journal of the Proceedings of the Linnean Society*. It's possible that Darwin, or someone else, possibly his son Francis, who edited his correspondence, carefully eliminated incriminating evidence of a conspiracy from his files. There is also the suggestion that Darwin "acquired" from Wallace the critical missing chunk of his theory of evolution, the so-called "Principle of Divergence", which accounts for the differentiation of species through time. The evidence brought together by Brooks and Brackman is certainly suggestive but remains circumstantial. That Darwin had for so long been engaged with the problem is sufficient for most commentators to accept his seniority in the Wallace–Darwin partnership. The rapidity with which Darwin was able to produce *On the Origin of Species* – it was published in November the following year, and, weighing in at nearly 500 pages, was a hefty tome – attests to this. Wallace's paper stimulated him to download twenty years of careful thought.[35]

The most remarkable aspect of the whole episode is that Darwin and his colleagues had the opportunity to intervene. It was highly unusual for Wallace to submit a paper in this way. His previous publications, after all, were submitted directly to journals or placed by Samuel Stevens, his agent in London. Admittedly Wallace knew from their correspondence following the publication of the "Sarawak Law" paper that Darwin was seriously interested in just the topics that he was himself addressing, and he may have hoped that Darwin's interest in the subject would prompt him to ensure that the paper would be published in a prominent venue. I suspect that Wallace was still smarting from the lack of attention afforded his "Sarawak Law" and was therefore making absolutely certain that the two influential people most likely to be interested, Darwin and Lyell, would see his new paper. The history of biology, however, would be very different had Wallace not serendipitously decided to send the manuscript to Darwin. Darwin would have awoken one day in late 1858 or early 1859 to find that his life's work had been scooped by an unknown bug collector marooned somewhere in the Moluccan Islands.

Wallace never complained in print about his treatment by

Darwin and his colleagues. He was unerringly deferential to Darwin, and indeed entitled his major work on evolution "Darwinism". He did complain a little about being deprived of the opportunity to polish the "Ternate Paper" prior to publication ("in reading it *now* it must be remembered that it was but a hasty first sketch, that I had no opportunity of revising it before it was printed in the journal of the Linnean Society"[36]). Perhaps the most sensitive barometer of Wallace's feelings towards Darwin is his private correspondence; here again we find no hint of resentment. An early response to news of the Linnean Society joint publication, in a letter to his mother, betrays only excitement at his elevation to the upper echelons of the Victorian biological world: "I have received letters from Mr. Darwin and Dr. Hooker, two of the most eminent naturalists in England, which has highly gratified me. I sent Mr. Darwin an essay on a subject on which he is now writing a great work. He showed it to Dr. Hooker and Sir C. Lyell, who thought so highly of it that they immediately read it before the Linnean Society. This assures me the acquaintance and assistance of these eminent men on my return home. . . ."[37] A month later, he wrote to his boyhood friend George Silk in similar terms: ". . . borrow the *Journal of Proceedings* [of the Linnean Society] for August last, and in the last article you will find some of my latest lucubrations, and also some complimentary remarks thereon by Sir Charles Lyell and Dr. Hooker, which (as I know neither of them) I am *a little* proud of."[38]

The publication of *On the Origin of Species* (in which Wallace is mentioned only fleetingly) occasioned more outpourings of Darwin admiration, also in a letter to George Silk:

I have read it [*On the Origin of Species*] through five or six times, each time with increasing admiration. It will live as long as the "Principia" of Newton. It shows that nature is, as I before remarked to you, a study that yields to none in grandeur and immensity. The cycles of astronomy or even the periods of geology will alone enable us to appreciate the vast depths of time we have to contemplate in the endeavour to understand the slow growth of life upon the earth. The most intricate effects of the law of gravitation, the mutual disturbances of all the bodies of the solar system, are simplicity itself compared with the intricate relations and complicated struggle which have determined what forms of life shall exist and in what proportions. Mr. Darwin has given the world a new science, and his name should, in my opinion, stand above that of every philosopher of ancient or modern times. The force of admiration can no further go!!!³⁹ [1860] ❧

Bates was also left in no doubt about Wallace's appreciation of *On the Origin of Species.*

❧ I know not how, or to whom, to express fully my admiration of Darwin's book. To *him* it would seem flattery, to others self-praise; but I do honestly believe that with however much patience I had worked and experimented on the subject, I could *never have approached* the completeness of his book, its vast accumulation of evidence, its overwhelming argument, and its admirable tone and spirit. I really feel thankful that it has *not* been left to me to give the theory to the world. Mr. Darwin has created a new science and a new philosophy; and I believe that never has such a complete illustration of a new branch of human knowledge been due to the labours and researches of a single man. Never have such vast masses of widely scattered and hitherto quite unconnected facts been combined into a system and brought to bear upon the establishment of such a grand and new and simple philosophy."⁴⁰ [1860]   ❧

A meeting was held at the Linnean Society in 1908 to mark the fiftieth anniversary of the Darwin–Wallace publication. Darwin had been dead twenty-six years, but Wallace was there to assess his and Darwin's contributions.

❧ Since the death of Darwin, in 1882, I have found myself in the somewhat unusual position of receiving credit and praise from popular writers under a complete misapprehension of what my share in Darwin's work really amounted to. It has been stated (not unfrequently) in the daily and weekly press, that Darwin and myself discovered "Natural Selection" simultaneously, while a more daring few have declared that I was the *first* to discover it, and I gave way to Darwin!

In order to avoid further errors of this kind (which this Celebration may possibly encourage), I think it will be well to give the actual facts as simply and clearly as possible.

The *one fact* that connects me with Darwin, and which, I am happy to say, has never been doubted, is that the idea of what is now termed "natural selection" or "survival of the fittest," together with its far-reaching consequences, occurred to us *independently*, and was first jointly announced before this Society fifty years ago.

But, what is often forgotten by the Press and the public is, that the idea occurred to Darwin in 1838, nearly twenty years earlier than to myself (in February, 1858); and that during the whole of that twenty years he had been laboriously collecting evidence from the vast mass of literature of biology, of horticulture, and of agriculture; as well as himself carrying out ingenious experiments and original observations, the extent of which is indicated by the range of subjects discussed in his "Origin of Species," and especially in that wonderful storehouse of knowledge, his "Animals and Plants under Domestication," almost the whole materials for which work had been collected, and to a large extent systematised, during that twenty years.

So far back as 1844, at a time when I had hardly thought of any serious study of nature, Darwin had written an outline of his views, which he communicated to his friends Sir Charles Lyell and Dr. (now Sir Joseph) Hooker. The former strongly urged him to publish an abstract of his theory as soon as possible, lest some other person might precede him; but he always refused till he had got together the whole of the materials for his intended great work. Then, at last, Lyell's prediction was fulfilled, and, without any apparent warning, my letter, with the enclosed Essay, came upon him, like a thunderbolt from a cloudless sky! This forced him to what he considered a premature publicity, and his two friends undertook to have our two papers read before this Society.

How different from this long study and preparation – this philosophical caution – this determination not to make known his fruitful conception till he could back it up by overwhelming proofs – was my own conduct.

The idea came to me as it had come to Darwin, in a sudden flash of insight; it was thought out in a few hours – was written down with such a sketch of its various applications and developments as occurred to me at the moment – then copied on thin letter paper and sent off to Darwin – all within one week. *I* was then (as often since) the "young man in a hurry": he, the painstaking and patient student seeking ever the full demonstration of the truth that he had discovered, rather than to achieve immediate personal fame.

Such being the actual facts of the case, I should have had no cause for complaint if the respective shares of Darwin and myself in regard to the elucidation of nature's method of organic development had been henceforth estimated as being, roughly, proportional to the time we had each bestowed upon it when it was thus first given to the world – that is to say, as twenty years is to one week. For, he had already made it his own. If the persuasion of his friends had prevailed with him, and he had published his theory after ten years' – fifteen years' – or even eighteen years' elaboration of it – *I* should have had no part in it whatever, and *he* would have been at once recognised as the sole and undisputed discoverer and patient investigator of this great law of "Natural Selection" in all its far-reaching consequences.

It was really a singular piece of good luck that gave to me any share whatever in the discovery. During the first half of the nineteenth century (and even earlier) many great biological thinkers and workers had been pondering over the problem and had even suggested ingenious but inadequate solutions. Some of these men were among the greatest intellects of our time, yet, till Darwin, all had failed; and it was only Darwin's extreme desire to perfect his work that allowed me to come in, as a very bad second, in the truly Olympian race in which all philosophical biologists, from Buffon[41] and Erasmus Darwin[42] to Richard Owen and Robert Chambers,[43] were more or less actively engaged.

And this brings me to the very interesting question: Why did so many of the greatest intellects fail, while Darwin and myself hit upon the solution of this problem – a solution which this Celebration proves to have been (and still to be) a satisfying one to a large number of those best able to form a judgment on its merits? As I have found what seems to me a good and precise answer to this question, and one which is of some psychological interest, I will, with your permission, briefly state what it is.

On a careful consideration, we find a curious series of correspondences, both in mind and in environment, which led Darwin and myself, alone among our contemporaries, to reach identically the same theory.

First (and most important, as I believe), in early life both Darwin and myself became ardent beetle-hunters. Now there is certainly no group of organisms that so impresses the collector by the almost infinite number of its specific forms, the endless modifications of structure, shape, colour, and surface-markings that distinguish them from each other, and their innumerable adaptations to diverse environments. These interesting features are exhibited almost as strikingly in temperate as in tropical regions, our own comparatively limited island-fauna possessing more than 3,000 species of this one order of insects.

Again, both Darwin and myself had, what he terms "the mere passion for collecting," not that of studying the minutiae of structure, either internal or external. I should describe it rather as an intense interest in the variety of living things – the variety that catches the eye of the observer even among those which are very much alike, but which are soon found to differ in several distinct characters.

Now it is this superficial and almost child-like interest in the outward forms of living things which, though often despised as unscientific, happened to be *the only one* which would lead us towards a solution of the problem of species. For nature herself distinguishes her species by just such characters – often exclusively so, always in some degree – very small changes in outline, or in the proportions of appendages, as give a quite distinct and recognisable *facies* to each, often aided by slight peculiarities in motion or habit; while in a larger number of cases differences of surface-texture, of colour, or in the details of the same general scheme of colour-pattern or of shading, give an unmistakable individuality to closely allied species.

It is the constant search for and detection of these often unexpected differences between very similar creatures, that gives such an intellectual charm and fascination to the mere collection of these insects; and when, as in the case of Darwin and myself, the collectors were of a speculative turn of mind, they were constantly led to think upon the "why" and the "how" of all this wonderful variety in nature – this overwhelming, and, at first sight, purposeless wealth of specific forms among the very humblest forms of life.

Then, a little later (and with both of us almost accidentally) we became travellers, collectors, and observers, in some of the richest and most interesting portions of the earth; and we thus had forced upon our attention all the strange phenomena of local and geographical distribution, with the numerous problems to which they give rise. Thenceforward our interest in the great mystery of *how* species came into existence was intensified, and – again to use Darwin's expression – "haunted" us.

Finally, both Darwin and myself, at the critical period when our minds were freshly stored with a considerable body of personal observation and reflection bearing upon the problem to be solved, had our attention directed to the system of *positive checks* as expounded by Malthus in his "Principles of Population." The effect of that was analogous to that of friction upon the specially prepared match, producing that flash of insight which led us immediately to the simple but universal law of the "survival of the fittest," as the long sought *effective* cause of the continuous modification and adaptations of living things.

It is an unimportant detail that Darwin read this book two years *after* his return from his voyage, while I read it *before* I went abroad, and it was a sudden recollection of its teachings that caused the solution to flash upon me. I attach much importance, however, to the large amount of solitude we both enjoyed during our travels, which, at the most impressionable period of our lives, gave us ample time for reflection on the phenomena we were daily observing.

This view, of the combination of certain mental faculties and external conditions that led Darwin and myself to an identical conception, also serves to explain why none of our precursors or contemporaries hit upon what is really so very simple a solution of the great problem. Such evolutionists as Robert Chambers, Herbert Spencer, and Huxley, though of great intellect, wide knowledge, and immense power of work, had none of them the special turn of mind that makes the collector and the species-man; while they all – as well as the equally great thinker on similar lines, Sir Charles Lyell – became in early life immersed in different lines of research which engaged their chief attention.[44] [1908]   ع

Evolution by Natural Selection

Wallace provided an accessible summary of the Darwin–Wallace theory in his final major work, *The World of Life* (1910). He broke it down into three basic axioms: heredity, variation, and a rate of population increase that exceeds the availability of resources.

ع   [HEREDITY]

Perhaps the most universal fact – sometimes termed "law" – of the organic world is, that like produces like – that offspring are like their parents. This is so common, so well known to everybody, so absolutely universal in ordinary experience, that we are only surprised when there seems to be any exception to it. In its widest sense as applied to species there are no exceptions. Not only does the acorn always produce an oak, the cat a kitten, which grows into a cat, the sheep a lamb, and so throughout all nature, but each different well-marked race also produces its like. We recognise Chinese and Negroes as being men of the same species as ourselves, but of different varieties or races, yet these varieties always produce their like, and no case has ever occurred of either race producing offspring in every respect like one of the other races, any more than there are cases of cart-horses producing racers or spaniels producing greyhounds.[45]

. . .

THE VARIATION OF SPECIES, ITS FREQUENCY AND ITS AMOUNT

Having now shown something of the nature of heredity, its universality and its limitations, we pass on to a rather fuller discussion of the nature and amount of those limitations, commonly known as the variability of species. It is this variability that constitutes the most important of the factors which bring about adaptation, and that peculiar change or modification of living things which we term distinct species. This change is often very small in amount, but it always extends to various parts or organs, and so pervades the whole structure as to modify to a perceptible extent the habits and mode of life, the actions and motions, so that we come to recognise each species as a complete entity distinct from all others.

There is no subject of such vital importance to an adequate conception of evolution, which is yet so frequently misapprehended, as variability. Perhaps owing to the long continued and inveterate belief in the immutability of species, the earlier naturalists came to look upon those conspicuous cases of variation which forced themselves upon their attention as something altogether abnormal and of no importance in the scheme of nature. Some of them went so far as to reject them altogether from their collections as interfering with the well-marked distinctness of species, which they considered to be a fundamental and certain fact of nature. Hence, perhaps, it was that Darwin himself, finding so little reference to variation among wild animals or plants in the works of the writers of his time, had no adequate conception of its universality or of its large general amount whenever extensive series of individuals were compared. He therefore always guarded himself against assuming its presence whenever required by using such expressions in regard to the power of natural selection as, "If they vary, for unless they do so, natural selection can effect nothing."

This was the more strange because wherever we look around us we find, in our own species, in our own race, in our own special section of that race, an amount of variation so large and so universal as to fully satisfy all the needs of the evolutionist for bringing about whatever changes in form, structure, habits or faculties that may be desired. By simply observing the people we daily meet in the street, in the railway carriage, at all public assemblages, among rich and poor, among lowly-born or high-born alike, variability stares us in the face. We see, for instance, not rarely, but almost daily and everywhere, short and tall men and women. We do not require to measure them or to be specially good judges of height to be able to observe this – the difference is not one of fractions of an inch only, but of whole inches, and even of several inches. We cannot go about much without constantly seeing short men who are about 5 feet 2 inches high, and tall men who are 6 feet 2 inches – a difference of a whole foot, while in almost every town of say 10,000 inhabitants, still greater differences are to be found.

But this special variation, so large and so frequent that it cannot be overlooked, is only one out of many which we may observe daily if we look for them. Some men have long legs and short bodies, others the reverse; some are long-armed, some are big-handed, some big-footed, and these differences are found in men differing little or nothing in height. Again we have big-headed and small-headed men, long-headed and round-headed, big-jawed, big-eared, big-eyed men, and the reverse; we see dark and light complexions, smooth or hairy faces; black, or brown, or red, or flaxen-haired men; slender or stout men, broad or narrow-chested, clumsy or graceful, energetic and active, or lazy and slow. Characters, too, vary just as much. Men are taciturn or talkative, cool or passionate, intelligent or stupid, poetical or prosy, witty or obtuse. And all these characteristics, whether physical or mental, are combined together in an infinite variety of ways, as if each of them varied independently with no constant or even usual associa- tion with any of the others; whence arises that wonderful diversity of appearance, attitudes, expression, ability, intellect, emotion, and what we term as a whole character, which adds so much to the possibilities and enjoyments of social life, and gives us in their higher develop- ments such mountain peaks of human nature as were manifested in Socrates and Plato, Homer and Virgil, Alexander and Phidias,[16] Bud- dha and Confucius in the older world; in Shakespeare and Newton, Michael Angelo [sic], Faraday,[47] and Darwin in more recent times.

And with all this endless variation wherever we look for it, we are told again and again in frequent reiteration, that variation is minute, is even infinitesimal, and only occurs at long intervals in single individuals, and that it is quite insufficient for natural selection to work with in the production of new species.

This blindness, no doubt, arose in some persons from the ingrained idea of man's special creation, at all events, and that it was almost impious to suppose that these variations could have had anything to do with his development from some lower forms. But among naturalists the idea long prevailed, as it does still to some extent, that in a state of nature there is little variation. Yet here, too, they might have found a clue in the fact, so often quoted, that a shepherd knows every individual sheep in his flock, and the huntsman every dog in his well-matched pack of hounds, and this notwithstanding that in both cases these animals are selected breeds in which all large deviations from the type form are usually rejected.[48]

. . .

POWERS OF INCREASE OF PLANTS AND ANIMALS

Of almost equal importance with ever-present variation is the power which all organisms possess of reproducing their kind so rapidly as to be able to take possession of any unoccupied spaces around them, and in many cases to expel other kinds by the vigour of their growth.

The rapidity of increase is most prominently seen among vegetables. These are capable, not only of a fivefold or tenfold annual increase, as among many of the higher animals, but one of many hundred or even thousandfold annually. A full-grown oak or beech tree is often laden with fruit on every branch, which must often reach 100,000, and sometimes perhaps a million in number, each acorn or nut being capable, under favourable conditions, of growing into a tree like its parent. Our wild cherries, hawthorns, and many other trees, are almost equally abundant fruitbearers, but in all these cases it is only rarely (in a state of nature) that any one seed grows to a fruit-bearing size, because, all having a superabundance of reproductive power, an equilibrium has been reached everywhere, and it is only when some vacancy occurs, as when a tempest uproots or destroys a number of trees, or some diminution of grazing animals allows more seedlings than usual to grow up, that any of the seeds of the various trees around have a chance of surviving; and the most vigorous of these will fill up the various gaps that have been produced.

But it is among the herbaceous plants that perhaps even greater powers of increase exist. Where our common foxglove luxuriates we often see its tall spikes densely packed with capsules, each crowded with hundreds of minute seeds, which are scattered by the wind over the surrounding fields, but only a few which are carried to especially favourable spots serve to keep up the supply of plants. Kerner,[49] in his *Natural History of Plants*, tells us that a crucifer, *Sisymbrium Sophia*, has been found to produce on an average 730,000 seeds, so that if vacant spaces of suitable land existed around it, one plant might, in three years only, cover an area equal to 2000 times that of the land-surface of the globe. A close ally of this, *Sisymbrium Irio*, is said to have sprung up abundantly among the ruins of London after the great fire of 1666. Yet it is not a common plant, and is a doubtful native, only occurring occasionally in English localities.

Turning to the animal kingdom, we still find the reproductive powers always large and often enormous. The slowest breeding of all is the elephant, which is supposed to rear one young one every 10 years; but, as it lives to more than 100 years, Darwin calculates that in 750 years (a few moments only in the geological history of the earth) each pair would, if all their offspring lived and bred, produce 19 millions of elephants.

The smaller mammals and most birds increase much more rapidly, as many of them produce two or more families every year. The rabbit is one of the most rapid, and Mr. Kearton[50] calculates that, under the most favourable conditions, a single pair might in 4 or 5 years increase to a million. In Australia, being favourable in climate, vegetation, and absence of enemies, they have so multiplied as to become a nuisance and almost a danger, and though their introduction was easy, it has so far been found impossible to get rid of them.[51]

. . .

[Thus through natural selection] Nature works to improve her stocks in the great world of life; and has been thus enabled not only to keep all in complete adaptation to an ever varying environment, but to fill up, as it were, every element, every different station, every crack and crevice in the earth's surface with wonderful and beautiful creatures which it is the privilege and delight of the naturalist to seek out, to study, and to marvel at.[52]

. . .

The facts outlined in the present chapter, of abundant and ever-present variability with enormous rapidity of increase, furnish a sufficient reply to those ill-informed writers who still keep up the parrot-cry that Darwinian theory is insufficient to explain the formation of new species by survival of the fittest.[53] [1910] ৯৯

Agreeing with Darwin

Wallace was one Darwin's staunchest defenders in the controversies that dogged him after the publication of the *Origin*. Indeed, Darwin paid homage to Wallace's rhetorical skills: "You certainly have the art of putting your ideas with remarkable force and clearness . . . it makes me almost envious."[54]

A particularly persistent critic was the Reverend Samuel Haughton,[55] who had supplied one of the very earliest responses to the Darwin–Wallace theory, many months before the publication of the *Origin*. He had addressed the Geological Society in Dublin in February 1859: "This speculation of Messrs Darwin and Wallace would not be worthy of notice were it not for the weight of authority [Lyell and Hooker] under whose auspices it has been brought forward. If it means what it says, it is a truism; if it means anything more, it is contrary to fact."[56]

Wallace records that:

~§ [i]n June, 1863, an article appeared in the *Annals and Magazine of Natural History* by the Rev. S. Haughton, entitled "On the Bee's Cell and the Origin of Species." At that time I was eager to enter the lists with anyone who attacked natural selection or Darwin's exposition of it. This article was full of the usual errors and misconceptions, some of the most absurd nature, but all set forth as if with the weight of authority in a scientific journal. I accordingly replied in the October number of the *Annals*, and criticized the critic rather severely.[57] [1905] §~

Wallace gave no quarter.

~§ My attention has been called to the paper in the "Annals" for June last on the above subjects, the author of which seems to me to have quite misunderstood and much misrepresented the facts and reasonings of Mr. Darwin on the question. As some of your readers may conclude, if it remains unanswered, that it is therefore unanswerable, I ask permission to make a few remarks on what seem to me its chief errors . . .

On the question of the "origin of species" Mr. Haughton enlarges considerably; but his chief arguments are reduced to the setting-up of "three unwarrantable assumptions," which he imputes to the Lamarckians and Darwinians, and then, to use his own words, "brings to the ground like a child's house of cards." The first of these is "*the indefinite variation of species continuously in the one direction.*" Now this is certainly never assumed by Mr. Darwin, whose argument is mainly grounded on the fact that variations occur in *every direction*. This is so obvious that it hardly needs insisting on. In every large family there is almost always one child taller, one darker, one thinner than the rest; one will have a larger nose, another a larger eye: they vary morally as well; some are more poetical, others more morose; one has a genius for numbers, another for painting. It is the same in animals: the puppies, or kittens, or rabbits of one litter differ in many ways from each other – in colour, in size, in disposition; so that, though they do not "*vary continuously in one direction,*" they do vary continuously in many directions; and thus there is always material for natural selection to act upon in *some* direction that may be advantageous.

In his remarks upon this "unwarrantable assumption" (which is altogether his own), Mr. Haughton has the following passage:– "In the writings of Darwin there is this singular inconsistency, that, while he shows the utmost effects of human breeding on domestic animals to be capable of production in ten or twenty years, he denies the right of his adversaries to appeal to the unaltered condition of the ass, the ostrich, and the cat for 3000 years," &c. The first part of this sentence is so completely out of the pale of grammatical construction, that I must conclude Mr. Haughton writes a very bad hand, and did not correct the proofs. But, so far from Mr. Darwin denying his opponents the use of the facts above alluded to, he himself offers them far stronger ones, in the many species of shells which have lived unchanged since the middle tertiary epochs, and of mammals whose remains are found in beds which testify that they have survived important changes of the earth's surface. No one who understands the theory of natural selection will imagine that these facts are in any way opposed to it.

The second supposed "unwarrantable assumption" is, "*That the causes of variation,* viz. natural advantage in the struggle for existence (Darwin), *are sufficient to account for the effects asserted to be produced.*" There certainly never was a more unwarrantable assertion made, than that Darwin assigned "natural advantage in the struggle for existence" as "the cause of variation." Darwin over and over again declares that the *cause* of variation is unknown (*Origin of Species*, pp. 8, 38), though the *fact* is certain and undeniable. Natural selection, acting through advantage in the struggle for existence, *accumulates* favourable variations, but in no sense *causes* them. This is the very foundation of Mr.

Darwin's theory; yet even this is misunderstood or misrepresented by Mr. Haughton.

The third "unwarrantable assumption" charged upon Mr. Darwin is, "*That succession implies causation*," "that the Palæozoic Cephalopoda [molluscs] produced the Red-Sandstone fishes," "that these in turn gave birth to the Liassic reptiles," &c., &c. Now those who have read the "Origin of Species" know that such absurd doctrines as these are nowhere taught there; and I can only say to those who have not read it that I challenge them or Mr. Haughton to produce any passages which will bear such a meaning.

In conclusion, it is asserted "that naturalists who have accepted by multitudes the new theory of the origin of species are, as a class, untrained in the use of the logical faculties, which, however, they may be charitably supposed to possess in common with other men." This is the judgment of the Rev. S. Haughton on such men as Lyell, Hooker, Lubbock,[58] Huxley, and Asa Gray.[59] A perusal of his paper, with the remarks I have now made upon it, will enable any one to judge how far Mr. Haughton himself possesses those "logical faculties" which he is half inclined to deny to the mass of British naturalists. There are several other minor points in his paper which might be alluded to; but it has already occupied as much space as it deserves, and I will only, in conclusion, quote from it a short paragraph which contains an important truth, but which may very fairly be applied in other quarters than those for which the author intended it:– "No progress in natural science is possible as long as men will take their rude guesses at truth for facts, and substitute the fancies of their imagination for the sober rules of reasoning."[60] [1863]

Following this article's publication, Wallace was taken to task by a colleague who insisted that "I ought not to have written in such a tone of ridicule of a man who was much older and more learned than myself."[61] Haughton, however, did not apparently bear a grudge, for Wallace reports that, in 1882, while on a visit to Dublin, he "had the great pleasure one morning of breakfasting with him and the other members of the managing committee at the Zoological Gardens, and of enjoying his instructive and witty conversation. The brilliant midsummer morning, the cosy room looking over the beautiful gardens, and the highly agreeable and friendly party assembled rendered this one of the many pleasant recollections of my life."[62]

Disagreeing with Darwin

Despite seeing eye to eye on the basics of evolution, Darwin and Wallace disagreed on many matters. Indeed, Darwin wrote in a letter to Wallace, "How lamentable it is that two men should take such widely different views, with the same facts before them; but this seems to be almost regularly our case, and much do I regret it."[63] Even natural selection – their joint product – was contentious. Wallace was inclined to attribute everything he saw in nature to natural selection whereas Darwin had a more pluralistic perspective on the factors responsible for evolutionary change. Wallace, perhaps a little ruefully, disclosed that "some of my critics declare that I am more Darwinian than Darwin himself, and in this, I admit, they are not far wrong".[64] Wallace dedicates a chunk of his autobiography to "the chief differences of opinion between Darwin and myself".[65]

Female Choice

The most significant area of disagreement was human evolution, which will be dealt with in a later section. A second area was Darwin's theory of female choice. Darwin had recognized that sexual selection – spurred by competition for access to mates – could drive evolutionary change in ways distinct from, but analogous to, natural selection. He identified two mechanisms. First, intra-sexual competition among, usually, males for access to females could promote the evolution of characters, like antlers in deer, involved in determining within-sex pecking order. Second, female choice for males with certain desirable traits could lead to the evolution of particularly exaggerated characteristics, like the peacock's tail. Wallace was dubious:

SEXUAL SELECTION THROUGH FEMALE CHOICE

Darwin's theory of sexual selection consists of two quite distinct parts – the combats of males so common among polygamous mammals and birds, and the choice of more musical or more ornamental male birds by the females. The first is an observed fact, and the development of weapons such as horns, canine teeth, spurs, etc., is a result of natural selection acting through such combats. The second is an inference from the observed facts of the display of the male plumage or ornaments; but the statement that ornaments have been developed by the female's choice of the most beautiful male *because he is the most beautiful,* is an inference supported by singularly little evidence. The first kind of sexual selection I hold as strongly and as thoroughly as Darwin himself; the latter I at first accepted, following Darwin's conclusions from what appeared to be strong evidence explicable in no other way; but I soon came to doubt the possibility of such an explanation, at first from considering the fact that in butterflies sexual differences are as strongly marked as in birds, and it was to me impossible to accept female choice in their case, while, as the whole question of colour came to be better understood, I saw equally valid reasons for its total rejection even in birds and mammalia.

But here my view really extends the influence of natural selection, because I show in how many unsuspected ways colour and marking is of use to its possessor. I first stated my objections to "female choice" in my review of the "Descent of Man" (1871),[66] and more fully developed it in my "Tropical Nature" (1878), while in my "Darwinism" (1889), I again discussed the whole subject, giving the results of more mature consideration. I had, however, already discussed the matter at some length with Darwin, and in a letter of September 18, 1869, I gave him my general argument as follows:

"I have a general and a special argument to submit.

1. Female birds and insects are usually exposed to more danger than the male, and in the case of insects their existence is necessary for a longer period. They therefore require, in some way or other, an increased amount of protection.
2. If the male and female were distinct species, with different habits and organizations, you would, I think, admit that a difference of colour, serving to make that one less conspicuous which evidently required more protection than the other, had been acquired by natural selection.
3. But you admit that variations appearing in one sex are (sometimes) transmitted to that sex only. There is, therefore, nothing to prevent natural selection acting on the two sexes as if they were two species.
4. Your objection that the same protection would, *to a certain extent,* be useful to the male seems to me quite unsound, and directly opposed to your own doctrine so convincingly urged in the

'Origin,' that natural selection *never improves an animal beyond its needs*. Admitting, therefore, abundant variation of colour in both sexes, it is impossible that the male can be brought by natural selection to resemble the female (unless such variations are always transmitted), because the difference in their colours is for the purpose of making up for their different organization and habits, and natural selection cannot give to the male more protection than he requires, which is less than in the female..."[67] [1905] ❧

Wallace especially valued his argument based on nesting habits of birds. He explained it more fully.

❧ CLASSIFICATION OF NESTS

For the purpose of this inquiry it is necessary to group nests into two great classes, without any regard to their most obvious differences or resemblances, but solely looking to the fact of whether the contents (eggs, young, or sitting bird) are hidden or exposed to view. In the first class we place all those in which the eggs and young are completely hidden, no matter whether this is effected by an elaborate covered structure, or by depositing the eggs in some hollow tree or burrow underground. In the second, we group all in which the eggs, young, and sitting bird are exposed to view, no matter whether there is the most beautifully formed nest or none at all. Kingfishers, which build almost invariably in holes in banks; woodpeckers and parrots, which build in hollow trees; the Icteridae [blackbirds etc.] of America, which all make beautiful covered and suspended nests; and our own wren, which builds a domed nest – are examples of the former; while our thrushes, warblers, and finches, as well as the crowshrikes, chatterers, and tanagers of the tropics, together with all raptorial birds and pigeons, and a vast number of others in every part of the world, all adopt the latter mode of building.

It will be seen that this division of birds, according to their nidification, bears little relation to the character of the nest itself. It is a functional not a structural classification. The most rude and the most perfect specimens of bird architecture are to be found in both sections. It has, however, a certain relation to natural affinities, for large groups of birds, undoubtedly allied, fall into one or the other division exclusively. The species of a genus or of a family are rarely divided between the two primary classes, although they are frequently divided between the two very distinct modes of nidification that exist in the first of them.

All the Scansorial or climbing, and most of the Fissirostral or wide-gaped birds, for example, build concealed nests; and in the latter group the two families which build open nests, the swifts and the goatsuckers, are undoubtedly very widely separated from the other families with which they are associated in our classifications. The tits vary much in their mode of nesting, some making open nests concealed in a hole, while others build domed or even pendulous covered nests, but they all come under the same class. Starlings vary in a similar way. The talking mynahs, like our own starlings, build in holes, the glossy starlings of the East (of the genus Calornis) form a hanging covered nest, while the genus Sturnopastor builds in a hollow tree. One of the most striking cases in which one family of birds is divided between the two classes is that of the finches; for while most of the European species build exposed nests, many of the Australian finches make them dome-shaped.

SEXUAL DIFFERENCES OF COLOUR IN BIRDS

Turning now from the nests to the creatures who make them, let us consider birds themselves from a somewhat unusual point of view, and form them into separate groups, according as both sexes, or the males only, are adorned with conspicuous colours.

The sexual differences of colour and plumage in birds are very remarkable, and have attracted much attention; and, in the case of polygamous birds, have been explained by Mr. Darwin's principle of sexual selection. We may, perhaps, understand how male pheasants and grouse have acquired their more brilliant plumage and greater size by the continual rivalry of the males both in strength and beauty – but this theory does not throw any light on the causes which have made the female toucan, bee-eater, parroquet, macaw, and tit in almost every case as gay and brilliant as the male, while the gorgeous chatterers, manakins, tanagers, and birds of paradise, as well as our own blackbird, have mates so dull and inconspicuous that they can hardly be recognised as belonging to the same species.

THE LAW WHICH CONNECTS THE COLOURS OF FEMALE BIRDS WITH THE MODE OF NIDIFICATION

The above-stated anomaly can, however, now be explained by the influence of the mode of nidification, since, with very few exceptions, I find it to be the rule – *that when both sexes are of strikingly gay and conspicuous colours the nest is of the first class, or such as to conceal the sitting birds; while, whenever the male is gay and conspicuous and the nest is open so as to expose the sitting bird to view, the female is of dull or obscure colours.*[68] [1868] ﷼

Despite his hypothesis to account for the drabness of (usually) female birds, Wallace never seems to have come up with a convincing explanation for the showiness of (usually) males. However, he never stopped doubting Darwin's theory: "There is so much still unknown that it will be very hard to convince me that there is no other possible explanation of the peacock's feather than the 'continued preference by females'. . . ."[69] [1889] Modern views are in line with Darwin's. While natural selection may indeed promote inconspicuous coloration in females, it is sexual selection through female choice that has typically driven the evolution of male ornamentation.

Genetics

Another disagreement with Darwin was over the inheritance of acquired characters, the underpinning of Lamarck's theory of evolution. Mendel published his experiments on the genetics of pea plants in 1866 but his work was ignored until its rediscovery at the turn of that century. Darwin therefore devised his own theory of inheritance, Pangenesis, predicated on circulating microscopic "gemmules" through which each tissue could contribute its characteristics to the germ cells. Wallace's initial response was enthusiastic:

﷼ I am reading Darwin's book ("Animals and Plants under Domestication"), and have read the "Pangenesis" chapter first, for I could not wait. The hypothesis is *sublime* in its simplicity and the wonderful manner in which it explains the most mysterious of the phenomena of life. To me it is *satisfying* in the extreme. I feel I can never give it up, unless it be *positively* disproved, which is impossible, or replaced by one which better explains the facts, which is highly improbable. Darwin has here decidedly gone ahead of Spencer in generalization. I consider it the most wonderful thing he has given us, but it will not be generally appreciated.[70] [1868] ॐ

Indeed it was not. Evidence against Pangenesis mounted up and Wallace's initial enthusiasm evaporated.

ॐ The difficulty of conceiving the actual operation of the theory of Pangenesis may be best illustrated by an example. Taking a bird, such as a peacock, the theory implies that not only every cell and fibre of bone, muscle, skin, and all internal organs gives off gemmules which all find their way into every one of the cells constituting the sperm or reproductive fluid, but that every one of the feathers also sends gemmules from each of the cells that build up its wonderfully complex structure, not only in the adult stage, but in the condition they assume in the young and adolescent birds; and further, that every detail of varying colour of the barbs of these feathers send off their gemmules, and that all this inconceivable number of gemmules must travel through the whole structure of the quill, and through all the tissues of the body, till they reach the reproductive organs, and every one of these gemmules must reach all or most of the sperm-cells, failing which there would be a corresponding deficiency in the offspring. But as important deficiencies of feathers, or of colour on the various feathers, which produce the beautiful patterns and ornaments of a bird's plumage only rarely occur, we must assume that the passage of the millions of gemmules from the ends of the feathers of a peacock's train through the whole length of the shaft, and then to the sperm-cells, is almost always successfully accomplished. In addition to the enormous difficulty, on any theory, of conceiving the processes of growth and development of the complex parts of living organisms, we have, on this theory, an equal or greater difficulty in the reverse process, by which the gemmules from every cell get back again to the sperm and germ cells. Without asserting that this process is impossible or inconceivable, it is well to endeavour to realise what it really is and its almost incredible complexity.[71] [1897] ॐ

ॐ PANGENESIS, AND THE HEREDITY OF
ACQUIRED CHARACTERS

Darwin always believed in the inheritance of acquired characters, such

as the effects of use and disuse of organs and of climate, food, etc.,
on the individual, as did almost every naturalist, and his theory of
pangenesis was invented to explain this among other affects of heredity. I therefore accepted pangenesis at first, because I have always
felt it a relief (as did Darwin) to have *some* hypothesis, however
provisional and improbable, that would serve to explain the facts; and
I told him that "I shall never be able to give it up till a better one
supplies its place." I never imagined that it could be directly disproved,
but Mr. F. Galton's[72] experiments of transfusing a large quantity of
the blood of rabbits into other individuals of quite different breeds,
and afterwards finding that the progeny was not in the slightest degree
altered, did seem to me to be very nearly a disproof, although Darwin
did not accept it as such. But when, at a much later period, Dr.
Weissman[73] showed that there is actually no valid evidence for the
transmission of such characters, and when he further set forth a mass
of evidence in support of his theory of the continuity of the germ-
plasm, the "better theory" was found, and I finally gave up pangenesis
as untenable. But this new theory really simplifies and strengthens the
fundamental doctrine of natural selection.[74] [1905]   ॐ

The most prominent Victorian champion of inheritance of
acquired characters was Herbert Spencer.

ॐ ARGUMENT FROM THE GIRAFFE

Mr. Spencer's argument is, briefly stated, that to develop such an
animal as the giraffe from some antelope-like ancestor requires many
co-incident and co-ordinate variations of different parts – each
increase in the length of the neck, of the head, of the fore or hind
limbs, or of either of the bones composing them, requiring corresponding increase of muscles, nerves, and blood-vessels, not only of
such as are immediately connected with the enlarged limb, but often
in remote parts of the body whose motions are necessarily co-
ordinated with it. He maintains that any increase of one part without
the adjusted increase of other parts would cause evil rather than good
– and that want of co-adaptation, even in a single muscle, would cause
fatal results when high speed had to be maintained while escaping
from an enemy. Then, again, not only the sizes but the shapes of the
bones have to be altered as the muscles are increased in size and the
motions of the various parts of the body change – and this introduces
fresh difficulties which are, again and again, declared to be insur-

mountable. And after elaborating all these alleged difficulties at great length, he arrives at the conclusion that, unless the increase or modification of parts due to use by the individual, is inherited, there can have been no evolution . . . .

Another tacit assumption is, that in nature all the individuals of a species have their parts so perfectly co-ordinated that any increase of one part only would disturb the harmonious adjustment and be a disadvantage. But this is totally to misconceive the situation. The adjustment of parts is a rough working adjustment, sufficient for the purpose of maintaining life, but capable of being improved (or deteriorated) by very many slight modifications of single parts. To illustrate this general adjustment, let us suppose we have before us for comparison all the county elevens of English cricketers. We shall have a body of some hundreds of picked men, all of whom are probably above the average as runners, are exceptionally quick with eye and hand, and are all more or less active and muscular. They vary, of course, in their special capacities, whether as batters, bowlers, fielders, or wicketkeepers, but it is certain that most of them would take a high place in almost any form of athleticism to which they chose to apply themselves. Yet these men would not resemble each other closely in stature or proportions. We should find among them tall and short, slender and stout; and among those of the same height proportions would differ, some being long-legged, others short-legged, and per-haps no two of the lot would be found to have exact the same proportions in all measurable parts of the body. We are thus shown that a high average result of strength and activity can be reached by very various combinations of the bones and muscles of the limbs and other variable parts, and we can hardly doubt that almost all of the men could be rendered still more efficient cricketers or athletes by some slight improvement in their organisation. One would run better if his legs were longer, another would throw and bowl better if his arms were shorter and more muscular; and such changes would be effective because these parts are now imperfectly co-ordinated with the rest of the body.

The considerations suggested by this illustrative case immensely increase the facilities for the improvement of any faculty required by natural selection, and they enable us to understand the process by which both natural and artificial selection have been able to modify the form and qualities of so many animals. It is not, as Mr. Spencer's argument assumes, by the selection of improvements in any special bone, or muscle, or limb that these modifications have been effected, but by *the selection of capacities or qualities resulting from the infinitely varied combination of variations that are always occurring.*[75] [1893]

Wallace's embrace of Weissman's ideas did not lead him to an appreciation of the significance of Mendel's work when it was rediscovered. Indeed, erroneously considering Mendel's ideas to constitute an alternative theory of evolution based on major changes, or "macromutations", Wallace shunned them. He wrote to E. B. Poulton[76] at Oxford, "Mendelism is something new, and within its very limited range, important, as leading to conceptions as to the causes and laws of heredity, but only misleading when adduced as the true origin of species in nature, as to which it seems to me to have no part."[77] Wallace, however, recognized that a theory of evolution could be robust even in the absence of an understanding of genetics.

Mr. Bennett[78] next returns to the laws of variation, and, because Mr. Darwin says that we are profoundly ignorant of these (although he himself has done so much to elucidate them), maintains that we cannot really know anything of the origin of species. As well might it be said that, because we are ignorant of the laws by which metals are produced and trees developed, we cannot know anything of the origin of steamships and railways. Spontaneous "variations" are but the materials out of which "species" are formed, and we do not require to know how the former are produced in order to learn the origin of the latter.[79] [1870]

Later, Wallace was willing to dismiss the entire problem of genetics with a rhetorical sleight of hand.

Does any one ask for a reason why no two gravel-stones or beach-pebbles, or even grains of sand, are absolutely identical in size, shape, surface, colour, and composition? When we trace back the complex series of causes and forces that have led to the production of these objects, do we not see that their absolute identity would be more remarkable than their diversity? So, when we consider how infinitely more complex have been the forces that have produced each individual animal or plant, and when we know that no two animals can possibly have been subject to identical conditions throughout the entire course of their development, we see that perfect identity in the result would be opposed to everything we know of natural agencies.

But variation is merely *the absence of identity*, and therefore requires no further explanation; neither do the diverse amounts of variation, for they correspond to the countless diversities of conditions to which animals have been exposed either during their own development or that of their ancestors.[80] [1880] ॐ

Name Selection

Wallace was never comfortable with Darwin's name for their joint discovery: for Wallace, "natural selection" misleadingly suggested the kind of conscious selection made by animal breeders. Ironically, however, the teleology Wallace saw as implicit in the term became an explicit part of his own theory as he grew older.

ॐ I have been so repeatedly struck by the utter inability of numbers of intelligent persons to see clearly, or at all, the self-acting and necessary effects of Natural Selection, that I am led to conclude that the term itself, and your mode of illustrating it, however clear and beautiful to many of us, are yet not the best adapted to impress it on the general naturalist public. The two last cases of this misunderstanding are (1) the article on "Darwin and His Teachings" in the last *Quarterly Journal of Science*, which, though very well written and on the whole appreciative, yet concludes with a charge of something like blindness, in your not seeing that Natural Selection requires the constant watching of an intelligent "chooser" like man's selection to which you so often compare it; and (2) in Janet's[81] recent work on the "Materialism of the Present Day," reviewed in last Saturday's *Reader*, by an extract from which I see that he considers your weak point to be that you do not see that "thought and direction are essential to the action of Natural Selection." The same objection has been made a score of times by your chief opponents, and I have heard it as often stated myself in conversation. Now, I think this arises almost entirely from your choice of the term Natural Selection, and so constantly comparing it in its effects to man's selection, and also to your so frequently personifying nature as "selecting," as "preferring," as "seeking only the good of the species," etc., etc. To the few this is as clear as daylight, and beautifully suggestive, but to many it is evidently a stumbling-block. I wish, therefore, to suggest to you the possibility of entirely avoiding this source of misconception in your great work (if not now too late), and also in any future editions of the "Origin," and I think it may be done without difficulty and very effectually by adopting Spencer's term (which he generally uses in preference to Natural Selection), viz. "Survival of the Fittest." This term is the plain expression of the *fact*; Natural Selection is a metaphorical expression of it, and to a certain degree *indirect* and *incorrect*, since, even personifying Nature, she does not so much select special variation as exterminate the most unfavourable ones.

Combined with the enormous multiplying powers of all organisms, and the "struggle for existence," leading to the constant destruction of by far the largest proportion – facts which no one of your opponents, as far as I am aware, has denied or misunderstood – "the survival of the fittest," rather than of those which were less fit, could not possibly be denied or misunderstood.[82] [1866] &

Replying, Darwin was sympathetic but intransigent: "The term Natural Selection has now been so largely used abroad and at home that I doubt whether it could be given up, and with all its faults I should be sorry to see the attempt made."[83]

Beyond Natural Selection

The "Sarawak Law" and "Ternate" papers only marked the beginning of Wallace's career in evolutionary biology. He made major contributions in a number of areas, most notably speciation and the study of the coloration of animals. Many of his insights are packed into a single remarkable paper on the swallow-tail butterflies of South-east Asia. Originally published in 1865, Wallace reprinted "the introductory portion of this . . . rather elaborate paper" in his 1870 volume of essays, "but in later editions it was omitted, as being rather too technical for general readers."[84] Excerpts of the 1870 version follow both to provide a flavour of Wallace's technical writing and to highlight some of his most impressive conceptual contributions to evolutionary biology. In this paper we see Wallace at his synthetic best, weaving together general statements ("superabundant population has been shown by Mr. Bates to be a general characteristic of all American groups and species which are objects of mimicry; and it is interesting to find his observations confirmed by examples on the other side of the Globe") and detailed personal observation ("it is curious that I captured one of these Papilios in the Aru Islands hovering along the ground, and settling on it occasionally, just as it is the habit of the Drusillas to do") to generate a comprehensive picture of the evolutionary factors affecting this group of butterflies. Note too Wallace's careful and accessible explanations, even here in his technical writing: his reference to "a blue-eyed, flaxen-haired Saxon man [who] had two wives, one a black-haired, red-skinned Indian squaw, the other a wooly-haired, sooty-skinned negress" is a marvellous way to illuminate the sex-limited genetic variation he was discussing.

In particular, this paper, with its careful parsing of biological variation, sees the introduction of the idea of "polymorphism", the coexistence of discrete forms within a species. In addition, Wallace's thinking on speciation was remarkably sophisticated. His definition – "Species are merely those strongly marked races or local forms which when in contact do not intermix, and when inhabiting distinct areas are generally believed to have had a separate origin, and to be incapable of producing a fertile hybrid offspring" – is extraordinarily close to today's most generally accepted definition: "Species are groups of actually or potentially inter-breeding populations that are reproductively isolated from other such groups." Students are taught that this "Biological Species Concept" was only introduced by evolutionary biologist Ernst Mayr in the 1940s, yet Wallace's definition contains its essence. Notice, however, that Wallace rapidly went on to point out that such a definition is effectively useless because of the great difficulty of actually applying it in practice: if individual butterflies from different Indonesian islands *looked* the same, he would put them in the same species, without ever checking that they could hybridize and produce fertile viable offspring.

Colour was a major biological preoccupation of Wallace's. His friend Bates had first described what is now known as "Batesian mimicry" in the Amazon: insects that are themselves perfectly palatable to birds have evolved to mimic other insects that are in some way unpalatable to birds. A bird, in learning to avoid the bad-tasting insects, simultaneously affords the mimics protection. Wallace realized that only one sex may be mimetic. In the examples he describes below the male is non-mimetic yet there are multiple mimetic female forms. So different are the males and females of these species that they had previously been classified as separate species.

THE MALAYAN PAPILIONIDAE OR SWALLOW-TAILED BUTTERFLIES, AS ILLUSTRATIVE OF THE THEORY OF NATURAL SELECTION

SPECIAL VALUE OF THE DIURNAL LEPIDOPTERA FOR ENQUIRIES OF THIS NATURE

When the naturalist studies the habits, the structure, or the affinities of animals, it matters little to which group he especially devotes himself; all alike offer him endless materials for observation and research. But, for the purpose of investigating the phenomena of geographical distribution and of local, sexual, or general variation, the several groups differ greatly in their value and importance. Some have too limited a range, others are not sufficiently varied in specific forms, while, what is of most importance, many groups have not received that amount of attention over the whole region they inhabit, which could furnish materials sufficiently approaching to complete-ness to enable us to arrive at any accurate conclusions as to the phenomena they present as a whole. It is in those groups which are, and have long been, favourites with collectors, that the student of distribution and variation will find his materials the most satisfactory, from their comparative completeness.

Pre-eminent among such groups are the diurnal Lepidoptera or Butterflies, whose extreme beauty and endless diversity have led to their having been assiduously collected in all parts of the world, and to the numerous species and varieties having been figured in a series of magnificent works, from those of Cramer,[85] the contemporary of Linnaeus, down to the inimitable productions of our own Hewitson.[86] But, besides their abundance, their universal distribution, and the great attention that has been paid to them, these insects have other qualities that especially adapt them to elucidate the branches of inquiry already alluded to. These are, the immense development and peculiar structure of the wings, which not only vary in form more than those of any other insects, but offer on both surfaces an end-less variety of pattern, colouring, and texture. The scales, with which they are more or less completely covered, imitate the rich hues and delicate surfaces of satin or of velvet, glitter with metallic lustre, or glow with the changeable tints of the opal. This delicately painted surface acts as a register of the minute differences of organization – a shade of colour, an additional streak or spot, a slight modification of outline continually recurring with the greatest regularity and fixity, while the body and all its other members exhibit no appreciable change. The wings of Butterflies, as Mr. Bates has well put it, "serve as a tablet on which Nature writes the story of the modifications of species;" they enable us to perceive changes that would otherwise be uncertain and difficult of observation, and exhibit to us on an enlarged scale the effects of the climatal and other physical conditions which influence more or less profoundly the organization of every living thing.

A proof that this greater sensibility to modifying causes is not imaginary may, I think, be drawn from the consideration, that while the Lepidoptera as a whole are of all insects the least essentially varied in form, structure, or habits, yet in the number of their specific forms they are not much inferior to those orders which range over a much wider field of nature, and exhibit more deeply seated structural modifications. The Lepidoptera are all vegetable-feeders in their larva-state, and suckers of juices or other liquids in their perfect form. In their most widely separated groups they differ but little from a common type, and offer comparatively unimportant modifications of structure or of habits. The Coleoptera, the Diptera, or the Hymenoptera, on the other hand, present far greater and more essential variations. In either of these orders we have both vegetable and animal feeders, aquatic, and terrestrial, and parasitic groups. Whole families are devoted to special departments in the economy of nature. Seeds, fruits, bones, carcases, excrement, bark, have each their special and dependent insect tribes from among them; whereas the Lepidoptera are, with but few exceptions, confined to the one function of devouring the foliage of living vegetation. We might therefore anticipate that their species-population would be only equal to that of sections of the other orders having a similar uniform mode of existence; and the fact that their numbers are at all comparable with those of entire orders, so much more varied in organization and habits, is, I think, a proof that they are in general highly susceptible of specific modification.[87]

DEFINITION OF THE WORD SPECIES

In estimating these numbers [of species] I have had the usual difficulty to encounter, of determining what to consider species and what varieties. The Malayan region, consisting of a large number of islands of generally great antiquity, possesses, compared to its actual area, a great number of distinct forms, often indeed distinguished by very slight characters, but in most cases so constant in large series of specimens, and so easily separable from each other, that I know not on what principle we can refuse to give them the name and rank of species. One of the best and most orthodox definitions is that of Pritchard,[88] the great ethnologist, who says, that "*separate origin and distinctness of race, evinced by a constant transmission of some characteristic peculiarity of organization,*" constitutes a species. Now leaving out the question of "origin," which we cannot determine, and taking only the proof of separate origin, "*the constant transmission of some characteristic peculiarity of organization,*" we have a definition which will compel us to neglect altogether the amount of difference between any two forms, and to consider only whether the differences that present themselves are *permanent.* The rule, therefore, I have endeavoured to adopt is, that when the difference between two forms inhabiting separate areas seems quite constant, when it can be defined in words, and when it is not confined to a single peculiarity only, I have considered such forms to be species. When, however, the individuals of each locality vary among themselves, so as to cause the distinctions between the two forms to become inconsiderable and indefinite, or where the differences, though constant, are confined to one particular only, such as size, tint, or a single point of difference in marking or in outline, I class one of the forms as a variety of the other.

I find as a general rule that the constancy of species is in an inverse ratio to their range. Those which are confined to one or two islands are generally very constant. When they extend to many islands, considerable variability appears; and when they have an extensive range over a large part of the Archipelago, the amount of unstable variation is very large. These facts are explicable on Mr. Darwin's principles. When a species exists over a wide area, it must have had, and probably still possesses, great powers of dispersion. Under the different conditions of existence in various portions of its area, different variations from the type would be selected, and, were they completely isolated, would soon become distinctly modified forms; but this process is checked by the dispersive powers of the whole species, which leads to the more or less frequent intermixture of the incipient varieties, which thus become irregular and unstable. Where, however, a species has a limited range, it indicates less active powers of dispersion, and the process of modification under changed conditions is less interfered with. The species will therefore exist under

one or more permanent forms according as portions of it have been
isolated at a more or less remote period.

LAWS AND MODES OF VARIATION

What is commonly called variation consists of several distinct phenomena which have been too often confounded. I shall proceed to
consider these under the heads of – 1st, simple variability; 2nd,
polymorphism; 3rd, local forms; 4th, co-existing varieties; 5th, races or
subspecies; and 6th, true species.

1. *Simple variability.* – Under this head I include all those cases in
which the specific form is to some extent unstable. Throughout the
whole range of the species, and even in the progeny of individuals,
there occur continual and uncertain differences of form, analogous
to that variability which is so characteristic of domestic breeds. It is
impossible usefully to define any of these forms, because there are
indefinite gradations to each other form. Species which possess these
characteristics have always a wide range, and are more frequently the
inhabitants of continents than of islands, though such cases are always
exceptional, it being far more common for specific forms to be fixed
within very narrow limits of variation. The only good example of this
kind of variability which occurs among the Malayan Papilionidae is in
Papilio Severus, a species inhabiting all the islands of the Moluccas
and New Guinea, and exhibiting in each of them a greater amount of
individual difference than often serves to distinguish well-marked
species. Almost equally remarkable are the variations exhibited in
most of the species of Ornithoptera,[89] which I have found in some
cases to extend even to the form of the wing and the arrangement of
the nervures [wing veins]. Closely allied, however, to these variable
species are others which, though differing slightly from them, are
constant and confined to limited areas. After satisfying oneself, by the
examination of numerous specimens captured in their native
countries, that the one set of individuals are variable and the others
are not, it becomes evident that by classing all alike as varieties of one
species we shall be obscuring an important fact in nature; and that
the only way to exhibit that fact in its true light is to treat the
invariable local form as a distinct species, even though it does not
offer better distinguishing characters than do the extreme forms of
the variable species. Cases of this kind are the Ornithoptera Priamus,
which is confined to the islands of Ceram and Amboyna, and is very
constant in both sexes, while the allied species inhabiting New Guinea

and the Papuan Islands is exceedingly variable; and in the island of Celebes is a species closely allied to the variable P. Severus, but which, being exceedingly constant, I have described as a distinct species under the name of Papilio Pertinax.

2. *Polymorphism or dimorphism.* – By this term I understand the co-existence in the same locality of two or more distinct forms, not connected by intermediate gradations, and all of which are occasion-ally produced from common parents. These distinct forms generally occur in the female sex only, and their offspring, instead of being hybrids, or like the two parents, appear to reproduce all the distinct forms in varying proportions. I believe it will be found that a con-siderable number of what have been classed as varieties are really cases of polymorphism. Albinoism and melanism are of this character, as well as most of those cases in which well-marked varieties occur in company with the parent species, but without any intermediate forms. If these distinct forms breed independently, and are never reproduced from a common parent, they must be considered as separate species, contact without intermixture being a good test of specific difference. On the other hand, intercrossing without produc-ing an intermediate race is a test of dimorphism. I consider, therefore, that under any circumstances the term "variety" is wrongly applied to such cases.

The Malayan Papilionidae exhibit some very curious instances of polymorphism, some of which have been recorded as varieties, others as distinct species; and they all occur in the female sex. Papilio Memnon is one of the most striking, as it exhibits the mixture of simple variability, local and polymorphic forms, all hitherto classed under the common title of varieties. The polymorphism is strikingly exhibited by the females, one set of which resembles the males in form, with a variable paler colouring; the others have a large spatulate tail to the hinder wings and a distinct style of colouring, which causes them closely to resemble P. Coon, a species having the two sexes alike and inhabiting the same countries, but with which they have no direct affinity. The tailless females exhibit simple variability, scarcely two being found exactly alike even in the same locality. The males of the island of Borneo exhibit constant differences of the under surface, and may therefore be distinguished as a local form, while the conti-nental specimens, as a whole, offer such large and constant differences from those of the islands, that I am inclined to separate them as a distinct species, to which the name P. Androgeus (Cramer) may be applied. We have here, therefore, distinct species, local forms, poly-

Polymorphic females of *Papilio memnon* (from Wallace's *The Malay Archipelago*)

morphism, and simple variability, which seem to me to be distinct phenomena, but which have been hitherto all classed together as varieties. I may mention that the fact of these distinct forms being one species is doubly proved. The males, the tailed and tailless females, have all been bred from a single group of the larvae, by Messrs. Payen and Bocarmé, in Java, and I myself captured, in Sumatra, a male P. Memnon, and a tailed female P. Achates, under circumstances which led me to class them as the same species.[90]

. . .

The phenomena of dimorphism and polymorphism may be well illustrated by supposing that a blue-eyed, flaxen-haired Saxon man had two wives, one a black-haired, red-skinned Indian squaw, the other a wooly-haired, sooty-skinned negress – and that instead of the children being mulattoes of brown or dusky tints, mingling the separate characteristics of their parents in varying degrees, all the boys should be pure Saxon boys like their father, while the girls should altogether resemble their mothers. This would be thought a wonderful fact; yet the phenomena here brought forward as existing in the insect-world are still more extraordinary; for each mother is capable not only of producing male offspring like the father, and female like herself, but also of producing other females exactly like her fellow-wife, and altogether differing from herself. If an island could be stocked with a colony of human beings having similar physiological idiosyncrasies with Papilio Pammon or Papilio Ormenus, we should see white men living with yellow, red, and black women, and their offspring always reproducing the same types; so that at the end of many generations the men would remain pure white, and the women of the same well-marked races as at the commencement.

The distinctive character therefore of dimorphism is this, that the union of these distinct forms does not produce intermediate varieties, but reproduces the distinct forms unchanged. In simple varieties, on the other band, as well as when distinct local forms or distinct species are crossed, the offspring never resembles either parent exactly, but is more or less intermediate between them. Dimorphism is thus seen to be a specialized result of variation, by which new physiological phenomena have been developed; the two should therefore, whenever possible, be kept separate.

3. *Local form, or variety.* – This is the first step in the transition from variety to species. It occurs in species of wide range, when groups of individuals have become partially isolated in several points of its area of distribution, in each of which a characteristic form has become more or less completely segregated. Such forms are very common in all parts of the world, and have often been classed by one author as varieties, by another as species. I restrict the term to those cases where the difference of the forms is very slight, or where the segregation is more or less imperfect. The best example in the present group is Papilio Agamemnon, a species which ranges over the greater part of tropical Asia, the whole of the Malay archipelago, and a portion of the Australian and Pacific regions. The modifications are principally of size and form, and, though slight, are tolerably constant in each locality. The steps, however, are so numerous and gradual that it would be impossible to define many of them, though the extreme forms are sufficiently distinct. Papilio Sarpedon presents somewhat similar but less numerous variations.

4. *Co-existing Variety.* – This is a somewhat doubtful case. It is when a slight but permanent and hereditary modification of form exists in company with the parent or typical form, without presenting those intermediate gradations which would constitute it a case of simple variability. It is evidently only by direct evidence of the two forms breeding separately that this can be distinguished from dimorphism. The difficulty occurs in Papilio Jason, and P. Evemon, which inhabit the same localities, and are almost exactly alike in form, size, and colouration, except that the latter always wants a very conspicuous red spot on the under surface, which is found not only in P. Jason, but in all the allied species. It is only by breeding the two insects that it can be determined whether this is a case of a co-existing variety or of dimorphism. In the former case, however, the difference being con-

stant and so very conspicuous and easily defined, I see not how we could escape considering it as a distinct species. A true case of co-existing forms would, I consider, be produced, if a slight variety had become fixed as a local form, and afterwards been brought into contact with the parent species, with little or no intermixture of the two; and such instances do very probably occur.

5. *Race or subspecies.* – These are local forms completely fixed and isolated; and there is no possible test but individual opinion to determine which of them shall be considered as species and which varieties. If stability of form and "*the constant transmission of some characteristic peculiarity of organization*" is the test of a species (and I can find no other test that is more certain than individual opinion) then every one of these fixed races, confined as they almost always are to distinct and limited areas, must be regarded as a species; and as such I have in most cases treated them. The various modifications of Papilio Ulysses, P. Peranthus, P. Codrus, P. Eurypilus, P. Helenus, &c., are excellent examples; for while some present great and well-marked, others offer slight and inconspicuous differences, yet in all cases these differences seem equally fixed and permanent. If, therefore, we call some of these forms species, and others varieties, we introduce a purely arbitrary distinction, and shall never be able to decide where to draw the line. The races of Papilio Ulysses, for example, vary in amount of modification from the scarcely differing New Guinea form to those of Woodlark Island and New Caledonia, but all seem equally constant; and as most of these had already been named and described as species, I have added the New Guinea form under the name of P. Autolycus. We thus get a little group of Ulyssine Papilios, the whole comprised within a very limited area, each one confined to a separate portion of that area, and, though differing in various amounts, each apparently constant.

Few naturalists will doubt that all these may and probably have been derived from a common stock, and therefore it seems desirable that there should be a unity in our method of treating them; either call them all varieties or all species. Varieties, however, continually get overlooked; in lists of species they are often altogether unrecorded; and thus we are in danger of neglecting the interesting phenomena of variation and distribution which they present. I think it advisable, therefore, to name all such forms; and those who will not accept them as species may consider them as subspecies or races.

6. *Species.* – Species are merely those strongly marked races or local forms which when in contact do not intermix, and when inhabiting distinct areas are generally believed to have had a separate origin, and

to be incapable of producing a fertile hybrid offspring. But as the test
of hybridity cannot be applied in one case in ten thousand, and even
if it could be applied would prove nothing, since it is founded on an
assumption of the very question to be decided – and as the test of
separate origin is in every case inapplicable – and as, further, the test
of non-intermixture is useless, except in those rare cases where the
most closely allied species are found inhabiting the same area, it will
be evident that we have no means whatever of distinguishing so-called
"true species" from the several modes of variation here pointed out,
and into which they so often pass by an insensible gradation. It is
quite true that in the great majority of cases, what we term "species"
are so well marked and definite that there is no difference in opinion
about them; but as the test of a true theory is, that it accounts for, or
at the very least is not inconsistent with, the whole of the phenomena
and apparent anomalies of the problem to be solved, it is reasonable
to ask that those who deny the origin of species by variation and
selection should grapple with the facts in detail, and show how the
doctrine of distinct origin and permanence of species will explain and
harmonize them. It has been recently asserted by Dr. J. E. Gray[91] (in
the Proceedings of the Zoological Society for 1863, page 134), that
the difficulty of limiting species is in proportion to our ignorance,
and that just as groups or countries are more accurately known and
studied in greater detail the limits of species become settled. This
statement has, like many other general assertions, its portion of both
truth and error. There is no doubt that many uncertain species,
founded on few or isolated specimens, have had their true nature
determined by the study of a good series of examples: they have been
thereby established as species or as varieties and the number of times
this has occurred is doubtless very great. But there are other, and
equally trustworthy cases, in which, not single species, but whole
groups have, by the study of a vast accumulation of materials, been
proved to have no definite specific limits.[92]

. . .

These few examples show, I think, that in every department of
nature there occur instances of the instability of specific form, which
the increase of materials aggravates rather than diminishes. And it
must be remembered that the naturalist is rarely likely to err on the
side of imputing greater indefiniteness to species than really exists.
There is a completeness and satisfaction to the mind in defining and
limiting and naming a species, which leads us all to do so whenever
we conscientiously can, and which we know has led many collectors to

reject vague intermediate forms as destroying the symmetry of their cabinets. We must therefore consider these cases of excessive variation and instability as being thoroughly well established and to the objection that, after all, these cases are but few compared with those in which species can be limited and defined, and are therefore merely exceptions to a general rule, I reply that a true law embraces all apparent exceptions, and that to the great laws of nature there are no real exceptions – that what appear to be such are equally results of law, and are often (perhaps indeed always) those very results which are most important as revealing the true nature and action of the law. It is for such reasons that naturalists now look upon the study of *varieties* as more important than that of well-fixed species. It is in the former that we see nature still at work, in the very act of producing those wonderful modifications of form, that endless variety of colour, and that complicated harmony of relations, which gratify every sense and give occupation to every faculty of the true lover of nature.[93]

. . .

MIMICRY

As in America, so in the Old World, species of Danaidae[94] are the objects which the other families most often imitate. But besides these, some genera of Morphidae and one section of the genus Papilio are also less frequently copied. Many species of Papilio mimic other species of these three groups so closely that they are undistinguishable when on the wing; and in every case the pairs which resemble each other inhabit the same locality.

The following list exhibits the most important and best marked cases of mimicry which occur among the Papilionidae of the Malayan region and India:

Mimickers	Species mimicked	Common habitat
	DANAIDAE	
1. Papilio paradoxa (male & female)	Euploea Midamus (male & female)	Sumatra, &c.
2. P. Caunus	E. Rhadamanthus	Borneo & Sumatra
3. P. Thule	Danais sobrina	New Guinea
4. P. Macareus	D. Aglaia	Malacca, Java
5. P. Agestor	Danais Tytia	Northern India

Mimickers	Species mimicked	Common habitat
6. P. Idaeoides	Hestia Leuconoe	Philippines
7. P. Delessertii	Ideopsis daos	Penang
MORPHIDAE		
8. P. Pandion (female)	Drusilla bioculata	New Guinea
PAPILIO (POLYDORUS-AND COON-GROUPS)		
9. P. Pammon (Romulus, female)	Papilio Hector	India
10. P. Theseus, var. (female)	P. Antiphus	Sumatra, Borneo
11. P. Theseus, var. (female)	P. Diphilus	Sumatra, Java
12. P. Memnon, var. (Achates, female)	P. Coon	Sumatra
13. P. Androgeus, var. (Achates, female)	P. Doubledayi	Northern India
14. P. Oenomaus (female)	P. Liris	Timor

We have, therefore, fourteen species or marked varieties of Papilio, which so closely resemble species of other groups in their respective localities, that it is not possible to impute the resemblance to accident. The first two in the list (Papilio paradoxa and P. Caunus) are so exactly like Euploea Midamus and E. Rhadamanthus on the wing, that although they fly very slowly, I was quite unable to distinguish them. The first is a very interesting case, because the male and female differ and each mimics the corresponding sex of the Euploea. A new species of Papilio which I discovered in New Guinea resembles Danais sobrina from the same country, just as Papilio Marcareus resembles Danais Aglaia in Malacca, and (according to Dr. Horsfield's[95] figure) still more closely in Java. The Indian Papilio Agestor closely imitates Danais Tytia, which has quite a different style of colouring from the preceding; and the extraordinary Papilio Idaeoides from the Philippine Islands, must, when on the wing, perfectly resemble the Hestia Leuconoe of the same region, as also does the Papilio Delessertii imitate the Ideopsis daos from Penang. Now in every one of these cases the Papilios are very scarce, while the Danaidae which they resemble are exceedingly abundant – most of them swarming so as to be a positive nuisance to the collecting entomologist by continually hovering before him when he is in search of newer and more varied captures. Every garden, every roadside, the suburbs of every village are full of them, indicating very clearly that their life is an easy one, and that they are free from persecution by the foes which keep down the population of less favoured races. This superabundant population has been shown by Mr. Bates to be a general characteristic of all American groups and species which are objects of mimicry; and it is interesting to find his observations confirmed by examples on the other side of the Globe.

The remarkable genus Drusilla, a group of pale coloured butterflies, more or less adorned with ocellate spots, is also the object of mimicry by three distinct genera (Melanitis, Hyantis, and Papilio). These insects, like the Danaidae, are abundant in individuals, have a very weak and slow flight, and do not seek concealment, or appear to have any means of protection from insectivorous creatures. It is natural to conclude, therefore, that they have some hidden property which saves them from attack; and it is easy to see that when any other insects, by what we call accidental variation, come more or less remotely to resemble them, the latter will share to some extent in their immunity. An extraordinary dimorphic form of the female of Papilio Ormenus has come to resemble the Drusillas sufficiently to be taken for one of that group at a little distance; and it is curious that I captured one of these Papilios in the Aru Islands hovering along the ground, and settling on it occasionally, just as it is the habit of the Drusillas to do. The resemblance in this case is only general; but this form of Papilio varies much, and there is therefore material for natural selection to act upon, so as ultimately to produce a copy as exact as in the other cases.

The eastern Papilios allied to Polydorus, Coon, and Philoxenus, form a natural section of the genus resembling, in many respects, the Aeneas-group of South America, which they may be said to represent in the East. Like them, they are forest insects, have a low and weak flight, and in their favourite localities are rather abundant in individuals; and like them, too, they are the objects of mimicry. We may conclude, therefore, that they possess some hidden means of protection, which makes it useful to other insects to be mistaken for them.

The Papilios which resemble them belong to a very distinct section of the genus, in which the sexes differ greatly; and it is those females only which differ most from the males, and which have already been alluded to as exhibiting instances of dimorphism, which resemble species of the other group.

The resemblance of P. Romulus to P. Hector is, in some specimens, very considerable, and has led to the two species being placed following each other in the British Museum Catalogues and by Mr. E. Doubleday. I have shown, however, that P. Romulus is probably a dimorphic form of the female P. Pammon, and belongs to a distinct section of the genus.

The next pair, Papilio Theseus, and P. Antiphus, have been united as one species both by De Haan[96] and in the British Museum Catalogues. The ordinary variety of P. Theseus found in Java almost as nearly resembles P. Diphilus, inhabiting the same country. The most interesting case, however, is the extreme female form of P. Memnon (figured by Cramer under the name of P. Achates), which has acquired the general form and markings of P. Coon, an insect which differs from the ordinary male P. Memnon, as much as any two species which can be chosen in this extensive and highly varied genus; and, as if to show that this resemblance is not accidental, but is the result of law, when in India we find a species closely allied to P. Coon, but with red instead of yellow spots (P. Doubledayi), the corresponding variety of P. Androgeus (P. Achates, Cramer, 182, A, B) has acquired exactly the same peculiarity of having red spots instead of yellow. Lastly, in the island of Timor, the female of P. Oenomaus (a species allied to P. Memnon) resembles so closely P. Liris (one of the Polydorus-group), that the two, which were often seen flying together, could only be distinguished by a minute comparison after being captured.

The last six cases of mimicry are especially instructive, because they seem to indicate one of the processes by which dimorphic forms have been produced. When, as in these cases, one sex differs much from the other, and varies greatly itself, it may happen that occasionally individual variations will occur having a distant resemblance to groups which are the objects of mimicry, and which it is therefore advantageous to resemble. Such a variety will have a better chance of preservation; the individuals possessing it will be multiplied; and their accidental likeness to the favoured group will be rendered permanent by hereditary transmission, and, each successive variation which increases the resemblance being preserved, and all variations departing from the favoured type having less chance of preservation, there will in time result those singular cases of two or more isolated and fixed forms, bound together by that intimate relationship which constitutes them the sexes of a single species. The reason why the females are more subject to this kind of modification than the males is, probably, that their slower flight, when laden with eggs, and their exposure to attack while in the act of depositing their eggs upon leaves, render it especially advantageous for them to have some additional protection. This they at once obtain by acquiring a resemblance to other species which, from whatever cause, enjoy a comparative immunity from persecution.

CONCLUDING REMARKS ON VARIATION IN LEPIDOPTERA

This summary of the more interesting phenomena of variation presented by the eastern Papilionidae is, I think, sufficient to substantiate my position, that the Lepidoptera are a group that offer especial facilities for such inquiries; and it will also show that they have undergone an amount of special adaptive modification rarely equalled among the more highly organized animals. And, among the Lepidoptera, the great and pre-eminently tropical families of Papilionidae and Danaidae seem to be those in which complicated adaptations to the surrounding organic and inorganic universe have been most completely developed, offering in this respect a striking analog to the equally extraordinary, though totally different, adaptations which present themselves in the Orchidae, the only family of plants in which mimicry of other organisms appears to play any important part, and the only one in which cases of conspicuous polymorphism occur; for as such we must class the male, female, and hermaphrodite forms of Catasetum tridentatum, which differ so greatly in form and structure that they were long considered to belong to three distinct genera.[97] [1870] ❧

   Darwin noted Wallace's interest in "protective coloration". He wrote to Wallace that "On Monday evening I called on Bates and put a difficulty before him, which he could not answer, and, as on some former similar occasion, his first suggestion was, 'You had better ask Wallace.' My difficulty is, why are caterpillars sometimes so beautifully and artistically coloured? Seeing that many are coloured to escape danger, I can hardly attribute their bright colour in other cases to mere physical conditions."[98] Wallace takes up the story in *My Life*:

❧   On reading this letter, I almost at once saw what seemed to be a very easy and probable explanation of the facts. I had then just been preparing for publication (in the *Westminster Review*) my rather elaborate paper on "Mimicry and Protective Colouring", and the numerous cases in which specially showy and slow-flying butterflies were known to have a peculiar odour and taste which protected them from the attacks of insect-eating birds and other animals, led me at once to suppose that the gaudily-coloured caterpillars must have a similar protection. I had just ascertained from Mr. Jenner Weir[99] that one of our common white moths (*Spilosoma menthrastri*) would not be eaten by most of the small birds in his aviary, nor by young turkeys. Now, as a white moth is as conspicuous in the dusk as a coloured caterpillar in the daylight, this case seemed to me so much on a par with the other that I felt almost sure my explanation would turn out correct. I at once wrote to Mr. Darwin to this effect, and his reply, dated February 26, is as follows:

"My Dear Wallace,

Bates was quite right; you are the man to apply to in a difficulty. I never heard anything more ingenious than your suggestion, and I hope you may be able to prove it true. That is a splendid fact about the white moths; it warms one's very blood to see a theory thus almost proved to be true."

The following week I brought the subject to the notice of The Fellows of the Entomological Society at their evening meeting (March 4), requesting that any of them who had the opportunity would make observations or experiments during the summer in accordance with Mr. Darwin's suggestion. I also wrote a letter to *The Field* newspaper which, as it explains my hypothesis in simple language, I here give entire:-

"Caterpillars and birds.

Sir,
May I be permitted to ask the co-operation of your readers in making some observations during the coming spring and summer which are of great interest to Mr. Darwin and myself? I will first state what observations are wanted, and then explain briefly why they are wanted. A number of our smaller birds devour quantities of caterpillars, but there is reason to suspect that they do not eat all alike. Now we want direct evidence as to which species they eat and which they reject. This may be obtained in two ways. Those who keep insectivorous birds, such as thrushes, robins, or any of the warblers (or any other that will eat caterpillars), may offer them all the kinds they can obtain,

and carefully note (1) which they eat, (2) which they refuse to touch, and (3) which they seize but reject. If the name of the caterpillar cannot be ascertained, a short description of its more prominent characters will do very well, such as whether it is hairy or smooth, and what are its chief colours, especially distinguishing such as are green or brown from such as are of bright and conspicuous colours, as yellow, red, or black. The food plant of the caterpillar should also be stated when known. Those who do not keep birds, but have a garden much frequented by birds, may put all the caterpillars they can find in a soup plate or other vessel, which must be placed in a larger vessel of water, so that the creatures cannot escape, and then after a few hours note which have been taken and which left. If the vessel could be placed where it might be watched from a window, so that the kind of birds which took them could also be noted, the experiment would be still more complete. A third set of observations might be made on young fowls [*sic.*], turkeys, guinea-fowls, pheasants, etc., in exactly the same manner.

Now the purport of these observations is to ascertain the law which had determined the coloration of caterpillars. The analogy of many other insects leads us to believe that all those which are green or brown, or of such speckled or mottled tints as to resemble closely the leaf or bark of the plant on which they feed, or the substance on which they usually repose, are thus to some degree protected from the attacks of birds and other enemies. We should expect, therefore, that all which are thus protected would be greedily eaten by birds whenever they can find them. But there are other caterpillars which seem coloured on purpose to be conspicuous, and it is very important to know whether they have another kind of protection, altogether independent of disguise, such as a disagreeable odour and taste. If they are thus protected, so that the majority of birds will never eat them, we can understand that to get the full benefit of this protection they should be easily recognized, should have some outward character by which birds would soon learn to know them and thus let them alone; because if birds could not tell the eatable from the uneatable till they had seized and tasted them, the protection would be of no avail, a growing caterpillar being so delicate that a wound is certain death. If, therefore, the eatable caterpillars derive a partial protection from their obscure and imitative colouring, then we can understand that it would be an advantage to the uneatable kinds to be well distinguished from them by bright and conspicuous colours.

I may add that this question has an important bearing on the whole theory of the origin of the colours of animals and especially of insects. I hope many of your readers may be thereby induced to make such observations as I have indicated, and if they will kindly send me their notes at the end of the summer, or earlier, I will undertake to compare and tabulate the whole, and to make known the results, whether they confirm or refute the theory here indicated.

Alfred R. Wallace.
9, St. Mark's Crescent, Regent's Park, N.W.,
March, 1867."

This letter brought me only one reply, from a gentleman in Cumberland, who informed me that the common "gooseberry" caterpillar, which is the larva of the magpie moth (*Abraxas grossulariata*), is refused by young pheasants, partridges, and wild ducks, as well as sparrows and finches, and that all birds to whom he offered it rejected it with evident dread and abhorrence. But in 1869 two entomologists, Mr. Jenner Weir and Mr. A. G. Butler,[100] gave an account of their two seasons' experiments and observations with several of our most gaily-coloured caterpillars, and with a considerable variety of birds, and also with lizards, frogs, and spiders, confirming my explanation in a most remarkable manner.[101] [1905] ﻉ

Darwinism

Wallace's magisterial summary of evolutionary biology, his *Darwinism*, came out in 1889, several years after Darwin's death. Wallace explained what provoked him to write it:

ﻉ Many of my correspondents, as well as persons I met in America, told me that they could not understand Darwin's "Origin of Species," but they did understand my lecture on "Darwinism;" and it therefore occurred to me that a popular exposition of the subject might be useful, not only as enabling the general reader to understand Darwin, but also to serve as an answer to the many articles and books professing to disprove the theory of natural selection. During the whole of the year 1888 I was engaged in writing this book, which, though largely following the lines of Darwin's work, contained a great many new features, and dwelt especially with those parts of the subject which had been most generally misunderstood.[102] [1905] ﻉ

Wallace has a knack of choosing the perfect example to make a point, as in this invocation of flightless insects on oceanic islands.

◄◊ In species which have a wide range the struggle for existence will often cause some individuals or groups of individuals to adopt new habits in order to seize upon vacant places in nature where the struggle is less severe. Some, living among extensive marshes, may adopt a more aquatic mode of life; others, living where forests abound, may become more arboreal. In either case we cannot doubt that the changes of structure needed to adapt them to their new habits would soon be brought about, because we know that variations in all the external organs and all their separate parts are very abundant and also considerable in amount. That such divergence of character has actually occurred we have some direct evidence. Mr. Darwin informs us that in the Catskill Mountains in the United States there are two varieties of wolves, one with light greyhound-like form which pursues deer, the other more bulky with shorter legs, which more frequently attacks sheep (*Origin of Species*, p. 71). Another good example is that of the insects of the island of Madeira, many of which have either lost their wings or have had them so much reduced as to be useless for flight, while the very same species on the continent of Europe possess fully developed wings. In other cases, the wingless Madeira species are distinct from, but closely allied to, winged species of Europe. The explanation of this change is, that Madeira, like many oceanic islands in the temperate zone, is much exposed to sudden gales of wind, and as most of the fertile land is on the coast, insects which flew much would be very liable to be blown out to sea and lost. Year after year, therefore, those individuals which had shorter wings, or which used them least, were preserved; and thus, in time, terrestrial, wingless, or imperfectly winged races or species have been produced. That this is the true explanation of this singular fact is proved by much corroborative evidence. There are some few flower-frequenting insects in Madeira to whom wings are essential, and in these the wings are somewhat larger than in the same species on the mainland. We thus see that there is no general tendency to the abortion of wings in Madeira, but that it is simply a case of adaptation to new conditions. Those insects to whom wings were not absolutely essential escaped a serious danger by not using them, and the wings therefore became reduced or were completely lost. But when they were essential they were enlarged and strengthened, so that the insect could battle against the winds and save itself from destruction at sea. Many flying insects, not varying fast enough, would be destroyed before they could establish themselves, and that we may explain the total absence from Madeira of several whole families of winged insects which must have had many opportunities of reaching the islands. Such are the large groups of the tiger-beetles (Cicindelidae), the chafers (Melolonthidae), the click-beetles (Elateridae), and many others.

But the most curious and striking confirmation of this portion of Mr. Darwin's theory is afforded by the case of Kerguelen Island. This island was visited by the *Transit of Venus* expedition. It is one of the stormiest places on the globe, being subject to almost perpetual gales, while, there being no wood, it is almost entirely without shelter. The Rev A. E. Eaton,[103] an experienced entomologist, was naturalist to the expedition, and he assiduously collected the few insects that were to be found. All were incapable of flight, and most of them entirely without wings. They included a moth, several flies, and numerous beetles. As these insects could hardly have reached the island in the wingless state, even if there were any other known land inhabited by them – which there is not – we must assume that, like the Madeiran insects, they were originally winged, and lost their power of flight because its possession was injurious to them.[104]
[1889] ছ

Biogeography

"There is no part of natural history more interesting or instructive than the study of the geographical distribution of animals."[1] Wallace early recognized that biogeographical and evolutionary arguments are essentially inseparable – the distribution of living things, after all, is largely determined by where they evolved. He first expressed this idea in his "Sarawak Law" and it remained a major preoccupation throughout his life. Indeed, like Wallace himself, the "Sarawak Law" defies disciplinary categorization: it could have been included in either the "evolution" or "biogeography" sections of this anthology. However, despite the overlap between Wallace's evolutionary and biogeographic interests, there remains a substantial body of specifically biogeographic work. Not only did he famously delineate the boundary between the Australasian and Oriental faunas ("Wallace's Line"), but he also set the standard for rigorous analysis of biogeographic problems in his monumental synthetic works on the subject, *Geographic Distribution of Animals* (1876) and *Island Life* (1880). In addition, Wallace's historical perspective on biological problems – again first outlined in the "Sarawak Law" – resolved the longstanding biogeographical problem of disjunct distributions, in which different members of a particular group occur in widely separated locations. Wallace pointed out that such a distribution may be a relic of a once continuous distribution, with extinction having eliminated the intervening populations. It is only proper that Wallace is regarded today as the father of the field of evolutionary biogeography.

In late 1847, Wallace spent a day in the British Museum's insect room. Impressed by the "overwhelming numbers of beetles and butterflies I was able to look over", he wrote to Bates, "I begin to feel rather dissatisfied with a mere local collection; little is to be learnt by it."[2] A local collection is a mere snapshot of biological diversity; Wallace realized that an understanding of the "origin of species" and of the factors affecting their distribution required a grander geographical quest. The remedy for that dissatisfaction was his trip with Bates to the Amazon.

The Amazon

Wallace and Bates were biological neophytes when they arrived in Brazil – new to serious scientific collecting and new to the Tropics – but they rapidly acquired exceptional skills as field biologists. Because they were in the business of collecting (to sell) as many exotic species as possible, they inevitably wanted reliable distributional data on their target species: there was no point travelling two weeks upriver in pursuit of a particular bird if, in fact, that bird was not present in the specified locality. It is hardly surprising, therefore, that Wallace and Bates acquired an acute interest in biogeography; for them, it simply addressed the practical matter of *where* to find the material that was their livelihood.

Biogeography can be thought of as the study of factors that limit the distribution of organisms. Wallace understood this clearly:

In each of these countries we find well-marked smaller districts, appearing to depend upon climate. The tropical and temperate parts of America and Africa have, generally speaking, distinct animals in each of them.

On a more minute acquaintance with the animals of any country, we shall find that they are broken up into yet smaller local groups, and that almost every district has peculiar animals found nowhere else. Great mountain-chains are found to separate countries possessing very distinct sets of animals. Those of the east and west of the Andes differ very remarkably. The Rocky Mountains also separate two distinct zoological districts; California and Oregon on the one side, possessing plants, birds, and insects, not found in any part of North America east of that range.

But there must be many other kinds of boundaries besides these, which, independently of climate, limit the range of animals. Places not more than fifty or a hundred miles apart often have species of insects and birds at the one, which are not found at the other. There must be some boundary which determines the range of each species; some external peculiarity to mark the line which each one does not pass.

These boundaries do not always form a barrier to the progress of the animal, for many birds have a limited range in a country where

A Victorian view of the Amazon forest (from Henry Bates, *Naturalist on the River Amazons*)

there is nothing to prevent them flying in every direction, – as in the case of the nightingale, which is quite unknown in some of our western counties. Rivers generally do not determine the distribution of species, because, when small, there are few animals which cannot pass them but in very large rivers the case is different, and they will, it is believed, be found to be the limits, determining the range of many animals of all orders.

With regard to the Amazon, and its larger tributaries, I have ascertained this to be the case.[3] [1853] ҉

He made his case for the importance of rivers in delimiting the distribution of species in one of his earliest scientific publications, "On the Monkeys of the Amazon". Note that he finished by taking to task an earlier naturalist who failed to live up to his own standards of record keeping.

҉ In the various works on natural history and in our museums, we have generally but the vaguest statements of locality. S. America, Brazil, Guiana, Peru, are among the most common; and if we have "River Amazon" or "Quito" attached to a specimen, we may think ourselves fortunate to get anything so definite: though both are on the boundary of two distinct zoological districts, and we have nothing to tell us whether the one came from the north or south of the Amazon, or the other from the east or the west of the Andes. Owing to this uncertainty of locality, and the additional confusion created by mistaking allied species from distant countries, there is scarcely an animal whose exact geographical limits we can mark out on the map.

On this accurate determination of an animal's range many interesting questions depend. Are very closely allied species ever separated by a wide interval of country? What physical features determine the boundaries of species and of genera? Do the isothermal lines ever accurately bound the range of species, or are they altogether independent of them? What are the circumstances which render certain rivers and certain mountain ranges the limits of numerous species, while others are not? None of these questions can be satisfactorily answered till we have the range of numerous species accurately determined.

During my residence in the Amazon district I took every opportunity of determining the limits of species, and I soon found that the Amazon, the Rio Negro and the Madeira formed the limits beyond which certain species never passed. The native hunters are perfectly acquainted with this fact, and always cross over the river when they want to procure particular animals, which are found even on the river's bank on one side, but never by any chance on the other. On approaching the sources of the rivers they cease to be a boundary, and most of the species are found on both sides of them. Thus several Guiana species come up to the Rio Negro and Amazon, but do not pass them; Brazilian species on the contrary reach but do not pass the Amazon to the north. Several Ecuador species from the east of the Andes reach down into the tongue of land between the Rio Negro and Upper Amazon, but pass neither of those rivers, and others from Peru are bounded on the north by the Upper Amazon, and on the east by the Madeira. Thus there are four districts, the Guiana, the Ecuador, the Peru and the Brazil districts, whose boundaries on one side are determined by the rivers I have mentioned.

In going up the Rio Negro the difference in the two sides of the river is very remarkable.

In the lower part of the river you will find on the north the *Jacchus bicolor* and the *Brachyurus couxiu*, and on the south the red-whiskered *Pithecia.* Higher up you will find on the north the *Ateles paniscus*, and on the south the new black *Jacchus* and the *Lagothrix Humboldtii.*

Spix,[4] in his work on the monkeys of Brazil, frequently gives "banks of the river Amazon" as a locality, not being aware apparently that the species found on one side very often do not occur on the other, though the fact is generally known to the natives. In these observations I have only referred to the monkeys, but the same phenomena occur both with birds and insects, as I have observed in many instances.[5] [1852] ટ≈

Ironically, Wallace found to his cost that he was guilty of the same crime as Spix. In a letter to Bates written in Sumatra he recalls,

≈ટ In a paper I read on "The Monkeys of the Lower Amazon and Rio Negro" I showed that the species were often different on the opposite sides of the river. Guayana species came up to the east bank, Columbian species to the west bank, and I stated that it was therefore important that travellers collecting on the banks of large rivers should note from which side every specimen came. Upon this Dr. Gray[6] came down upon me with a regular floorer. "Why," said he, "we have specimens collected by Mr. Wallace himself marked 'Rio Negro' only." I do not think I answered him properly at the time, that those specimens were sent from Barra [Manaus] before I had the slightest idea myself that the species were different on the two banks.[7] [1861] ટ≈

South-east Asia: "Wallace's Line"

Wallace's eight years in what he called the "Malay Archipelago" were the "central and controlling incident" of his life.[8] Able to capitalize on his hard-won experience as a field naturalist in Brazil, his South-east Asia expedition was a biological *tour de force.* He chose the region because

≈ટ During my constant attendance [after returning from Brazil] at the meetings of the Zoological and Entomological Societies, and visits to the insect and bird departments of the British Museum, I had obtained sufficient information to satisfy me that the very finest field for an exploring and collecting naturalist was to be found in the Malay Archipelago, of which just sufficient is known to prove its wonderful richness, while no part of it, with the one exception of the island of Java, had been well explored as regards its natural history.[9] [1905] ટ≈

The area was also likely to illuminate interesting biogeographic questions: not only does it comprise a huge number of islands that vary both in size and in the extent of their isolation from each other, but it also bridges the gap from Australia to Asia, from kangaroos to monkeys.

◄§ Nowhere does the ancient doctrine – that differences or similarities in the various forms of life that inhabit different countries are due to corresponding physical differences or similarities in the countries themselves – meet with so direct and palpable a contradiction. Borneo and New Guinea, as alike physically as two distinct countries can be, are zoologically wide as the poles asunder; while Australia, with its dry winds, its open plains, its stony deserts, and its temperate climate, yet produces birds and quadrupeds which are closely related to those inhabiting the hot damp luxuriant forests, which everywhere clothe the plains and mountains of New Guinea.[10] [1869]   §►

In a letter to Bates written from the field, Wallace refers to the discontinuity between Australasian and Asian forms.

◄§ In this Archipelago there are two distinct faunas rigidly circumscribed, which differ as much as those of South America and Africa, and more than those of Europe and North America: yet there is nothing on the map or on the face of the islands to mark their limits. The boundary line often passes between islands closer than others in the same group. I believe the western part to be a separated portion of continental Asia, the eastern the fragmentary prolongation of a former Pacific continent. In mammalia and birds the distinction is marked by genera, families, and even orders confined to one region; in *insects* by a number of *genera* and little groups of peculiar species, the *families* of insects having generally a universal distribution.[11] [1858]   §►

Wallace's discovery was serendipitous.

◄§ Having been unable to find a vessel direct to Macassar [Ujung Pandang, Sulawesi], I took passage to Lombok, whence I was assured I should easily reach my destination. By this delay, which seemed to me at the time a misfortune, I was enabled to make some very interesting collections in Bali and Lombok, two islands which I should otherwise never have seen. I was thus enabled to determine the exact boundary between two of the primary zoological regions, the Oriental and the Australian.[12] [1905]

Wallace published in 1860 a formal account of his discovery.

ON THE ZOOLOGICAL GEOGRAPHY OF THE MALAY ARCHIPELAGO

In Mr. Sclater's paper on the Geographical Distribution of Birds, read before the Linnean Society, and published in the "Proceedings" for February 1858,[13] he has pointed out that the western islands of the Archipelago belong to the Indian, and the eastern to the Australian region of Ornithology. My researches in these countries lead me to believe that the same division will hold good in every branch of Zoology; and the object of my present communication is to mark out the precise limits of each region, and to call attention to some inferences of great general importance as regards the study of the laws of organic distribution.

The Australian and Indian regions of Zoology are very strongly contrasted. In one the Marsupial order constitutes the great mass of the mammalia, – in the other not a solitary marsupial animal exists. Marsupials of at least two genera (*Cuscus* and *Belideus*) are found all over the Moluccas and in Celebes [Sulawesi]; but none have been detected in the adjacent islands of Java and Borneo. Of all the varied forms of *Quadrumana, Carnivora, Insectivora,* and *Ruminantia* which abound in the western half of the Archipelago, the only genera found in the Moluccas are *Paradoxurus* and *Cervus.* The *Sciuridæ,* so numerous in the western islands, are represented in Celebes by only two or three species, while not one is found further east. Birds furnish equally remarkable illustrations. The Australian region is the richest in the world in Parrots; the Asiatic is (of tropical regions) the poorest. Three entire families of the Psittacine order are peculiar to the former region, and two of them, the Cockatoos and the Lories, extend up to its extreme limits, without a solitary species passing into the Indian islands of the Archipelago. The genus *Palæornis* is, on the other hand, confined with equal strictness to the Indian region. In the Rasorial order [birds obtaining their food by scratching the ground], the *Phasianidæ* are Indian, the *Megapodiidæ* Australian; but in this case one species of each family just passes the limits into the adjacent region. The genus *Tropidorhynchus,* highly characteristic of the Australian region, and everywhere abundant as well in the Moluccas and New

Guinea as in Australia, is quite unknown in Java and Borneo. On the other hand, the entire families of *Bucconidæ*, *Trogonidæ* and *Phyllorni-thidæ*, and the genera *Pericrocotus*, *Picnonotus*, *Trichophorus*, *Ixos*, in fact, almost all the vast family of Thrushes and a host of other genera, cease abruptly at the eastern side of Borneo, Java, and Bali. All these groups are *common birds* in the great Indian islands; they abound everywhere; they are the characteristic features of the ornithology; and it is most striking to a naturalist, on passing the narrow straits of Macassar and Lombok, suddenly to miss them entirely, together with the *Quadrumana* and *Felidæ*, the *Insectivora* and *Rodentia*, whose varied species people the forests of Sumatra, Java, and Borneo.

To define exactly the limits of the two regions where they are (geographically) most intimately connected, I may mention that during a few days' stay in the island of Bali I found birds of the genera *Copsychus*, *Megalaima*, *Tiga*, *Ploceus*, and *Sturnopastor*, all characteristic of the Indian region and abundant in Malacca [Peninsular Malaysia], Java, and Borneo; while on crossing over to Lombock, during three months collecting there, not one of them was ever seen; neither have they occurred in Celebes nor in any of the more eastern islands I have visited. Taking this in connexion with the fact of *Cacatua*, *Tropidorhyn-chus*, and *Megapodius* having their western limit in Lombock, we may consider it established that the Strait of Lombok (only 15 miles wide) marks the limits and abruptly separates two of the great Zoological regions of the globe. The Philippine Islands are in some respects of doubtful location, resembling and differing from both regions. They are deficient in the varied Mammals of Borneo, but they contain no Marsupials. The Psittaci are scarce, as in the Indian region; the Lories are altogether absent, but there is one representative of the Cocka-toos. Woodpeckers, Trogons, and the genera *Ixos*, *Copsychus*, and *Ploceus* are highly characteristic of India. *Tanysiptera* and *Megapodius*, again, are Australian forms, but these seem represented by only solitary species. The islands possess also a few peculiar genera. We must on the whole place the Philippine Islands in the Indian region, but with the remark that they are deficient in some of its most striking features. They possess several isolated forms of the Australian region, but by no means sufficient to constitute a real transition thereto.

Leaving the Philippines out of the question for the present, the western and eastern islands of the Archipelago, as here divided, belong to regions more distinct and contrasted than any other of the great zoological divisions of the globe. South America and Africa, separated by the Atlantic, do not differ so widely as Asia and Australia:

Asia with its abundance and variety of large Mammals and no Marsupials, and Australia with scarcely anything but Marsupials; Asia with its gorgeous *Phasianidæ*, Australia with its dull-coloured *Megapodiidæ*; Asia the poorest tropical region in Parrots, Australia the richest: and all these striking characteristics are almost unimpaired at the very limits of their respective districts; so that in a few hours we may experience an amount of zoological difference which only weeks or even months of travel will give us in any other part of the world!

Moreover there is nothing in the aspect or physical character of the islands to lead us to expect such a difference; their physical and geological differences do not coincide with the zoological differences. There is a striking homogeneity in the two *halves* of the Archipelago. The great volcanic chain runs through both parts; Borneo is the counterpart of New Guinea; the Philippines closely resemble the equally fertile and equally volcanic Moluccas; while in eastern Java begins to be felt the more arid climate of Timor and Australia. But these resemblances are accompanied by an extreme zoological diversity, the Asiatic and Australian regions finding in Borneo and New Guinea respectively their highest development.

But it may be said: "The separation between these two regions is not so absolute. There *is* some transition. There *are* species and genera common to the eastern and western islands." This is true, yet (in my opinion) proves no transition in the proper sense of the word; and the nature and amount of the resemblance only shows more strongly the absolute and original distinctness of the two divisions. The exception here clearly proves the rule.[14] [1860]   ès

In 1868 T. H. Huxley referred to the South-east Asian biogeographic discontinuity as "Wallace's Line" and the name has stuck. Its precise location, however, has been controversial. Even Wallace could not make up his mind, placing Celebes (Sulawesi) on one side of the line to start with and switching it later to the other side.

ès   Its fauna presents the most puzzling relations, showing affinities to Java, to the Philippines, to the Moluccas, to New Guinea, to continental India, and even to Africa; so that it is almost impossible to decide whether to place it in the Oriental or the Australian region. On the whole the preponderance of its relations appears to be with the latter, though it is undoubtedly very anomalous, and may, with almost as much propriety, be classed with the former.[15] [1876]   ès

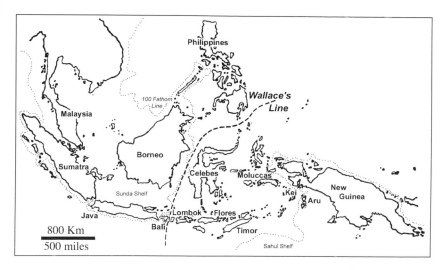

Map showing "Wallace's Line", marking the boundary between the Australasian and Oriental faunas (modified from Wilma George, *Biologist Philosopher*)

Remarkably, modern geophysical studies that track the continental drift of land masses through time reveal a historical pattern consistent with Wallace's deductions from the distributions of birds and mammals. South-east Asia is the site of a slow motion geological collision between the ancient super-continents of Gondwanaland and Laurasia. Wallace's Line is the biological legacy of continental drift. Indeed, we now understand that Wallace found Sulawesi so hard to fit into his scheme because it is an ancient fusion of two distinct land masses, and is therefore part Gondwana- and part Laurasia-derived. It is hardly surprising that it should prove biologically anomalous – its animals and plants represent a mix of two long-separated groups.

Synthesis

Wallace's two major works on biogeography were, as noted above, *The Geographical Distribution of Animals* (1876) and *Island Life* (1880). *Island Life*, Wallace wrote to Charles Darwin, "will form a sort of supplement to my former work [*The Geographical Distribution of Animals*], and will, I trust, be more readable and popular."[16] Darwin was happy to give it his endorsement: "I have now read your book, and it has interested me deeply. It is quite excellent, and seems to me the best book which you have ever published;

but this may be merely because I have read it last."[17] Darwin also
sent along detailed notes on Wallace's arguments. He objected in
particular to Wallace's claims about an Arctic element in the
floras of southern temperate regions, complaining, "This is rather
too speculative for my old noddle."[18] This excerpt is from the
very beginning of the book, in which Wallace lays out the basic
problems of biogeography:

❧   When an Englishman travels by the nearest sea-route from Great
Britain to Northern Japan he passes by countries very unlike his own,
both in aspect and natural productions. The sunny isles of the
Mediterranean, the sands and date-palms of Egypt, the arid rocks of
Aden, the cocoa groves of Ceylon [Sri Lanka], the tiger-haunted
jungles of Malacca [Peninsular Malaysia] and Singapore, the fertile
plains and volcanic peaks of Luzon [Philippines], the forest-clad
mountains of Formosa [Taiwan], and the bare hills of China, pass
successively in review; till after a circuitous journey of thirteen
thousand miles he finds himself at Hakodadi [Hokkaido] in Japan.
He is now separated from his starting-point by the whole width of
Europe and Northern Asia, and by an almost endless succession of
plains and mountains, arid deserts or icy plateaus, yet when he visits
the interior of the country he sees so many familiar natural objects
that he can hardly help fancying that he is close to his home. He
finds the woods and fields tenanted by tits, hedge-sparrows, wrens,
wagtails, larks, redbreasts, thrushes, buntings, and house-sparrows,
some absolutely identical to our own feathered friends, others so
closely resembling them that it requires a practised ornithologist to
tell the difference. If he is fond of insects, he notices many butterflies
and a host of beetles which, though on close examination they are
found to be distinct from ours, are yet of the same general aspect,
and seem just what might be expected in any part of Europe. There
are also of course many birds and insects which are quite new and
peculiar, but these are by no means so numerous or conspicuous as
to remove the general impression of a wonderful resemblance
between the productions of such remote islands as Britain and Yesso
[Hokkaido].
    Now let an inhabitant of Australia sail to New Zealand, a distance
of less than thirteen hundred miles, and he will find himself in a
country whose productions are totally unlike those of his own. Kangaroos and wombats there are none, the birds are almost all entirely
new, insects are very scarce and quite unlike the handsome or strange

Australian forms, while even the vegetation is all changed, and no gum-tree, or wattle, or grass-tree meets the traveller's eye.

But there are some more striking cases even than this, of the diversity of the productions of countries not far apart. In the Malay Archipelago there are two islands named Bali and Lombok, each about as large as Corsica, and separated by a strait only fifteen miles wide at its narrowest part. Yet these islands differ far more from each other in their birds and quadrupeds than do England and Japan. The birds of the one are extremely *unlike* those of the other, the difference being such as to strike even the most ordinary observer. Bali has red and green woodpeckers, barbets, weaver-birds, and black-and-white magpie-robins, none of which are found in Lombok, where, however, we find screaming cockatoos and friar-birds, and the strange mound-building megapodes, which are all equally unknown in Bali. Many of the kingfishers, crow-shrikes, and other birds, though of the same general form, are of very distinct species; and though a considerable number of birds are the same in both islands the difference is none the less remarkable – as proving that mere distance is one of the least important of the causes which have determined the likeness or unlikeness in the animals of different countries.

In the western hemisphere we find equally striking examples. The Eastern United States possess very peculiar and interesting plants and animals, the vegetation becoming more luxuriant as we go south but not altering in essential character, so that when we reach Alabama or Florida we still find ourselves in the midst of pines, oaks, sumachs, magnolias, vines, and other characteristic forms of the temperate flora; while the birds, insects, and land-shells are of the same general character with those found further north. But if we now cross over the narrow strait, about fifty miles wide, which separates Florida from the Bahama Islands, we find ourselves in a totally different country, surrounded by a vegetation which is essentially tropical, and generally identical with that of Cuba. The change is most striking, because there is little difference of climate, of soil, or apparently of position, to account for it; and when we find that the birds, the insects, and especially the land-shells of the Bahamas are almost all West Indian, while the North American types of plants and animals have almost completely disappeared, we shall be convinced that such differences and resemblances cannot be due to existing conditions, but must depend upon laws and causes to which mere proximity of position offers no clue.

Hardly less uncertain and irregular are the effects of climate. Hot countries usually differ widely from cold ones in all their organic forms; but the difference is by no means constant, nor does it bear any proportion to the difference of temperature. Between frigid Canada and sub-tropical Florida there are less marked differences in the animal productions than between Florida and Cuba or Yucatan, so much more alike in climate and so much nearer together. So the differences between the birds and quadrupeds of temperate Tasmania and tropical North Australia are slight and unimportant as compared with the enormous differences we find when we pass from the latter country to equally tropical Java. If we compare corresponding portions of different continents, we find no indication that the almost perfect similarity of climate and general conditions has any tendency to produce similarity in the animal world. The equatorial parts of Brazil and of the West Coast of Africa are almost identical in climate and in luxuriance of vegetation, but their animal life is totally diverse. In the former we have tapirs, sloths, and prehensile-tailed monkeys; in the latter elephants, antelopes, and man-like apes; while among birds, the toucans, chatterers, and humming-birds of Brazil are replaced by the plantain-eaters, bee-eaters, and sun-birds of Africa. Parts of South-temperate America, South Africa, and South Australia, correspond closely in climate; yet the birds and quadrupeds of these three districts are as completely unlike each other as those of any parts of the world that can be named.

If we visit any of the great islands of the globe, we find that they present similar anomalies in their animal productions, for while some exactly resemble the nearest continents others are widely different. Thus the quadrupeds, birds and insects of Borneo correspond closely to those of the Asiatic continent, while those of Madagascar are extremely unlike African forms, although the distance from the continent is less in the latter case than in the former. And if we compare the three great islands Sumatra, Borneo, and Celebes [Sulawesi] – lying as it were side by side in the same ocean – we find that the two former, although furthest apart, have almost identical productions, while the two latter, though closer together, are more unlike than Britain and Japan situated in different oceans and separated by the largest of the great continents.[19] [1880]   ❧

Wallace had "accepted and supported Dr. P. L. Sclater's division of the earth's surface into six great zoological regions,[20] founded upon a detailed examination of the distribution of birds, but equally applicable to mammalia, reptiles, and several other great divisions, and best serving to illustrate and explain the diversities and apparent contradictions in the distribution of all land mammals".[21] Though the precise boundaries of the biogeographic regions remain controversial, *The Geographical Distribution of Animals*, Wallace's massive two-volume compendium that integrated for the first time geological, palaeontological, and contemporary distributional data, effectively established Sclater's doctrine. In 1879, Wallace published a synopsis, emphasizing how fossils can aid in explaining an apparently anomalous disjunct distribution in which a group or species is found in two or more widely separated locations.

Returning to the general question of zoological distribution and its anomalies, it has been shown, I trust, that the only mode of explaining the existing distribution of living things is by a constant reference to those comparatively slight but often important changes of sea and land, which the most recent researches show to be alone probable; and, what is still more important, by recognising the undoubted fact that every group of animals whose distribution is discontinuous is now more or less in a fragmentary condition, and has, in all probability, once had a much more extensive range, to which its present distribution may offer no clue whatever. Who would ever have imagined, for example, that the horse tribe, now confined to Africa and Asia, formerly ranged over the entire American continent, north and south, in great abundance and variety; or that the camel tribe, now confined to Central Asia and the Andean region of South America, formerly abounded in North America, whence in fact our existing camels were almost certainly derived? How easy it is to imagine that analogous causes to those which have so recently exterminated the horses of America and Europe might have acted in a somewhat different direction, and have led to the survival of horses in South America and Africa, and their extermination elsewhere. Had this been the case, how strong would have been the argument for a former union of these two continents; yet we now know that these widely separated species would be merely the relics of a once dominant group which had occupied and become extinct in all the northern continents.

Discoveries of extinct forms remote from the countries they now inhabit, are continually furnishing us with new proofs that the great northern continents of the two hemispheres were really the birthplace of almost if not quite all the chief forms of animal life upon the globe;

while change of climate, culminating in the glacial epoch, seems to have been the motive power which has driven many of these forms into the tropical lands where they now alone exist.

If we give full weight to these various considerations, and at the same time bear constantly in mind the extreme imperfection of our knowledge of extinct land animals, we shall, I believe, have no difficulty in explaining most of the apparent anomalies in zoological distribution, and in imagining a possible and even probable solution for those extreme cases of difficulty which the facts at our command do not yet permit us to explain in detail.

Let us now briefly summarise the general principles on which the solution of problems in zoological distribution depends.

During the evolution of existing forms of animal life, we may picture to ourselves the production of successive types, each in turn increasing in variety of species and genera, spreading over more or less extensive regions of the earth's surface, and then, after arriving at a maximum of development, passing through various stages of decay, dwindling to a single genus or a single species, and finally becoming extinct. While the forms of life are thus, each in turn, moving on from birth to maturity and from maturity to decay and death, the earth's surface will be undergoing important physical changes, which will sometimes unite and sometimes separate contiguous continents or islands, leading now to the intermingling, now to the isolation, of the progressing or diminishing groups of animals. Again, we know that climates have often changed over a considerable portion of the earth, so that what was at one time an almost tropical region has become at another time temperate, and then even arctic; and these changes have, it is believed, been many times repeated, leading each time to important changes, migrations, and extinctions of animal and vegetable life.

It is by the combined effect of these three distinct sets of causes, acting and reacting on each other in various complex ways, that have produced those curious examples of erratic distribution of species and genera which have been so long a puzzle to the naturalist, but which have now, it is believed, been shown to be the natural and inevitable results of the process of animal development [evolution], combined with constant changes in the geography and in the climate of the earth.[22] [1879] ❧

Wallace's combination of field experience and analytical expertise made him in effect one of the very early ecologists. Here he alludes to two important ecological ideas: the "interme-

diate disturbance hypothesis", which postulates that maximal biological diversity is maintained in situations in which dominant
species are prevented from taking over; and "succession",
whereby a predictable sequence of plants colonize a previously
cleared area.

But besides these inorganic causes – soil, climate, aspect, etc. –
which seem primarily to determine the distribution of plants, and,
through them, of many animals, there are other and often more
powerful causes in the organic environment which acts in a variety of
ways. Thus, it has been noticed that over fields or heaths where cattle
and horses have free access seedling trees and shrubs are so constantly
eaten down that none ever grow to maturity, even although there may
be plenty of trees and woods around. But if a portion of this very same
land is enclosed and all herbivorous quadrupeds excluded, it very
quickly becomes covered with a dense vegetation of trees and shrubs.
Again, it has been noticed that on turfy banks constantly cropped by
sheep a very large variety of dwarf plants are to be found. But if these
animals are kept out and the vegetation allowed to grow freely, many
of the dwarfer and more delicate plants disappear owing to the rapid
growth of grasses, sedges, or shrubby plants, which, by keeping off the
sun and air and exhausting the soil, prevent the former kinds from
producing seed, so that in a few years they die out and the vegetation
becomes more uniform.

A modified form of the same general law is seen when any ground
is cleared of all vegetation, perhaps cultivated for a year or two, and
then left fallow. A large crop of weeds then grows up (the seeds of
which must have been brought by the wind or by birds, or have lain
dormant in the ground); but in the second and third years these
change their proportions, some disappear, while a few new ones
arrive, and this change goes on till a stable form of vegetation is
formed, often very different from that of the surrounding country.[23]
[1910]

Wallace was also well aware of the latitudinal trend in species
richness: polar regions have few species, temperate ones have
more, and tropical ones still more. As with many of the issues
that he touched on, explanations remain controversial, but he
would still find plenty of support among contemporary ecologists
for his theory of climatic stability as the major factor in promoting
high tropical species richness.

⋙   No less indicative of delicate response to variation of temperature, and therefore of close adaptation to the whole modified environment, is the continuous increase in the number of species with every important change in latitude. Although this increase is but slight for moderate changes and is therefore liable to be masked by other favourable or adverse conditions of the environment, it yet makes itself visible in every continent; and in the comparison between the north or mid-temperate and the tropical zones is so pronounced that in fairly comparable areas the tropical species are often (and probably on the average) double those of the temperate zones. This seems to be the case among the higher animals, as well as among all the vascular plants.

Now all this is indicative of long and minute adjustment to the special inorganic as well as the organic conditions; and the reason why the tropics as a whole far surpass the temperate zones in the number of their specific forms, is, not the greater amount of heat alone, but rather the much greater uniformity of climatical conditions generally, during long periods – perhaps during the whole range of geological time. Whatever changes have occurred through astronomical causes, such as greater excentricity of the earth's orbit, must necessarily have produced extremes of climate towards the poles, while the equatorial regions would remain almost unaffected, except by a slight and very slow rise or fall of the average temperature, which we know to be of little importance to vegetation so long as other conditions remain tolerably uniform and favourable.

It is this long-continued uniformity of favourable conditions within the tropics, or more properly within the great equatorial belt about 2000 miles in width, that has permitted and greatly favoured ever-increasing delicacy of adjustments of the various species to their whole environment. Thus has arisen that multiplicity of species intermingled in the same areas, none being able, as in the temperate zone, to secure such a superior position as to monopolise large areas to the exclusion of others.[24] [1910]   ⋙

Natural History and Conservation

Wallace combined a naturalist's enthusiasm for the plants and animals he was seeing with both the observational acuteness of the professional collector and rare writing skills. Not surprisingly, his works – especially his books of travel writing – are rich in memorable biological descriptions. Looking back when writing his autobiography, he realized that the naturalist's basic excitement at finding something he had never before seen was the same whether in the depths of the Amazon rainforest or on a more mundane Welsh hillside.

&ε But I soon found that by merely identifying the plants I found in my walks [around Neath, Wales] I lost much time in gathering the same species several times, and even then not being always quite sure that I had found the same plant before. I therefore began to form a herbarium, collecting good specimens and drying them carefully between drying papers and a couple of boards weighted with books or stones. My brother, however, did not approve of my devotion to this study, even though I had absolutely nothing else to do, nor did he suggest any way in which I could employ my leisure more profitably. He said very little to me on the subject beyond a casual remark, but a letter from my mother showed me that he thought I was wasting my time. Neither he nor I could foresee that it would have any effect on my future life, and I myself only looked upon it as an intensely interesting occupation for time that would be otherwise wasted. Even when we were busy I had Sundays perfectly free, and used then to take long walks over the mountains with my collecting box, which I brought home full of treasures. I first named the species as nearly as I could do so, and then laid them out to be pressed and dried. At such times I experienced the joy which every discovery of a new form of life gives to the lover of nature, almost equal to those raptures which I afterwards felt at every capture of new butterflies on the Amazon, or at the constant stream of new species of birds, beetles, and butterflies in Borneo, the Moluccas, and the Aru Islands.[1] [1905] &ε

Most of Wallace's fieldwork, needless to say, was carried out a long way from Wales.

✑ GENERAL FEATURES OF THE EQUATORIAL FORESTS

It is not easy to fix upon the most distinctive features of these virgin forests, which nevertheless impress themselves upon the beholder as something quite unlike those of temperate lands, and as possessing a grandeur and sublimity altogether their own. Amid the countless modifications in detail which these forests present, we shall endeavour to point out the chief peculiarities as well as the more interesting phenomena which generally characterise them.

The observer new to the scene would perhaps be first struck by the varied yet symmetrical trunks, which rise up with perfect straightness to a great height without a branch, and which, being placed at a considerable average distance apart, give an impression similar to that produced by the columns of some enormous building. Overhead, at a height, perhaps, of a hundred and fifty feet, is an almost unbroken canopy of foliage formed by the meeting together of these great trees and their interlacing branches; and this canopy is usually so dense that but an indistinct glimmer of the sky is to be seen, and even the intense tropical sunlight only penetrates to the ground subdued and broken up into scattered fragments. There is a weird gloom and a solemn silence, which combine to produce a sense of the vast – the primeval – almost of the infinite. It is a world in which man seems an intruder, and where he feels overwhelmed by the contemplation of the ever-acting forces which, from the simple elements of the atmosphere, build up the great mass of vegetation which overshadows and almost seems to oppress the earth.[2] [1891]   ✑

> Wallace never lost his capacity for being awed by nature, and he complained that his scientific colleagues too readily exchanged that sense of wonder for the myopic pedantry of detailed research. The reductionism of modern biology – that unwavering focus on DNA – makes Wallace's comment even more relevant today than it was in his.

✑   The modern scientific morphologists seem so wholly occupied in tracing out the mechanism of organisms that they hardly seem to appreciate the overwhelming marvel of the powers of life, which result in such infinitely varied structures and such strange habits and so-called instincts. The older I grow the more marvellous seem to me the mere variety of form and habit in plants and animals, and the unerring certitude with which from a minute germ the whole complex organism is built up, true to the type of its kind in all the infinitude of details! It is this which gives such a charm to the watching of plants growing, and of kittens so rapidly developing their senses and habitudes![3] [1892] ε►

The Amazon

Wallace's account of his Amazon trip was inevitably compromised by the loss of so much of his material on the way home. Nevertheless, his Amazon book is peppered with evocative descriptions of the sights and sounds of the Brazilian rainforest.

Rainforest Cacophony

◄§ Every night, while in the upper part of the river, we had a concert of frogs, which made most extraordinary noises. There are three kinds, which can frequently be all heard at once. One of these makes a noise something like what one would expect a frog to make, namely a dismal croak, but the sounds uttered by the others were like no animal noise that I ever heard before. A distant railway-train approaching, and a blacksmith hammering on his anvil, are what they exactly resemble. They are such true imitations, that when lying half-dozing in the canoe I have often fancied myself at home, hearing the familiar sounds of the approaching mail-train, and the hammering of the boiler-makers at the iron-works. Then we often had the "guarhibas," or howling monkeys, with their terrific noises, the shrill grating whistle of the cicadas and locusts, and the peculiar notes of the suacáras and other aquatic birds; add to these the loud unpleasant hum of the mosquito in your immediate vicinity, and you have a pretty good idea of our nightly concert on the Tocantins.[4] [1853] ε►

Umbrella Bird

◄§ The next morning my hunter arrived, and immediately went out in his canoe among the islands, where the umbrella-birds are found. In the evening after dark he returned, bringing one fine specimen. This singular bird is about the size of a raven, and is of a similar colour, but its feathers have a more scaly appearance, from being

Engraving from Wallace's great biogeographic work *The Geographic Distribution of Animals*, showing "a forest scene on the Upper Amazon, with some characteristic birds". An umbrella bird is in left foreground.

margined with a different shade of glossy blue. It is also allied to the crows in its structure, being very similar to them in its feet and bill. On its head it bears a crest, different from that of any other bird. It is formed of feathers more than two inches long, very thickly set, and with hairy plumes curving over at the end. These can be laid back so as to be hardly visible, or can be erected and spread out on every side, forming a hemispherical, or rather a hemi-ellipsoidal dome, completely covering the head, and even reaching beyond the point of the beak: the individual feathers then stand out something like the down-bearing seeds of the dandelion. Besides this, there is another ornamental appendage on the breast, formed by a fleshy tubercle, as thick as a quill and an inch and a half long, which hangs down from the neck, and is thickly covered with glossy feathers, forming a large pendent plume or tassel. This also the bird can either press to its breast, so as to be scarcely visible, or can swell out, so as almost to conceal the forepart of its body. In the female the crest and the neck-plume are less developed, and she is altogether a smaller and much less handsome bird. It inhabits the flooded islands of the Rio Negro and the Solimoes, never appearing on the mainland. It feeds on fruits, and utters a loud, hoarse cry, like some deep musical instrument; whence its Indian name, *Ueramimbé*, "trumpet-bird." The whole of the neck, where the plume of feathers springs from, is covered internally with a thick coat of hard, muscular fat, very difficult to be cleaned away, – which, in preparing the skins, must be done, as it would putrefy, and cause the feathers to drop off. The birds are tolerably abundant, but are shy, and perch on the highest trees, and, being very muscular, will not fall unless severely wounded. My hunter worked very perseveringly to get them, going out before daylight and often not returning till nine or ten at night, yet he never brought me more than two at a time, generally only one, and sometimes none.[5]

. . .

Having remained here [three days' journey up the Rio Negro from Manaus] a month, and obtained twenty-five specimens of the umbrella-bird, I prepared to return to Barra [Manaus]. On the last day my hunter went out he brought me a fine male bird alive. It had been wounded slightly on the head, just behind the eye, and had fallen to the ground stunned, for in a short time it became very active, and when he brought it to me was as strong and fierce as if it was quite uninjured. I put it in a large wicker basket, but as it would take no food during two days I fed it by thrusting pieces of banana down its throat; this I continued for several days, with much difficulty, as its claws were very sharp and powerful. On our way to Barra I found by the river-side a small fruit which it ate readily; this fruit was about the size of a cherry, of an acid taste, and was swallowed whole. The bird arrived safely in the city, and lived a fortnight; when one day it suddenly fell off its perch and died. On skinning it, I found the shot had broken the skull and entered to the brain, though it seems surprising that it should have remained so long apparently in perfect health. I had had, however, an excellent opportunity of observing its habits, and its method of expanding and closing its beautiful crest and neck-plume.[6] [1853]   &

Sloth

Another interesting little animal was a young sloth, which Antonio, an Indian boy, who had enlisted himself in our service, brought alive from the forest. It was not larger than a rabbit, was covered with coarse grey and brown hair, and had a little round head and face resembling the human countenance quite as much as a monkey's, but with a very sad and melancholy expression. It could scarcely crawl along the ground, but appeared quite at home on a chair, hanging on the back, legs, or rails. It was a most quiet, harmless little animal, submitting to any kind of examination with no other manifestation of

Engraving from *The Geographic Distribution of Animals*, showing "a Brazilian forest, with characteristic mammalia". Sloths can be seen bottom left.

displeasure than a melancholy whine. It slept hanging with its back downwards and its head between its fore-feet. Its favourite food is the leaf of the *Cecropia peltata*, of which it sometimes ate a little from a branch we furnished it with.[7] [1853]   ৵

Jaguar

৵ As I was walking quietly along I saw a large jet-black animal come out of the forest about twenty yards before me, which took me so much by surprise that I did not at first imagine what it was. As it moved slowly on, and its whole body and long curving tail came into full view in the middle of the road, I saw that it was a fine black jaguar. I involuntarily raised my gun to my shoulder, but remembering that both barrels were loaded with small shot, and that to fire would exasperate without killing him, I stood silently gazing. In the middle of the road he turned his head, and for an instant paused and gazed at me, but having, I suppose, other business of his own to attend to, walked steadily on, and disappeared in the thicket. As he advanced, I heard the scampering of small animals, and the whizzing flight of ground birds, clearing the path for their dreaded enemy.

This encounter pleased me much. I was too much surprised, and occupied too much with admiration, to feel fear. I had at length had a full view, in his native wilds, of the rarest variety of the most powerful

and dangerous animal inhabiting the American continent. I was, however, by no means desirous of a second meeting, and, as it was near sunset, thought it most prudent to turn back towards the village.[8]
[1853] ❧

Pet Parrots

❧ The only live animals I had with me were a couple of parrots, which were a never-failing source of amusement. One was a little "Marianna," or Macaí of the Indians, a small black-headed, white-breasted, orange-neck and thighed parrot; the other, an Anacá, a most beautiful bird, banded on the breast and belly with blue and red, and the back of the neck and head covered with long bright red feathers margined with blue, which it would elevate when angry, forming a handsome crest somewhat similar to that of the harpy eagle; its ornithological name is *Derotypus accipitrinus*, the hawk-headed parrot. There was a remarkable difference in the characters of these birds. The Anacá was of a rather solemn, morose, and irritable disposition; while the Mariánna was a lively little creature, inquisitive as a monkey, and playful as a kitten. It was never quiet, running over the whole canoe, climbing into every crack and cranny, diving into all the baskets, pans, and pots it could discover, and tasting everything they contained. It was a most omnivorous feeder, eating rice, farinha [cassava flour], every kind of fruit, fish, meat, and vegetable, and drinking coffee too as well as myself; and as soon as it saw me with basin in hand, would climb up to the edge, and not be quiet without having a share, which it would lick up with the greatest satisfaction, stopping now and then, and looking knowingly round, as much as to say, "This coffee is very good," and then sipping again with increased gusto. The bird evidently liked the true flavour of the coffee, and not that of the sugar, for it would climb up to the edge of the coffee-pot, and hanging on the rim plunge boldly down till only its little tail appeared above, and then drink the coffee grounds for five minutes together. The Indians in the canoe delighted to imitate its pretty clear whistle, making it reply and stare about, in a vain search after its companions. Whenever we landed to cook, the Mariánna was one of the first on shore, – not with any view to an escape, but merely to climb up some bush or tree and whistle enjoyment of its elevated position, for as soon as eating commenced, it came down for a share of fish or coffee. The more sober Anacá would generally remain quietly in the canoe, till, lured by the cries and whistles of its lively

little companion, it would venture out to join it; for, notwithstanding
their difference of disposition, they were great friends, and would sit
for hours side by side, scratching each other's heads, or playing
together just like a cat and a kitten; the Mariánna sometimes so
exasperating the Anacá by scratches and peckings, and by jumping
down upon it, that a regular fight would ensue, which, however, soon
terminated, when they would return to their former state of brotherhood. I intended them as presents to two friends in Barra, but was
almost sorry to part them.[9] [1853]   ⇚

South-east Asia

Orang-Utan

Wallace was one of the first scientists to make systematic observations on one of our closest relatives.

⇚   The Orangutan is known to inhabit Sumatra and Borneo, and
there is every reason to believe that it is confined to these two great
islands, in the former of which, however, it seems to be much more
rare. In Borneo it has a wide range, inhabiting many districts on the
southwest, southeast, northeast, and northwest coasts, but appears to
be chiefly confined to the low and swampy forests. It seems, at first
sight, very inexplicable that the Mias[10] should be quite unknown in
the Sarawak valley, while it is abundant in Sambas, on the west, and
Sadong, on the east. But when we know the habits and mode of life
of the animal, we see a sufficient reason for this apparent anomaly in
the physical features of the Sarawak district. In the Sadong, where I
observed it, the Mias is only found when the country is low level and
swampy, and at the same time covered with a lofty virgin forest. From
these swamps rise many isolated mountains, on some of which the
Dyaks[11] have settled and covered with plantations of fruit trees. These
are a great attraction to the Mias, which comes to feed on the unripe
fruits, but always retires to the swamp at night. Where the country
becomes slightly elevated, and the soil dry, the Mias is no longer to be
found. For example, in all the lower part of the Sadong valley it
abounds, but as soon as we ascend above the limits of the tides, where
the country, though still flat, is high enough to be dry, it disappears.
Now the Sarawak valley has this peculiarity – the lower portion though
swampy is not covered with a continuous lofty forest, but is principally
occupied by the Nipa palm; and near the town of Sarawak [Kuching]

where the country becomes dry, it is greatly undulated in many parts, and covered with small patches of virgin forest, and much second-growth jungle on the ground, which has once been cultivated by the Malays or Dyaks.

Now it seems probable to me that a wide extent of unbroken and equally lofty virgin forest is necessary to the comfortable existence of these animals. Such forests form their open country, where they can roam in every direction with as much facility as the Indian on the prairie, or the Arab on the desert, passing from tree-top to tree-top without ever being obliged to descend upon the earth. The elevated and the drier districts are more frequented by man, more cut up by clearings and low second-growth jungle – not adapted to its peculiar mode of progression, and where it would therefore be more exposed to danger, and more frequently obliged to descend upon the earth. There is probably also a greater variety of fruit in the Mias district, the small mountains which rise like islands out of it serving as gardens or plantations of a sort, where the trees of the uplands are to be found in the very midst of the swampy plains.

It is a singular and very interesting sight to watch a Mias making his way leisurely through the forest. He walks deliberately along some of the larger branches in the semi-erect attitude which the great length of his arms and the shortness of his legs cause him naturally to assume; and the disproportion between these limbs is increased by his walking on his knuckles, not on the palm of the hand, as we should do. He seems always to choose those branches which intermingle with an adjoining tree, on approaching which he stretches out his long arms, and seizing the opposing boughs, grasps them together with both hands, seems to try their strength, and then deliberately swings himself across to the next branch, on which he walks along as before. He never jumps or springs, or even appears to hurry himself, and yet manages to get along almost as quickly as a person can run through the forest beneath. The long and powerful arms are of the greatest use to the animal, enabling it to climb easily up the loftiest trees, to seize fruits and young leaves from slender boughs which will not bear its weight, and to gather leaves and branches with which to form its nest. I have already described how it forms a nest when wounded, but it uses a similar one to sleep on almost every night. This is placed low down, however, on a small tree not more than from twenty to fifty feet from the ground, probably because it is warmer and less exposed to wind than higher up. Each Mias is said to make a fresh one for himself every night; but I should think that is hardly probable, or their

A female orang (from *The Malay Archipelago*)

remains would be much more abundant; for though I saw several about the coal-mines, there must have been many Orangs about every day, and in a year their deserted nests would become very numerous. The Dyaks say that, when it is very wet, the Mias covers himself over with leaves of pandanus, or large ferns, which has perhaps led to the story of his making a hut in the trees.

The Orang does not leave his bed until the sun has well risen and has dried up the dew upon the leaves. He feeds all through the middle of the day, but seldom returns to the same tree two days running. They do not seem much alarmed at man, as they often stared down upon me for several minutes, and then only moved away slowly to an adjacent tree. After seeing one, I have often had to go half a mile or more to fetch my gun, and in nearly every case have found it on the same tree, or within a hundred yards, when I returned. I never saw two full-grown animals together, but both males and females are sometimes accompanied by half-grown young ones, while, at other times, three or four young ones were seen in company. Their food consists almost exclusively of fruit, with occasionally leaves, buds, and young shoots. They seem to prefer unripe fruits, some of which were very sour, others intensely bitter, particularly the large red, fleshy arillus [fruit] of one which seemed an especial favourite. In other cases they eat only the small seed of a large fruit, and they almost always waste and destroy more than they eat, so that there is a continual rain of rejected portions below the tree they are feeding on. The Durian is an especial favourite, and quantities of this delicious fruit are destroyed wherever it grows surrounded by forest, but they will not cross clearings to get at them. It seems wonderful how the animal can tear open this fruit, the outer covering of which is so thick and tough, and closely covered with strong conical spines, It probably bites off a few of these first, and then, making a small hole, tears open the fruit with its powerful fingers.[12]

The Mias rarely descends to the ground, except when pressed by hunger, it seeks succulent shoots by the riverside; or, in very dry weather, has to search after water, of which it generally finds sufficient in the hollows of leaves. Only once I saw two half-grown Orangs on the ground in a dry hollow at the foot of the Simunjon hill. They were playing together, standing erect, and grasping each other by the arms. It may be safely stated, however, that the Orang never walks erect, unless when using its hands to support itself by branches overhead or when attacked. Representations of its walking with a stick are entirely imaginary.

The Dyaks all declare that the Mias is never attacked by any animal in the forest, with two rare exceptions [pythons and crocodiles]; and the accounts I received of these are so curious that I give them nearly in the words of my informants, old Dyak chiefs, who had lived all their lives in the places where the animal is most abundant. The first of whom I inquired said: "No animal is strong enough to hurt the Mias, and the only creature he ever fights with is the crocodile. When there is no fruit in the jungle, he goes to seek food on the banks of the river where there are plenty of young shoots that he likes, and fruits that grow close to the water. Then the crocodile sometimes tries to seize him, but the Mias gets upon him, and beats him with his hands and feet, and tears him and kills him." He added that he had once seen such a fight, and that he believes that the Mias is always the victor.[13] [1869] &

The earlier version of this account published as an article in 1856 – prior to Wallace's Ternate discovery – finished with a hint that Wallace's encounter with the orang had started him thinking about the evolution of humans.

It is a remarkable circumstance, that an animal so large, so peculiar, and of such a high type of form as the Orang-Utan, should yet be confined to such a limited district, – to two islands, and those almost at the limits of the range of the higher mammalia; for, eastward of Borneo and Celebes, the Quadrumana and most of the higher mammalia almost disappear. One cannot help speculating on a former condition of this part of the world which should give a wider range to these strange creatures, which at once resemble and mock the "human form divine," – which so closely approach us in structure, and yet differ so widely from us in many points of their external form. And when we consider that almost all other animals have in previous ages been represented by allied, yet distinct forms, – that the bears and tigers, the deer, the horses, and cattle of the tertiary period were distinct from those which now exist, with what intense interest, with what anxious expectation must we look forward to the time when the progress of civilization in those hitherto wild countries may lay open the monuments of a former world, and enable us to ascertain approximately the period when the present species of Orangs first made their appearance, and perhaps prove the former existence of allied species still more gigantic in their dimensions, and more or less human in their form and structure! Some such discoveries we may not unreasonably anticipate, after the wonders that geology has already made known to us. Animals the most isolated in existing nature have been shown to be but the last of a series of allied species which have lived and died upon the earth. Every class and every order has furnished some examples, from which we may conclude, that all isolations in nature are apparent only, and that whether we discover their remains or no, every animal now existing has had its representatives in past geological epochs.[14] [1856]   ❧

Although unquestionably an acute and patient observer, Wallace was first and foremost a *collector*. Not for him the modern mantra of "take nothing but photographs": his main tool for collecting vertebrates was his gun. Having once shot a nursing orang mother, Wallace suddenly found that he had an orang orphan on his hands. He wrote this whimsical letter home about his newfound responsibilities.

❧ I must now tell you of the addition to my household of an orphan baby, a curious little half-nigger baby, which I have nursed now more than a month. I will tell you presently how I came to get it, but must first relate my inventive skill as a nurse. The little innocent was not weaned, and I had nothing proper to feed it with, so was obliged to give it rice-water. I got a large-mouthed bottle, making two holes in the cork, through one of which I inserted a large quill so that the baby could suck. I fitted up a box for a cradle with a mat for it to lie upon, which I had washed and changed every day. I feed it four times a day, and wash it and brush its hair every day, which it likes very much, only crying when it is hungry or dirty. In about a week I gave it the rice-water a little thicker, and always sweetened it to make it nice. I am afraid you would call it an ugly baby, for it has a dark brown skin and red hair, a very large mouth, but very pretty little hands and feet. It has now cut its two lower front teeth, and the uppers are coming. At first it would not sleep alone at night, but cried very much; so I made it a pillow of an old stocking, which it likes to hug, and now sleeps very soundly. It has powerful lungs, and sometimes screams tremendously, so I hope it will live.

But I must now tell you how I came to take charge of it. Don't be alarmed; I was the cause of its mother's death. It happened as follows: – I was out shooting in the jungle and saw something up a tree which I thought was a large monkey or orang-utan, so I fired at it, and down fell this little baby – in its mother's arms. What she did up in the tree of course I can't imagine, but as she ran about the branches quite easily, I presume she was a wild "woman of the woods"; so I have preserved her skin and skeleton, and am trying to bring up her only daughter, and hope some day to introduce her to fashionable society at the Zoological Gardens. When its poor mother fell mortally wounded, the baby was plunged head over ears in a swamp about the consistence of peasoup, and when I got it out looked very pitiful. It clung to me very hard when I carried it home, and having got its little hands unawares into my beard, it clutched so tight that I had great difficulty in extricating myself. Its mother, poor creature, had very long hair, and while she was running about the trees like a mad woman, the little baby had to hold fast to prevent itself from falling, which accounts for the remarkable strength of its little fingers and toes, which catch hold of anything with the firmness of a vice. About a week ago I bought a little monkey with a long tail, and as the baby was very lonely while we were out in the daytime, I put the little monkey into the cradle to keep it warm. Perhaps you will say that this was not proper. "How could you do such a thing?" But, I assure you, the baby likes it exceedingly, and they are excellent friends. When the monkey wants to run away, as he often does, the baby clutches him by the tail or ears and drags him back; and if the monkey does succeed in escaping, screams violently till he is brought back again. Of course, baby cannot walk yet, but I let it crawl about on the floor to exercise its limbs; but it is the most wonderful baby I ever saw, and has such strength in its arms that it will catch hold of my trousers as I sit at work, and hang under my legs for a quarter of an hour at a time without being the least tired, all the time trying to suck, thinking, no doubt, it has got hold of its poor dear mother. When it finds no milk is to be had, there comes another scream, and I have to put it back in its cradle and give it "Toby" – the little monkey – to hug, which quiets it immediately. From this short account you will see that my baby is no common baby, and I can safely say, what so many have said before with much less truth, "There never was such a baby as my baby," and I am sure nobody ever had such a dear little duck of a darling of a little brown hairy baby before.[15] [1855]   ⇒

Sad to say, Wallace's baby died. "I much regretted the loss of my little pet, which I had at one time looked forward to bringing up to years of maturity, and taking home to England. For several months it had afforded me daily amusement by its curious ways and the inimitably ludicrous expressions of its little countenance."[16] Sentimentality apart, however, Wallace's mission was to collect specimens for commercial purposes. The baby orang's corpse fetched £6 from the British Museum.

Birds of Paradise

Wallace was disappointed that he was not able to see and collect more birds of paradise than he did. "I fear the somewhat scanty results of my exertions may have led to the opinion that they failed for want of judgement or perseverance."[17] As he recounts, however, circumstances conspired against him to limit his collecting of these birds:

⇒ Nature seems to have taken every precaution that these, her choicest treasures, may not lose value by being too easily obtained. First we find an open, harbourless, inhospitable coast, exposed to the full swell of the Pacific Ocean; next, a rugged and mountainous country, covered with dense forests, offering in its swamps and precipices and serrated ridges an almost impassable barrier to the central regions; and lastly, a race of the most savage and ruthless character, in the very lowest stage of civilization. In such a country and among such a people are found these wonderful productions of nature. In those trackless wilds do they display that exquisite beauty and that marvellous development of plumage, calculated to excite admiration and astonishment among the most civilized and most intellectual races of man. A feather is itself a wonderful and a beautiful thing. A bird clothed with feathers is almost necessarily a beautiful creature. How much, then, must we wonder at and admire the modification of

Engraving from *The Geographic Distribution of Animals*, showing "a Malayan forest, with some of its peculiar birds"

simple feathers into the rigid, polished, wavy ribbons which adorn *P. rubra*, the mass of airy plumes on *P. apoda*, the tufts and wires of *Seleucides alba*, or the golden buds borne upon airy stems that spring from the tail of *Cicinnurus regius*; while gems and polished metals can alone compare with the tints that adorn the breast of *Parotia sexsetacea* and *Astrapia nigra*, and the immensely developed shoulder-plumes of *Epimachus magnus*.[18] [1862] ❧

❧ Just as I got home I overtook Ali [Wallace's Malay assistant] returning from shooting with some birch hanging from his belt. He seemed much pleased, and said, "Look here, sir, what a curious bird," holding out what at first completely puzzled me. I saw a bird with a mass of splendid green feathers on its breast, elongated into two glittering tufts; but, what I could not understand was a pair of long white feathers, which stuck straight out from each shoulder. Ali assured me that the bird stuck them out this way itself, when fluttering its wings, and that they had remained so without his touching them. I now saw

Wallace's Standard Wing bird of
paradise, *Semioptera wallacii* (from
The Malay Archipelago)

that I had got a great prize, no less than a completely new form of the
Bird of Paradise, differing most remarkably from every other known
bird. The general plumage is very sober, being a pure ashy olive, with a
purplish tinge on the back; the crown of the head is beautifully glossed
with pale metallic violet, and the feathers of the front extend as much
over the beak as in most of the family. The neck and breast are scaled
with fine metallic green, and the feathers on the lower part are
elongated on each side, so as to form a two-pointed gorget [patch of
throat colour], which can be folded beneath the wings, or partially
erected and spread out in the same way as the side plumes of most of
the birds of paradise. The four long white plumes which give the bird
its altogether unique character, spring from little tubercles close to the
upper edge of the shoulder or bend of the wing; they are narrow,
gentle curved, and equally webbed on both sides, of a pure creamy
white colour. They are about six inches long, equalling the wing, and
can be raised at right angles to it, or laid along the body at the pleasure
of the bird. The bill is horn colour, the legs yellow, and the iris pale
olive. This striking novelty has been named by Mr. G. R. Gray[19] of the
British Museum, *Semioptera wallacii*, or "Wallace's Standard wing."[20] ❧

Thus one of my objects in coming to the far east was accomplished. I had obtained a specimen of the King Bird of Paradise (*Paradisea regia*), which had been described by Linnaeus from skins preserved in a mutilated state by the natives. I knew how few Europeans had ever beheld the perfect little organism I now gazed upon, and how very imperfectly it was still known in Europe. The emotions excited in the mind of a naturalist, who has long desired to see the actual thing which he has hitherto known only by description, drawing, or badly-preserved external covering – especially when that thing is of surpassing rarity and beauty – require the poetic faculty fully to express them. The remote island in which I found myself situated, in an almost unvisited sea, far from the tracks of merchant fleets and navies; the wild luxuriant tropical forest, which stretched far away on every side; the rude uncultured savages who gathered round me – all had their influence in determining the emotions with which I gazed upon this "thing of beauty." I thought of the long ages of the past, during which the successive generations of this little creature had run their course – year by year being born, and living and dying amid these dark and gloomy woods, with no intelligent eye to gaze upon their loveliness; to all appearance such a wanton waste of beauty. Such ideas excite a feeling of melancholy. It seems sad, that on the one hand such exquisite creatures should live out their lives and exhibit their charms only in these wild inhospitable regions, doomed for ages yet to come to hopeless barbarism; while on the other hand, should civilized man ever reach these distant lands, and bring moral, intellectual, and physical light into the recesses of these virgin forests, we may be sure that he will so disturb the nicely-balanced relations of organic and inorganic nature as to cause the disappearance, and finally the extinction, of these very beings whose wonderful structure and beauty he alone is fitted to appreciate and enjoy. This consideration must surely tell us that all living things were *not* made for man. Many of them have no relation to him. The cycle of their existence has gone on independently of his, and is disturbed or broken by every advance in man's intellectual development; and their happiness and enjoyment, their loves and hates, their struggles for existence, their vigorous life and early death, would seem to be immediately related to their own well-being and perpetuation alone, limited only by the equal well-being and perpetuation of the numberless other organisms with which each is more or less intimately connected.[21]

. . .

But what I valued almost as much as the birds [Great Bird of Paradise] themselves was the knowledge of their habits, which I was daily obtaining both from the accounts of my hunters, and from the conversation of the natives. The birds had now commenced what the people here call their "sácaleli," or dancing-parties, in certain trees in the forest, which are not fruit trees as I at first imagined, but which have an immense head of spreading branches and large but scattered leaves, giving a clear space for the birds to play and exhibit their plumes. On one of these trees a dozen or twenty full-plumaged male birds assemble together, raise up their wings, stretch out their necks, and elevate their exquisite plumes, keeping them in a continual vibration. Between whiles they fly across from branch to branch in great excitement, so that the whole tree is filled with waving plumes in every variety of attitude and motion. The bird itself is nearly as large as a crow, and is of a rich coffee brown colour. The head and neck is of a pure straw yellow above and rich metallic green beneath. The long plumy tufts of golden orange feathers spring from the sides beneath each wing, and when the bird is in repose are partly concealed by them. At the time of its excitement, however, the wings are raised vertically over the back, the head is bent down and stretched out, and the long plumes are raised up and expanded till they form two magnificent golden fans, striped with deep red at the base, and fading off into the pale brown tint of the finely divided and softly waving points. The whole bird is then overshadowed by them, the crouching body, yellow head, and emerald green throat forming but the foundation and setting to the golden glory which waves above. When seen in this attitude, the Bird of Paradise really deserves its name, and must be ranked as one of the most beautiful and most wonderful of living things.[22]

. . .

The Red Birds of Paradise are not shot with blunt arrows, as in the Aru Islands and some parts of New Guinea, but are snared in a very ingenious manner. A large climbing Arum bears a red reticulated fruit, of which the birds are very fond. The hunters fasten this fruit on a stout forked stick, and provide themselves with a fine but strong cord. They then seek out some tree in the forest on which these birds are accustomed to perch, and climbing up it fasten the stick to a branch and arrange the cord in a noose so ingeniously, that when the bird comes to eat the fruit its legs are caught, and by pulling the end of the cord, which hangs down to the ground, it comes free from the branch and brings down the bird. Sometimes, when food is abundant elsewhere, the hunter sits from morning till night under his tree with the cord in his hand, and even for two or three whole days in succession, without even getting a bite; while, on the other hand, if very lucky, he may get two or three birds in a day. There are only eight or ten men at Bessir [on Waigeo, off the western tip of New Guinea] who practise this art, which is unknown anywhere else in the island. I determined, therefore, to stay as long as possible, as my only chance of getting a good series of specimens; and although I was nearly starved, everything eatable by civilized man being scarce or altogether absent, I finally succeeded.[23]

. . .

The true Paradise Birds are omnivorous, feeding on fruits and insects – of the former preferring the small figs; of the latter, grass-hoppers, locusts, and phasmas [stick insects], as well as cockroaches and caterpillars. When I returned home, in 1862, I was so fortunate as to find two adult males of this species in Singapore; and as they seemed healthy, and fed voraciously on rice, bananas, and cock-roaches, I determined on giving the very high price asked for them – £100 – and to bring them to England by the overland route under my own care. On my way home I stayed a week at Bombay, to break the journey, and to lay in a fresh stock of bananas for my birds. I had great difficulty, however, in supplying them with insect food, for in the Peninsular and Oriental steamers cockroaches were scarce, and it was only by setting traps in the store-rooms, and by hunting an hour every night in the forecastle, that I could secure a few dozen of these creatures, – scarcely enough for a single meal. At Malta, where I stayed a fortnight, I got plenty of cockroaches from a bake-house, and when I left, took with me several biscuit-tins' full, as provision for the voyage home. We came through the Mediterranean in March, with a very cold wind; and the only place on board the mail-steamer where their large cage could be accommodated was exposed to a strong current of air down a hatchway which stood open day and night, yet the birds never seemed to feel the cold. During the night journey from Mar-seilles to Paris it was a sharp frost; yet they arrived in London in perfect health, and lived in the Zoological Gardens for one, and two years, often displaying their beautiful plumes to the admiration of the spectators.[24] [1869]

Birdwing Butterflies

During my very first walk into the forest at Batchian [Bacan], I had seen sitting on a leaf out of reach, an immense butterfly of a dark colour marked with white and yellow spots. I could not capture it as it flew away high up into the forest, but I at once saw that it was a female of a new species of Ornithoptera or "bird-winged butterfly," the pride of the Eastern tropics. I was very anxious to get it and to find the male, which in this genus is always of extreme beauty. During the two succeeding months I only saw it once again, and shortly afterwards I saw the male flying high in the air at the mining village. I had begun to despair of ever getting a specimen, as it seemed so rare and wild; till one day, about the beginning of January, I found a beautiful shrub with large white leafy bracts and yellow flowers, a species of Mussaenda, and saw one of these noble insects hovering over it, but it was too quick for me, and flew away. The next day I went again to the same shrub and succeeded in catching a female, and the day after a fine male. I found it to be as I had expected, a perfectly new and most magnificent species, and one of the most gorgeously coloured butterflies in the world. Fine specimens of the male are more than seven inches across the wings, which are velvety black and fiery orange, the latter colour replacing the green of the allied species. The beauty and brilliancy of this insect are indescribable, and none but a naturalist can understand the intense excitement I experienced when I at length captured it. On taking it out of my net and opening the glorious wings, my heart began to beat violently, the blood rushed to my head, and I felt much more like fainting than I have done when in apprehension of immediate death. I had a headache the rest of the day, so great was the excitement produced by what will appear to most people a very inadequate cause.

I had decided to return to Ternate in a week or two more, but this grand capture determined me to stay on till I obtained a good series of the new butterfly, which I have since named *Ornithoptera croesus*.[25]

. . .

The next two days were so wet and windy that there was no going out; but on the succeeding one the sun shone brightly, and I had the good fortune to capture one of the most magnificent insects the world contains, the great bird-winged butterfly, *Ornithoptera poseidon*. I trembled with excitement as I saw it coming majestically towards me, and could hardly believe I had really succeeded in my stroke till I had taken it out of the net and was gazing, lost in admiration, at the velvet

Male bird-wing butterfly, *Trogonoptera brookiana*, named by Wallace in honour of James Brooke, "White Rajah of Sarawak" (For more on Brooke, see pp. 32 and 374 in this volume)

black and brilliant green of its wings, seven inches across, its bolder body, and crimson breast. It is true I had seen similar insects in cabinets at home, but it is quite another thing to capture such oneself – to feel it struggling between one's fingers, and to gaze upon its fresh and living beauty, a bright gem shining out amid the silent gloom of a dark and tangled forest. The village of Dobbo held that evening at least one contented man.[26] [1869] ẽ

Insect Disguise

The perfection of insect camouflage – one of his longstanding interests – never failed to excite Wallace's admiration.

ẽ By far the most singular and most perfect disguise I have ever met with in a Lepidopterous insect is that of a common Indian butterfly, *Kallima inachis*, and its Malayan ally *Kallima paralekta*. I had the satisfaction of observing the habits of the latter in Sumatra, where it is rather plentiful at the end of the dry season. It is a large and showy insect when on the wing, the upper surface being glossed with blue and purple, and the fore wings crossed obliquely by a broad band of rich orange. The under surface of the wings is totally different, and is seen at a glance to resemble a dead leaf. The hind wings terminate in a little tail, which forms the stalk of the leaf, and from this to the apex is a slightly curved dark brown line representing the midrib. The transverse striae which cross the discoidal cell in many butterflies are here continued so as to form lateral veins, and the usual submarginal striae, on the hind wings, slightly modified, represent others towards the base of the wing. But it is only when the habits of the insect are observed that the disguise becomes manifested in all its perfection. This butterfly, like many others, has the habit of resting only upon a nearly vertical twig or branch, with the wings closed together so as completely to conceal the upper surface. In this position, the little tail of the hind wings exactly touches the branch, and we now see why it is always curved inwards a little; for if it were quite straight, it would hang clear of the branch, and thus fail to represent an attached leaf. There is a little scallop or hollow on the margin of the fore wings at the base, which serves to conceal the head of the butterfly, which is very small for its size, and the long antennae are carried back and hidden between the folded wings. When sitting on a twig in the manner described, the insect is to all appearance a perfect dry leaf, – yet it is evident that its chances of escape would be much increased if it were surrounded by real dry leaves instead of by green ones, for if, when pursued, it took shelter in a growing bush, it could hardly fail to be still a conspicuous object. Marvellous to relate, it does possess the habit of almost invariably entering a bush loaded with dead leaves, and is so instantly lost to sight, owing to its close resemblance to all the surrounding objects, that I doubt if the most vigilant fly-catcher could detect it. I have myself often been utterly puzzled. I have watched it settle, apparently in a very conspicuous situation, a few yards off, but on crawling carefully up to the spot have been quite unable to detect any living thing. Sometimes, while gazing intently, a butterfly would start out from just before my eyes, and again enter another dead bush a few yards off, again to be lost in the same manner. Once or twice only was I able to detect it sitting, and admire the wonderful disguise which a most strange combination of colour, form, and habits enabled it instantaneously to assume. But there is yet another peculiarity which adds to the concealment of this species. Scarcely two of the specimens are alike in colour on the under side, but vary through all the shades of pale buff, yellow, brown, and deep rusty orange which dried leaves assume. Others are speckled over with little black dots like mildewed leaves, or have clusters of spots or irregular blotches, like the minute fungi that attack dead leaves; so that a dozen of these insects might settle on a perfectly bare spray, and clothe it at once with withered foliage not distinguishable from that of the surrounding branches![27] [1867]   ẹ≈

Conservation

Perhaps more than anyone else of his generation, Wallace understood the value – and the vulnerability – of the natural world. He recognized that part of the problem is ignorance of what species exist in threatened habitats: we cannot, after all, plan to conserve something if we do not know that it is there. He therefore wrote passionately on the importance of training biologists to carry on the descriptive work he and others had begun.

◄§ It is for such [biogeographic] inquiries the modern naturalist collects his materials; it is for this that he still wants to add to the apparently boundless treasures of our national museums, and will never rest satisfied as long as the native country, the geographical distribution, and the amount of variation of any living thing remains imperfectly known. He looks upon every species of animal and plant now living as the individual letters which go to make up one of the volumes of our earth's history; and, as a few lost letters may make a sentence unintelligible, so the extinction of the numerous forms of life which the progress of cultivation invariably entails will necessarily obscure this invaluable record of the past. It is, therefore, an important object, which governments and scientific institutions should immediately take steps to secure, that in all tropical countries colonised by Europeans the most perfect collections possible in every branch of natural history should be made and deposited in national museums, where they may be available for study and interpretation.

If this is not done, future ages will certainly look back upon us as a people so immersed in the pursuit of wealth as to be blind to higher considerations. They will charge us with having culpably allowed the destruction of some of those records of Creation which we had it in our power to preserve; and while professing to regard every living thing as the direct handiwork and best evidence of a Creator, yet, with a strange inconsistency, seeing many of them perish irrecoverably from the face of the earth, uncared for and unknown.[28] [1863]   §►

◄§ It is really deplorable that in so many of our tropical dependencies no attempt has been made to preserve for posterity any *adequate* portions of the native vegetation, especially of the virgin forests. As an example, the island of Singapore was wholly covered with grand virgin forest at the beginning of the last [the nineteenth] century. When I was there in 1854 the greater part of it was still forest, but timber-cutting and clearing for gambir [plant used for tanning] and other plantations has gone on without restriction till there is now hardly any true virgin forest left; and quite recently the finest portion left has been allowed to be destroyed by a contractor in order to get granite for harbour works, which might almost as easily have been obtained elsewhere. The grand forest trees were actually burnt to make way for the granite diggers!

Surely, before it is too late, our Minister for the Colonies should be urged without delay to give stringent orders that in all the protected Malay States, in British Guiana [Guyana], Trinidad, Jamaica, Ceylon [Sri Lanka], Burma, etc., a suitable provision shall be made of forest or mountain "reserves," not for the purpose of forestry and timber-cutting only, but in order to preserve adequate and even abundant examples of those most glorious and entrancing features of our earth, its native forests, woods, mountain slopes, and alpine pastures in every country under our control. It is not only our duty to posterity that such reserves should be made for the purpose of enjoyment and study by future generations, but it is absolutely necessary in order to prevent further deterioration of the climate and destruction of the fertility of the soil, which has already taken place in Ceylon and some parts of India to a most deplorable extent. For this end not only must timber-producing forests of an ample size be secured, but on all mountain slopes continuous belts of at least 400 or 500 yards wide should be reserved wherever forests still exist or where they have been already lost be reproduced as soon as possible, so as to form retainers of moisture by the surface vegetation, checks to evaporation by the shade of the trees, guards against torrential rains, mud slides, snow slides where such are prevalent, and protection against winds. On level or nearly level ground, where such varied uses would not be required, similar belts at greater distances apart should be saved for local uses and amelioration of climate, besides "botanical reserves" of adequate extent to give a representation of each type of vegetation in the country.

I would also strongly urge that, in all countries where there are still vast areas of tropical forests, as in British Guiana, Burma, etc., all future sales or concessions of land for any purpose should be limited to belts of moderate breadth, say half a mile or less, to be followed by a belt of forest of the same width; and further, that at every mile or half-mile, and especially where streams cross the belts, transverse patches of forest, from one to two furlongs wide, shall be reserved, to remain public property and to be utilised in the public interest. Thus only can the salubrity and general amenity of such countries be handed on to our successors. Of course the general position of these belts and clearings should be determined by local conditions; but there should be no exception to the rule that all rivers and streams except the very smallest should be reserved as public property and absolutely secured against pollution; while all natural features of especial interest or beauty should also be maintained for public use and enjoyment. . . .

There must be hundreds of young botanists in Europe and America who would be glad to go to collect, say for three years, in any of these islands [South-east Asia] if their expenses were paid. There would be work for fifty of them, and if they were properly distributed over the islands from Sumatra to New Guinea in places decided upon by a committee of botanists who knew the country, with instructions to limit their work to a small area which they could examine thoroughly, to make forest trees their main object, but obtain all other flowering plants they met with, a more thorough and useful botanical explora-tion would be the result than the labours of all other collectors in the same area have accomplished, or are likely to accomplish, during the next century. And if each of these collectors had a moderate salary for another three years in order to describe and publish the results of their combined work on a uniform plan, and in a cheap form, the total expense for all the nations of Europe combined would be a mere trifle. Here is a great opportunity for some of our millionaires to carry out this important scientific exploration before these glorious forests are recklessly diminished or destroyed – a work which would be sure to lead to the discovery of great numbers of plants of utility or beauty, and would besides form a basis of knowledge from which it would be possible to approach the various great governments urging the estab-lishment, as a permanent possession for humanity, of an adequate number of such botanical, or rather biological, "reserves" as I have here suggested in every part of the world.[29] [1910] ಹ

> Wallace's plea for conservation was not confined to the Tropics. In Britain, too, he saw evidence of an environmental disaster, and even worried about the possibility of anthropogenic climate change.

ಹ I would call attention to the way in which the lavish production of minerals disfigures the country, diminishes vegetable and animal life, and destroys the fertility (for perhaps hundreds of generations) of large tracts of valuable land. It would be interesting to have a survey made of the number of acres of land covered by slag-heaps and cinder-tips at our iron and copper works, and by the waste and refuse mounds at our various mines and slate quarries, together with the land destroyed or seriously injured by smoke and deleterious gases in

those "black countries" which it pains the lover of nature to travel
through. The extent of once fertile land thus rendered more or less
permanently barren would, I believe, astonish and affright us. How
strikingly contrasted, both in their motive and results, are those noble
works of planting or of irrigation which permanently increase both
the beauty and productiveness of a country, and carry down their
blessings to succeeding generations![30] [1873] ॐ

ॐ Some years back one of our gardening periodicals obtained from
gardeners of forty or fifty years' experience a body of facts clearly
indicating a comparatively recent change of climate. It was stated that
in many parts of the country, especially in the north, fruits were
formerly grown successfully and of good quality in gardens where they
cannot be grown now; and this occurred in places sufficiently removed
from manufacturing centres to be unaffected by any direct deleterious
influence of smoke. But an increase of cloud, and consequent diminution of sunshine, would produce just such a result; and this increase
is almost certain to have occurred, owing to the enormously increased
amount of dust thrown into the atmosphere as our country has
become more densely populated, and especially owing to the vast
increase of our smoke-producing manufactories. It seems highly probable, therefore, that to increase the wealth of our capitalist-manufac-
turers we are allowing the climate of our whole country to be greatly
deteriorated in a way which diminishes both its productiveness and its
beauty, thus injuriously affecting the enjoyment and the health of the
whole population, since sunshine is itself an essential condition of
healthy life. When this fact is thoroughly realized we shall surely put a
stop to such a reckless and wholly unnecessary production of injurious
smoke and dust.[31] [1898] ॐ

In Wallace's opinion, Americans had butchered their natural
environment even more than had Europeans.

ॐ What most impresses the nature-loving Englishman while travelling in America is, the newness and rawness of the country, and the
almost universal absence of that harmonious interblending of wild
nature with human cultivation, which is so charming over a large part
of England. In these North-Eastern States, the native forests have been
so ruthlessly destroyed that fine trees are comparatively rare, and such
noble elms, beeches, oaks, and sycamores as are to be found arching
over the lanes and shading the farmhouses and cottages in a thousand

English villages, are only to be seen near a few of the towns in the older settled States, or as isolated specimens which are regarded as something remarkable. Instead of the old hedgerows with tall elms, spreading oaks, and an occasional beech, hornbeam, birch, or holly, we see everywhere the ugly snake-fence of split rails, or the still more unsightly boundary of barbed wire. Owing to the country being mostly cut up into one-mile square sections, subdivided into quarters, along the outer boundaries of which is the only right-of-way for access to the different farms, the chief country roads or tracks zig-zag along these section-lines without any regard to the contours of the land. It is probably owing to the cost of labour and the necessity of bringing large areas under cultivation as quickly as possible, that our system of fencing by live hedges, growing on a bank, with a ditch on one side for drainage, seems to be absolutely unknown in America; and hence the constant references of English writers on rural scenery and customs to "the ditch," or "the hedge," are unintelligible to most Americans.

The extreme rapidity with which the land has been cleared of its original forest seems to have favoured the spread of imported weeds, many of which are specially adapted to seize upon and monopolise newly exposed or loosened soil; and this has prevented the native plants, which might have adapted themselves to the new conditions had the change gone on very slowly, from gaining a footing. Hence it is that the cultivated fields and the artificial grass lands are less flowery than our hedge-bordered fields and old pastures, while the railway banks never exhibit such displays of floral beauty as they often do with us. An American writer in *The Century* for June, 1887, summarises the general result of these varied causes, with a severe truthfulness that would hardly be courteous in a stranger, in the following words:

> "A whole huge continent has been so touched by human hands, that over a large part of its surface it has been reduced to a state of unkempt, sordid ugliness; and it can be brought back into a state of beauty only by further touches of the same hands, more intelligently applied."[32]

Let us hope that intelligence of this kind will soon be cultivated as an essential part of education in all American schools. This alone will, however, have no effect so long as the fierce competition of great capitalists, farmers and manufacturers, reduces the actual cultivator of the soil, whether owner, tenant, or labourer, to a condition of sordid poverty, and a life of grinding labour which leaves neither leisure nor desire for the creation or preservation of natural beauty in his surroundings.[33] [1891] ಎ

Wallace was particularly moved by the plight of the California giant redwood trees.

ಎ Neither the thundering waters of Niagara, nor the sublime precipices and cascades of Yosemite, nor the vast expanse of the prairies, nor the exquisite delight of the alpine flora of the Rocky Mountains – none of these seem to me so unique in their grandeur, so impressive in their display of the organic forces of nature, as the . . . magnificent "big trees" of California. Unfortunately these alone are within the power of man totally to destroy, as they have been already partially destroyed. Let us hope that the progress of true education will so develop the love and admiration of nature that the possession of these altogether unequalled trees will be looked upon as a trust for all future generations, and that care will be taken, before it is too late, to preserve not only one or two small patches, but some more extensive tracts of forest, in which they may continue to flourish, in their fullest perfection and beauty, for thousands of years to come, as they have flourished in the past, in all probability for millions of years and over a far wider area.[34] [1891] ಎ

Having been one of the first Europeans to study birds of paradise, it must have been painful for Wallace to observe the Victorian fashion of wearing feathers in hats. Eventually, measures were taken to stem the tide of avian slaughter. Wallace wrote in 1908 to Lord Avebury (the naturalist Sir John Lubbock): "Allow me to wish every success to your Bill for preserving beautiful birds from destruction. To stop the import is the only way – short of the still more drastic method of heavily fining everyone who wears feathers in public, with imprisonment for a second offence. But we are not ripe for that yet."[35] [1908]

Wallace understood nature's delicate ecological balance, and just how easily it could be upset. We are now highly familiar with the damage that an introduced species can do in an ecosystem. However, Wallace was writing about it at a time when European colonists still commonly (and blithely) brought with them animals and plants from home and rejoiced when these introductions became established, devastating the indigenous environment in the process.[36] Again, Wallace's ecological acumen put him well

ahead of his time: today he would have been at the vanguard of
those sounding the alarm at the impact of "bioinvasions" and at
the threat of the homogenization – the "McDonaldsization" – of
global ecosystems.

Every change becomes the centre of an ever-widening circle of
effects. The different members of the organic world are so bound
together by complex relations, that any one change generally involves
numerous other changes, often of the most unexpected kind. We
know comparatively little of the way in which one animal or plant is
bound up with others, but we know enough to assure us that groups
the most apparently disconnected are often dependent on each other.
We know, for example, that the introduction of goats into St. Helena
utterly destroyed a whole flora of forest trees; and with them all the
insects, mollusca, and perhaps birds directly or indirectly dependent
on them.[37] [1876]

Geography, Geology, and Glaciology

Like Darwin, Wallace was an accomplished geologist and physical geographer. Victorian scientists were not as discipline-bound as we are today: the underlying structure of the landscape was as much a feature of the environment to Darwin and Wallace as were the animals and plants that populate it.

ᐃ I trust that the reader who has followed me throughout [*Island Life*] will be imbued with the conviction that ever presses upon myself, of the complete interdependence of organic and inorganic nature. Not only does the marvellous structure of each organised being involve the whole past history of the earth, but such apparently unimportant facts as the presence of certain types of plants or animals in one island rather than in another, are now shown to be dependent on the long series of past geological changes – on those marvellous astronomical revolutions which cause a periodic variation of terrestrial climates – on the apparently fortuitous action of storms and currents in the conveyance of germs – and on the endlessly varied actions and reactions of organised beings on each other.[1] [1880] ᐁ

Wallace's first encounters with geography, however, were inauspicious.

ᐃ Next to Latin grammar the most painful subject I learned was geography, which ought to have been the most interesting. It consisted almost entirely in learning by heart the names of the chief towns, rivers, and mountains of the various countries from, I think, Pinnock's "School Geography," which gave the minimum of useful or interesting information. It was something like learning the multiplication table both in the painfulness of the process and the permanence of the results. The incessant grinding in both, week after week, and year after year, resulted in my knowing both the product of any two numbers up to twelve, and the chief towns of any English county so thoroughly, that the result was automatic, and the name of

Staffordshire brought into my memory Stafford, Litchfield, Leek, as surely and rapidly as eight times seven brought fifty-six. The labour and mental effort to one who like myself had little verbal memory was very painful, and though the result has been a somewhat useful acquisition during life, I cannot think but that the same amount of mental exertion wisely directed might have produced far greater and more generally useful results. When I had to learn the chief towns of the provinces of Poland, Russia, Asia Minor, and other parts of Western Asia, with their almost unpronounceable names, I dreaded the approaching hour, as I was sure to be kept in for inability to repeat them, and it was sometimes only by several repetitions that I could attain even an approximate knowledge of them. No interesting facts were ever given in connection with these names, no accounts of the country by travellers were ever read, no good maps ever given us, nothing but the horrid stream of unintelligible place-names, to be learned in their due order as belonging to a certain country.[2] [1905] ঌ

Wallace's practical experience as a surveyor made up for the shortcomings of Hertford Grammar School. One of his major undertakings while exploring the remote River Uapés during his Amazon journey was to map the river, which proved no simple matter.

ঌ Being now in a part of the country that no European traveller had ever before visited, I exceedingly regretted my want of instruments to determine the latitude, longitude, and height above the sea. The two last I had no means whatever of ascertaining, having broken my boiling-point thermometer, and lost my smaller one, without having been able to replace either. I once thought of sealing up a flask of air, by accurately weighing which on my return, the density of air at that particular time would be obtained, and the height at which a barometer would have stood might be deduced. But, besides that this would only give a result equal to that of a single barometer observation, there were insuperable difficulties in the way of sealing up the bottle, for whether sealing-wax or pitch were used, or even should the bottle be hermetically sealed, heat must be applied, and at the moment of application would, of course, rarefy the air within the bottle, and so produce in such a delicate operation very erroneous results. My observations, however, on the heights of the falls we passed, would give their sum as about two hundred and fifty feet; now if we add fifty for the fall of the river between them, we shall obtain three hundred feet, as the probable height of the point I reached above the mouth of the river; and, as I have every reason to believe that that is not five hundred feet above the sea, we shall obtain eight hundred feet as the probable limit of the height of the river above the sea-level, at the point I reached. Nothing, however, can accurately determine this fact, but a series of barometer or "boiling-point" observations; and to determine this height above the next great fall, and ascertain the true course and sources of this little-known but interesting and important river, would be an object worth the danger and expense of the voyage.[3] [1853] ે≫

Geology

Wallace's geological education was derived from Charles Lyell's great work *The Principles of Geology* (1830–33), and Wallace regarded himself as a lifelong apostle of Lyell. In particular, he promulgated Lyell's uniformitarianism, whereby the geological upheavals of the past can be explained by the protracted action of processes observable today. He concludes his essay on "inaccessible valleys" (like Yosemite) with a plug for Lyell:

≫ It was for the purpose of bringing clearly before non-geologic readers the total inaccuracy of the popular view – that every rock-walled valley or steep Alpine gorge has had its origin in some great "convulsion of nature" – and to impress upon such readers the grand but simple theory, which we owe mainly to the late Sir Charles Lyell, of the efficiency of causes now in action in producing the varied contours of the earth's surface, that this account of some of the most remarkable of known valleys has been written.[4] [1893] ે≫

Given that Wallace's major work was biogeographic, he was inevitably considerably hamstrung by the lack, at that time, of a theory of continental drift. German meteorologist Alfred Wegener first published his theory – which was to remain controversial for more than fifty years – in 1915,[5] just two years after Wallace's death. Wallace's vision of the surface of the planet was nevertheless a dynamic one: rather than *horizontal* continental drift, he envisioned a kind of *vertical* drift in which the land masses are thrust up or down.

◄§ THE MOVEMENTS OF CONTINENTS

As we find these stratified rocks of different periods spread over almost the whole of existing continents where not occupied by igneous or metamorphic rocks, it follows that at one period or another each part of the continent has been under the sea, but at the same time not far from the shore. Geologists now recognize two kinds of movements by which the deposits so formed have been elevated into dry land – in the one case the strata remain almost level and undisturbed, in the other they are contorted and crumpled, often to an enormous extent. The former often prevails in plains and plateaus, while the latter is almost always found in the great mountain ranges. We are thus led to picture the land of the globe as a flexible area in a state of slow but incessant change; the changes consisting of low undulations which creep over the surface so as to elevate and depress limited portions in succession without perceptibly affecting their nearly horizontal position; and also of intense lateral compression, supposed to be produced by partial subsidence along certain lines of weakness in the earth's crust, the effect of which is to crumple the strata and force up certain areas in great contorted masses, which, when carved out by subaerial denudation [erosion] into peaks and valleys, constitute our great mountain systems. In this way every part of a continent may again and again have sunk beneath the sea, and yet as a whole may never have ceased to exist as a continent or a vast continental archipelago. And, as subsidence will always be accompanied by deposition, piles of marine strata many thousand feet thick may have been formed in a sea which was never very deep, by means of a slow depression either continuous or intermittent, or through alternate subsidences and elevations, each of moderate amount.[6] [1880]   §►

With his particular interest in the evolution of humans, Wallace paid special attention to palaeontological studies of human material. Here, in *The Wonderful Century*, he uses the human record to illustrate the reluctance of scientists to accept new data until the occurence of what today would be termed a "paradigm shift". Wallace the scientist memorably concluded that "whenever the scientific men of any age disbelieve other men's careful observations without inquiry, the scientific men are *always* wrong."

❧ THE ANTIQUITY OF MAN

Following the general acceptance of a glacial epoch by about twenty years, but to some extent connected with it, came the recognition that man had existed in Northern Europe along with numerous animals which no longer live there – the mammoth, the woolly rhinoceros, the wild horse, the cave-bear, the lion, the sabre-toothed tiger, and many others – and that he had left behind him, in an abundance of rude flint implements, the record of his presence. Before that time geologists, as well as the whole educated world, had accepted the dogma that man only appeared upon the earth when both its physical features and its animal and vegetable forms were exactly as we find them to-day; and this belief, resting solely on negative evidence, was so strongly and irrationally maintained that the earlier discoveries could not get a hearing. A careful but enthusiastic French observer, M. Boucher de Perthes,[7] had for many years collected with his own hands, from the great deposits of old river gravels in the valley of the Somme near Amiens, abundance of large and well-formed flint implements. In 1847 he published an account of them, but nobody believed his statements, till, ten years later, Dr. Falconer,[8] and shortly afterward, Professor Prestwich[9] and Mr. John Evans,[10] examined the collections and the places where they were found, and were at once convinced of their importance; and their testimony led to the general acceptance of the doctrine of the great antiquity of the human race. From that time researches on this subject have been carried on by many earnest students, and have opened up a number of altogether new chapters in human history.

So soon as the main facts were established, many old records of similar discoveries were called to mind, all of which had been ignored or explained away on account of the strong prepossession in favour of the very recent origin of man. In 1715 flint weapons had been found in excavations near Gray's Inn Lane [central London], along with the skeleton of an elephant. In 1800 another discovery was made in Suffolk of flint weapons and the remains of extinct animals in the same deposits. In 1825 Mr. McEnery, of Torquay, discovered worked flints along with the bones and teeth of extinct animals in Kent's cavern.[11] In 1840 a good geologist confirmed these discoveries, and sent an account of them to the Geological Society of London, but the paper was rejected as being too improbable for publication! All these discoveries were laughed at or explained away, as the glacial striae and grooves so beautifully exhibited in the Vale of Llanberris [Snowdonia,

Wales] were at first endeavoured to be explained as the wheel-ruts caused by the chariots of the ancient Britons! These, combined with numerous other cases of the denial of facts on *à priori* grounds, have led me to the conclusion that, whenever the scientific men of any age disbelieve other men's careful observations without inquiry, the scientific men are *always* wrong.

Even after these evidences of man's great antiquity were admitted, strenuous efforts were made to minimise the time as measured by years; and it was maintained that man, although undoubtedly old, was entirely post-glacial. But evidence has been steadily accumulating of his existence at the time of the glacial epoch, and even before it; while two discoveries of recent date seem to carry back his age far into pre-glacial times. These are, first, the human cranium, bones, and works of art which have been found more than a hundred feet deep in the gold-bearing gravels of California, associated with abundant vegetable remains of extinct species, and overlaid by four successive lava streams from long extinct volcanoes. The other case is that of rude stone implements discovered by a geologist of the Indian Survey in Burma in deposits which are admitted to be of at least Pliocene age. In both these cases the evidence is disputed by some geologists, who seem to think that there is something unscientific, or even wrong, in admitting evidence that would prove the Pliocene age of any other animal to be equally valid in the case of man. There is assumed to be a great improbability of his existence earlier than the very end of the Tertiary epoch. But all the indications drawn from his relations to the anthropoid apes point to an origin far back in Tertiary time. For each one of the great apes – the gorilla, the chimpanzee, the orang, and even the gibbon – resemble man in certain features more than do their allies, while in other points they are less like him. Now, if man has been developed from a lower animal form, we must seek his ancestors not in the direct line between him and any of the apes, but in a line toward a common ancestor to them all; and this common ancestor must certainly date back to the early part of the Tertiary epoch, because in the Miocene period anthropoid apes not very different from living forms have been found fossil.[12] [1898]   ❧

One of the many early storms that the Darwin–Wallace theory had to weather was geophysical. Whereas evolution required earth to be geologically ancient, the thermodynamics-based calculations of William Thomson, Lord Kelvin,[13] suggested in 1864 that it was relatively young. In retrospect, we know Kelvin's estimate to have been a vast underestimate because he was unaware of radioactivity as a potential source of heat. However, to the late Victorian Darwinians, Kelvin's challenge was extremely significant: in a letter to Wallace, Darwin complained "but then comes Sir W. Thomson like an odious spectre."[14] Wallace gave his ingenious response in *Island Life* (1880). In cramming all evolution into Kelvin's short time-span, Wallace's argument is, of course, as wrong as Kelvin's, but it illustrates well his capacity for constructing plausible scenarios from the most unpromising of materials and his willingness to stretch a point (in this case Lyell's uniformitarianism) when it suited him.

❦ But, according to the physicists, no such periods as are here contemplated can be granted. From a consideration of the possible sources of the heat of the sun, as well as from calculations of the period during which the earth can have been cooling to bring about the present rate of increase of temperature as we descend beneath the surface, Lord Kelvin has concluded that the crust of the earth cannot have been solidified much longer than 100 million years (the maximum possible being 400 millions), and this conclusion is held by Dr. Croll[15] and other men of eminence to be almost indisputable. It will therefore be well to consider on what data the calculations of geologists have been founded, and how far the views here set forth, as to frequent changes of climate throughout all geological time, may affect the rate of biological change.[16]

. . .

THE RATE OF GEOLOGICAL CHANGE PROBABLY GREATER IN VERY REMOTE TIMES

The opinion that denudation and deposition went on more rapidly in earlier times owing to the frequent occurrence of vast convulsions and cataclysms was strenuously opposed by Sir Charles Lyell, who so well showed that causes of the very same nature as those now in action were sufficient to account for all the phenomena presented by the rocks throughout the whole series of geological formations. But while upholding the soundness of the views of the "uniformitarians" as opposed to the "convulsionists," we must yet admit that there is reason for believing in a gradually increasing intensity of all telluric action as we go back into past time. This subject has been well treated by Mr. W. J. Sollas,[17] who shows that, if, as all physicists maintain, the Sun gave out perceptibly more heat in past ages than now, this alone would cause an increase in almost all the forces that have brought about geological phenomena. With greater heat there would be a more extensive aqueous atmosphere, and, perhaps, a greater difference between equatorial and polar temperatures; hence more violent winds, heavier rains and snows, and more powerful oceanic currents, all producing more rapid denudation. At the same time, the initial heat of the earth being greater, it would be cooling more rapidly, and thus the forces of contraction – which cause the upheaving of mountains, the eruption of volcanoes, and the subsidence of extensive areas – would be more powerful and would still further aid the process of denudation. Yet again, the earth's rotation was certainly more rapid in very remote times, and this would cause more impetuous tides and still further add to the denuding power of the ocean. It thus appears that, if we go back into the past, *all* the forces tending to the continued destruction and renewal of the earth's surface would be in more powerful action, and must therefore tend to reduce the time required for the deposition and upheaval of the various geological formations. It may be true, as many geologists assert, that the chances here indicated are so slow that they would produce comparatively little effect within the time occupied by the known sedimentary rocks, yet, whatever effect they did produce would certainly be in the direction here indicated, and as several causes are acting together, their combined effects may have been by no means unimportant. It must also be remembered that such an increase of the primary forces on which all geologic change depends would act with great effect in still further intensifying those alternations of cold and warm periods in each hemisphere, or, more frequently, of excessive and equable seasons, which have been shown to be the result of astronomical, combined with geographical, revolutions; and this would again increase the rapidity of denudation and deposition, and thus still further reduce the time required for the production of the known sedimentary rocks. It is evident therefore that these various considerations all combine to prove that, in supposing that the rate of denudation has been on the average only what it is now, we are almost certainly over-estimating, the time required to have produced the whole series of formations from the Cambrian upwards.[18]

. . .

ORGANIC MODIFICATION DEPENDENT ON CHANGE OF CONDITIONS

Having thus shown that the physical changes of the earth's surface may have gone on much more rapidly and occupied much less time than has generally been supposed, we have now to inquire whether there are any considerations which lead to the conclusion that organic changes may have gone on with corresponding rapidity.

There is no part of the theory of natural selection which is more clear and satisfactory than that which connects changes of specific forms with changes of external conditions or environment. If the external world remains for a moderate period unchanged, the organic world soon reaches a state of equilibrium through the struggle for existence; each species occupies its place in nature, and there is then no inherent tendency to change. But almost any change whatever in the external world disturbs this equilibrium, and may set in motion a whole series of organic revolutions before it is restored. A change of climate in any direction will be sure to injure some and benefit other species. The one will consequently diminish, the other increase in number; and the former may even become extinct. But the extinction of a species will certainly affect other species which it either preyed upon, or competed with, or served for food; while the increase of any one animal may soon lead to the extinction of some other to which it was inimical. These changes will in their turn bring other changes; and before an equilibrium is again established, the proportions, ranges, and numbers, of the species inhabiting the country may be materially altered. . . .

Geographical changes would be still more important, and it is almost impossible to exaggerate the modifications of the organic world that might result from them. A subsidence of land separating a large island from a continent would affect the animals and plants in a variety of ways. It would at once modify the climate, and so produce a series of changes from this cause alone; but more important would be its effect by isolating small groups of individuals of many species and thus altering their relations to the rest of the organic world. Many of these would at once be exterminated, while others, being relieved from competition, might flourish and become modified into new species. Even more striking would be the effects when two continents, or any two land areas which had been long separated, were united by an upheaval of the strait which divided them. Numbers of animals would now be brought into competition for the first time. New enemies and new competitors would appear in every part of the country; and a struggle would commence which, after many fluctuations, would certainly result in the extinction of some species, the modification of others, and a considerable alteration in the proportionate numbers and the geographical distribution of almost all.

Any other changes which led to the intermingling of species whose ranges were usually separate would produce corresponding results. Thus, increased severity of winter or summer temperature, causing southward migrations and the crowding together of the productions of distinct regions, must inevitably produce a struggle for existence, which would lead to many changes both in the characters and the distribution of animals. Slow elevations of the land would produce another set of changes, by affording an extended area in which the more dominant species might increase their numbers; and by a greater range and variety of alpine climates and mountain stations, affording room for the development of new forms of life.

GEOGRAPHICAL MUTATIONS AS A MOTIVE POWER IN BRINGING ABOUT ORGANIC CHANGES

Now, if we consider the various geographical changes which, as we have seen, there is good reason to believe have ever been going on in the world, we shall find that the motive power to initiate and urge on organic changes has never been wanting. In the first place, every continent, though permanent in a general sense, has been ever subject to innumerable physical and geographical modifications. At one time the total area has increased, and at another has diminished; great plateaus have gradually risen up, and have been eaten out by denudation into mountain and valley; volcanoes have burst forth, and, after accumulating vast masses of eruptive matter, have sunk down beneath the ocean, to be covered up with sedimentary rocks, and at a subsequent period again raised above the surface; and the loci of all these grand revolutions of the earth's surface have changed their position age after age, so that each portion of every continent has again and again been sunk under the ocean waves, formed the bed of some inland sea, or risen high into plateaus and mountain ranges. How great must have been the effects of such changes on every form of organic life! And it is to such as these we may perhaps trace those great changes of the animal world which have seemed to revolutionize it, and have led us to class one geological period as the age of reptiles, another as the age of fishes, and a third as the age of mammals.

But such changes as these must necessarily have led to repeated unions and separations of the land masses of the globe, joining together continents which were before divided, and breaking up others into great islands or extensive archipelagoes. Such alterations of the means of transit would probably affect the organic world even more profoundly than the changes of area, of altitude, or of climate, since they afforded the means, at long intervals, of bringing the most diverse forms into competition, and of spreading all the great animal and vegetable types widely over the globe. But the isolation of considerable masses of land for long periods also afforded the means of preservation to many of the lower types, which thus had time to become modified into a variety of distinct forms, some of which became so well adapted to special modes of life that they have continued to exist to the present day, thus affording us examples of the life of early ages which would probably long since have become extinct had they been always subject to the competition of the more highly organized animals. As examples of such excessively archaic forms, we may mention the mud-fishes and the ganoids [a group of primitive bony fish], confined to limited fresh-water areas; the frogs and toads, which still maintain themselves vigorously in competition with higher forms; and among mammals the Ornithorhynchus [duck-billed platypus, a monotreme (egg-laying mammal)] and Echidna [spiny ant-eater, a monotreme] of Australia; the whole order of Marsupials – which, out of Australia, where they are quite free from competition, only exist abundantly in South America, which was certainly long isolated from the northern continents; the Insectivora, which, though widely scattered, are generally nocturnal or subterranean in their habits; and the Lemurs, which are most abundant in Madagascar, where they have long been isolated, and almost removed from the competition of higher forms.[19] [1880] ❧

Glaciology

The Ice Age was a hot topic in Victorian scientific circles. In 1840, Swiss naturalist and glaciologist Louis Agassiz[20] both published his *Études sur les Glaciers* and, with English geologist and clergyman William Buckland,[21] toured Britain, where he found more evidence of past glaciation. Wallace wrote extensively on glaciology, and contributed theoretically with his modification of James Croll's astronomical theory[22] of the cause of Ice Ages, and with technical comments on the factors causing the movement of

glaciers and the formation of geomorphological features like
mountain lakes. Here is Wallace's summary of the Ice Age's
impact:

≈ᶾ   The complete similarity of the conclusions reached by four different sets of observers in four different areas – Switzerland, northwest-
ern Europe, the British Isles, and North America – after fifty years of
continuous research, and after every other less startling theory had
been put forth and rejected as wholly inconsistent with the phenomena to be explained, renders it as certain as any conclusion from
indirect evidence can be, that a large portion of the north temperate
zone, now enjoying a favourable climate and occupied by the most
civilized nations of the world, was, at a very recent epoch, geologically
speaking, completely buried in ice, just as Greenland is now. How
recently the ice has passed away is shown by the perfect preservation
of innumerable moraines, perched blocks, erratics, and glaciated rock-
surfaces, showing that but little denudation has occurred to modify
the surface; while undoubted relics of man found in glacial or
interglacial deposits prove that it occurred during the human period.
It is clear that man could not have lived in any area while it was
actually covered by the ice-sheet, while any indications of his presence
at an earlier period would almost certainly be destroyed by the
enormous abrading and grinding power of the ice.

   Besides the areas above referred to, there are widespread indications of glaciation in parts of the world where a temperate climate
now prevails. In the Pyrenees, Caucasus, Lebanon, and Himalayas
glacial moraines are found far below the lower limits they now attain.
In the Southern Hemisphere similar indications are found in New
Zealand, Tasmania, and the southern portion of the Andes; but
whether this cold period was coincident with that of the Northern
Hemisphere we have at present no means of determining, nor even
whether they were coincident among themselves, since it is quite
conceivable that they may have been due to local causes, such as
greater elevation of the land, and not to any general cause acting
throughout the south temperate zone.

   In the north temperate zone, however, the phenomena are so
widespread and so similar in character, with only such modifications
as are readily explained by proximity to, or remoteness from, the
ocean, that we are almost sure they must have been simultaneous, and
have been due to the same general causes, though perhaps modified
by local changes in altitude and consequent modification of winds or

ocean-currents. The time that has elapsed since the glaciation of the Northern Hemisphere passed away is, geologically, very small indeed, and has been variously estimated at from 20,000 to 100,000 years. At present the smaller period is most favoured by geologists, but the duration of the ice age itself, including probably one or more inter-glacial mild periods, is admitted to be much longer, and probably to approach the higher figure above given.

The undoubted fact, however, that a large part of the north temperate zone has been recently subjected to so marvellous a change of climate, is of immense interest from many points of view. It teaches us in an impressive way how delicate is the balance of forces which renders what are now the most densely peopled areas habitable by man. We can hardly suppose that even the tremendously severe ice age of which we have evidence is the utmost that can possibly occur; and, on the other hand, we may anticipate that the condition of things which in earlier geological times rendered even the polar regions adapted for a luxuriant woody vegetation may again recur, and thus vastly extend the area of our globe which is adapted to support human life in abundance and comfort. In the endeavour to account for the change of climate and of physical geography which brought about so vast a change, and then, after a period certainly approaching, and perhaps greatly exceeding, a hundred thousand years, caused it to pass away, some of the most acute and powerful intellects of our day have exerted their ingenuity; but, so far as obtaining general acceptance for the views of any one of them, altogether in vain. There seems reason to believe, however, that the problem is not an insoluble one; and when the true cause is reached, it will probably carry with it the long-sought datum from which to calculate with some rough degree of accuracy the duration of geological periods. But, whether we can solve the problem of its cause or no, the demonstration of the recent occurrence of a Glacial Epoch or Great Ice Age, with the determination of its main features over the Northern Hemisphere, will ever rank as one of the great scientific achievements of the Nineteenth Century.[23] [1898] ಆ

Humans

Wallace made many distinguished contributions to anthropology in which he combined his experience of "savage peoples" from his years in the field with his analytical and theoretical expertise as a biologist. However, many of his writings on the subject, especially later ones, go beyond straightforward academic discussion: unlike Darwin, for whom humans were merely evolutionary extrapolations from the Great Ape lineage, Wallace cheerfully threw biological, metaphysical, and political elements into his analysis of human problems. As a result, his conclusions were sometimes startling – and he invariably succeeded in upsetting someone (often Darwin) with every new article – but they were also always cogently and carefully argued. Wallace's insights have aged well: they remain today as provocative and refreshing as when they first appeared.

"Uncivilised people"

Wallace's journeys exposed him to indigenous "savage" peoples. Although many of the other leading scientists of his day had also spent considerable periods of time overseas, none could boast knowledge of indigenous peoples as intimate as his. In both the Amazon and South-east Asia he was typically travelling alone (Bates and Wallace went their separate ways after travelling together for probably a little less than the first year of their trip), or with a few local companions, and he was very much living off the land, dependent upon the people he met for food and lodging. In contrast, the journeys of Darwin, Huxley, and Hooker were all in association with official government expeditions, and they were most of the time surrounded by their own compatriots, only observing local people from a distance. Even when Darwin was not on board the *Beagle*, he typically travelled in some style, the guest of local colonials. Wallace's perspectives on indigenous people were therefore much more informed than those of his colleagues. He wrote from Borneo to an unknown correspondent, "The more I see of uncivilised people, the better I think of human nature on the whole, and the essential differences between so-called civilised and savage man seem to disappear."[1]

*[The] most unexpected sensation of surprise and delight was my first meeting and living with man in a state of nature – with absolute uncontaminated savages! This was on the Uuapés river [tributary to the Rio Negro], and the surprise of it was that I did not in the least expect to be so surprised. I had already been two years in the country, always among Indians of many tribes; but these were all what are called tame Indians, they wore at least trousers and shirt; they had been (nominally) converted to Christianity, and were under the government of the nearest authorities; and all of them spoke either Portuguese or the common language, called "Lingoa-Geral."

But these true wild Indians of the Uaupés were at once seen to be something totally different. They had nothing that we call clothes; they had peculiar ornaments, tribal marks, etc.; they all carried weapons or tools of their own manufacture; they are living in a large house, many families together, quite unlike the hut of the tame Indians; but, more than all, their whole aspect and manner were different – they were all going about their own work or pleasure which had nothing to do with white men or their ways; they walked with the free step of the independent forest-dweller, and, except the few that were known to my companion, paid no attention whatever to us, mere strangers of an alien race. In every detail they were original and self-sustaining as are the wild animals of the forests, absolutely independent of civilization, and who could and did live their own lives in their own way, as they had done for countless generations before America was discovered. I could not have believed that there would be so much difference in the aspect of the same people in their native state and when living under European supervision. The true denizen of the Amazonian forests, like the forest itself, is unique and not to be forgotten.[2] [1905]   ह

Despite retaining many of the paternalistic attitudes of his day – he continued, for example, to refer to the "inferior races" – Wallace came to recognize that the impact of colonial rule was often detrimental.

ह   For nearly twelve years I travelled and lived mostly among uncivilised or completely savage races, and I became convinced that they all possessed good qualities, some of them in a very remarkable degree, and that in all the great characteristics of humanity they are wonderfully like ourselves. Some, indeed, among the brown Polynesians especially, are declared by numerous independent and unprejudiced observers, to be both physically, morally, and intellectually our equals, if not our superiors; and it has always seemed to me one of the disgraces of our civilisation that these fine people have not in a single case been protected from contamination by the vices and follies of our more degraded classes, and allowed to develop their own social and political organism under the advice of some of our best and wisest men and the protection of our world-wide power. That would have been indeed a worthy trophy of our civilisation. What we have actually done, and left undone, resulting in the degradation and lingering extermination of so fine a people, is one of the most pathetic of its tragedies.[3] [1906]   ह

Slavery, abolished in Britain and its colonies in 1833 (the slave trade had been halted in 1807), was still practised in Brazil when Wallace arrived in 1848. Still a young man when writing this account in his Amazon book, Wallace's condemnation is much milder than it would have been had he been writing later in life.

⋘ But looking at it in this, its most favourable light, can we say that slavery is good or justifiable? Can it be right to keep a number of our fellow-creatures in a state of adult infancy – of unthinking childhood? It is the responsibility and self-dependence of manhood that calls forth the highest powers and energies of our race. It is the struggle for existence, the "battle of life," which exercises the moral faculties and calls forth the latent sparks of genius. The hope of gain, the love of power, the desire of fame and approbation, excite to noble deeds, and call into action all those faculties which are the distinctive attributes of man.

Childhood is the animal part of man's existence, manhood the intellectual; and when the weakness and imbecility of childhood remain, without its simplicity and pureness, its grace and beauty, how degrading is the spectacle! And this is the state of the slave when slavery is the best that it can be. He has no care of providing food for his family, no provision to make for old age. He has nothing to incite him to labour but the fear of punishment, no hope of bettering his condition, no future to look forward to of a brighter aspect. Everything he receives is a favour; he has no rights, – what can he know therefore of duties? Every desire beyond the narrow circle of his daily labours is shut out from his acquisition. He has no intellectual pleasures, and, could he have education and taste them, they would assuredly embitter his life; for what hope of increased knowledge, what chance of any further acquaintance with the wonders of nature or the triumphs of art, than the mere hearing of them, can exist for one who is the property of another, and can never hope for the liberty of working for his own living in the manner that may be most agreeable to him?[4] [1853] ⋙

Wallace the naturalist brought his observational skills to bear on his fellow humans.

⋘ On entering the great malocca [long house] a most extraordinary and novel scene presented itself. Some two hundred men, women, and children were scattered about the house, lying in the maqueiras,

squatting on the ground, or sitting on the small painted stools, which
are made only by the inhabitants of this river. Almost all were naked
and painted, and wearing their various feathers and other ornaments.
Some were walking or conversing, and others were dancing, or playing
small fifes and whistles. The regular festa [*sic*] had been broken up
that morning; the chiefs and principal men had put off their feather
head-dresses, but as caxirí [local alcohol] still remained, the young
men and women continued dancing. They were painted over their
whole bodies in regular patterns of a diamond or diagonal character,
with black, red, and yellow colours; the former, a purple or blue
black, predominating. The face was ornamented in various styles,
generally with bright red in bold stripes or spots, a large quantity of
the colour being applied to each ear, and running down on the sides
of the cheeks and neck, producing a very fearful and sanguinary
appearance. The grass in the ears was now decorated with a little tuft
of white downy feathers, and some in addition had three little strings
of beads from a hole pierced in the lower lip. All wore the garters,
which were now generally painted yellow. Most of the young women
who danced had besides a small apron of beads of about eight inches
by six inches, arranged in diagonal patterns with much taste; besides
this, the paint on their naked bodies was their only ornament; they
had not even the comb in their hair, which the men are never
without.

The men and boys appropriated all the ornaments, thus reversing
the custom of civilised countries and imitating nature, who invariably
decorates the male sex with the most brilliant colours and most
remarkable ornaments. On the head all wore a coronet of bright red
and yellow toucans' feathers, set in a circlet of plaited straw. The
comb in the hair was ornamented with feathers, and frequently a
bunch of white heron's plumes attached to it fell gracefully down the
back. Round the neck or over one shoulder were large necklaces of
many folds of white or red beads, as well as the white cylindrical stone
hung on the middle of a string of some black shining seeds.

The ends of the monkey-hair cords which tied the hair were
ornamented with little plumes, and from the arm hung a bunch of
curiously-shaped seeds, ornamented with bright coloured feathers
attached by strings of monkeys' hair. Round the waist was one of their
most valued ornaments, possessed by comparatively few, – the girdle
of oncas' [jaguar] teeth. And lastly, tied round the ankles were large
bunches of a curious hard fruit, which produce a rattling sound in
the dance. In their hands some carried a bow and a bundle of curabís,

or war arrows; others a murucú, or spear of hard polished wood, or an oval painted gourd, filled with small stones and attached to a handle, which, being shaken at regular intervals in the dance, produced a rattling accompaniment to the leg ornaments and the song.

The wild and strange appearance of these handsome, naked, painted Indians, with their curious ornaments and weapons, the stamp and song and rattle which accompanies the dance, the hum of conversation in a strange language, the music of fifes and flutes and other instruments of reed, bone, and turtles' shells, the large calabashes of caxirí constantly carried about, and the great smoke-blackened gloomy house, produced an effect to which no description can do justice, and of which the sight of half-a-dozen Indians going through their dances for show, gives but a very faint idea.

I stayed looking on a considerable time, highly delighted at such an opportunity of seeing these interesting people in their most character-istic festivals. I was myself a great object of admiration, principally on account of my spectacles, which they saw for the first time and could not at all understand. A hundred bright pairs of eyes were continually directed on me from all sides, and I was doubtless the great subject of conversation.[5] [1853] ಒ

Wallace went beyond mere descriptive anthropology, however. In the extract below he applies the biogeographic logic of his "Wallace's Line" to humans in South-east Asia.

ಒ CONTRASTS OF RACES

Before I had arrived at the conviction that the eastern and western halves of the [Malay] Archipelago belonged to distinct primary regions of the earth, I had been led to group the natives of the Archipelago under two radically distinct races. In this I differed from most ethnologists who had before written on the subject; for it had been the almost universal custom to follow William von Humboldt[6] and Prichard,[7] in classing all the Oceanic races as modifications of one type. Observation soon showed me, however, that Malays and Papuans differed radically in every physical, mental, and moral char-acter; and more detailed research, continued for eight years, satisfied me that under these two forms, as types, the whole of the peoples of the Malay Archipelago and Polynesia could be classified. On drawing the line which separates these races, it is found to come near to that which divides the zoological regions, but somewhat eastward of it; a circumstance which appears to me very significant of the same causes having influenced the distribution of mankind that have determined the range of other animal forms.

The reason why exactly the same line does not limit both is sufficiently intelligible. Man has means of traversing the sea which animals do not possess; and a superior race has power to press out or assimilate an inferior one. The maritime enterprise and higher civilization of the Malay races have enabled them to overrun a portion of the adjacent region, in which they have entirely supplanted the indigenous inhabitants if it ever possessed any; and to spread much of their language, their domestic animals, and their customs far over the Pacific, into islands where they have but slightly, or not at all, modified the physical or moral characteristics of the people.

I believe, therefore, that all the peoples of the various islands can be grouped either with the Malays or the Papuans; and that these two have no traceable affinity to each other. I believe, further, that all the races east of the line I have drawn have more affinity for each other than they have for any of the races west of that line; that, in fact, the Asiatic races include the Malays, and all have a continental origin, while the Pacific races, including all to the east of the former (except perhaps some in the Northern Pacific), are derived, not from any existing continent, but from lands which now exist or have recently existed in the Pacific Ocean.[8] [1869] ❧

"A being apart": Human Evolution

Wallace's biggest disagreement with Darwin was over human evolution. He explained their differences succinctly in his autobiography.

◄ THE ORIGIN OF MAN AS AN INTELLECTUAL AND MORAL BEING

On this great problem the belief and teaching of Darwin was, that man's whole nature – physical, mental, intellectual, and moral – was developed from the lower animals by means of the same laws of variation and survival; and, as a consequence of this belief, that there was no difference in *kind* between man's nature and animal nature, but only one of degree. My view, on the other hand, was, and is, that there is a difference in kind, intellectually and morally, between man and other animals; and that while his body was undoubtedly developed by the continuous modification of some ancestral animal form, some different agency, analogous to that which first produced organic *life*, and then originated *consciousness*, came into play in order to develop the higher intellectual and spiritual nature of man. . . .

These views caused much distress of mind to Darwin, but, as I have shown, they do not in the least affect the general doctrine of natural selection. It might be as well urged that because man has produced the pouter-pigeon, the bull-dog, and the dray-horse, none of which could have been produced by natural selection alone, therefore the agency of natural selection is weakened or disproved. Neither, I urge, is it weakened or disproved if my theory of the origin of man is the true one.[1] [1905] ►

The subject was not originally such a bone of contention between Darwin and Wallace. Darwin had famously avoided addressing the issue of human evolution in the *Origin* and would only do so with the publication in 1871 of his *Descent of Man*. Wallace, though, had no such inhibitions, and within two years of his return to Britain from South-east Asia had presented a major paper on the subject to the Anthropological Society of London. Darwin was impressed: "The great leading idea is quite new to me, viz. that during late ages the mind will have been modified more than the body."[2]

It is a remarkable paper,[3] both for the reasoning its first part contains, and for the soaring utopian rhetoric with which it finishes. (Wallace modified the ending in subsequent versions.) Wallace entered the anthropological debate between those who considered all humans to belong to one species – the mono-genists – and those who considered each race a separate species – the polygenists – and attempted a reconciliation by suggesting two phases of human evolution. First, proto-humans evolved in physical form, adapting to their local environments (hence races), but then "there came into existence a being in whom that subtle force we term *mind*, became of greater importance than his mere bodily structure." The human brain permits us to respond to environmental challenges – "the mere capacity of clothing" ourselves has, for example, allowed us to inhabit cold regions without the corresponding evolution of cold-related adaptations. Thus natural selection no longer plays a role in the evolution of the human body: it solely exerts its effect in our mental and moral attributes. Racial differences are therefore a frozen snapshot of an earlier phase of our evolution.

THE ORIGIN OF HUMAN RACES AND THE ANTIQUITY OF MAN DEDUCED FROM THE THEORY OF "NATURAL SELECTION"

Among the most advanced students of man, there exists a wide difference of opinion on some of the most vital questions respecting his nature and origin. Anthropologists are now, indeed, pretty well agreed that man is not a recent introduction into the earth. All who have studied the question now admit that his antiquity is very great; and that, though we have to some extent ascertained the minimum of time during which he *must* have existed, we have made no approximation towards determining that far greater period during which he *may* have, and probably *has*, existed. We can with tolerable certainty affirm that man must have inhabited the earth a thousand centuries ago, but we cannot assert that he positively did not exist, or that there is any good evidence against his having existed, for a period of a hundred thousand centuries. We know positively that he was contemporaneous with many now extinct animals, and has survived changes of the earth's surface fifty or a hundred times greater than any that have occurred during the historical period; but we cannot place any definite limit to the number of species he may have outlived, or to the amount of terrestrial change he may have witnessed.

But while on this question of man's antiquity there is a very general agreement, – and all are waiting eagerly for fresh evidence to clear up those points which all admit to be full of doubt, – on other and not less obscure and difficult questions a considerable amount of dogmatism is exhibited; doctrines are put forward as established truth, no doubt or hesitation is admitted, and it seems to be supposed that no further evidence is required, or that any new facts can modify our convictions. This is especially the case when we inquire, *Are the various forms under which man now exists primitive, or derived from preexisting forms; in other words, is man of one or many species?* To this question we immediately obtain distinct answers diametrically opposed to each other: the one party positively maintaining that man is a *species* and is essentially *one* – that all differences are but local and temporary variations, produced by the different physical and moral conditions by which he is surrounded; the other party maintaining with equal confidence that man is a genus of *many species,* each of which is practically unchangeable, and has ever been as distinct, or even more distinct, than we now behold them. This difference of opinion is somewhat remarkable, when we consider that both parties are well acquainted with the subject; both use the same vast accumulation of facts; both reject those early traditions of mankind which profess to give an account of his origin; and both declare that they are seeking fearlessly after truth alone. I believe, however, it will be found to be the old story over again of the shield – gold on one side and silver on the other – about which the knights disputed; each party will persist in looking only at the portion of truth on his own side of the question, and at the error which is mingled with his opponent's doctrine. It is my wish to show how the two opposing views can be combined so as to eliminate the error and retain the truth in each, and it is by means of Mr. Darwin's celebrated theory of "Natural Selection" that I hope to do this, and thus to harmonise the conflicting theories of modern anthropologists.

Let us first see what each party has to say for itself. In favour of the unity of mankind it is argued that there are no races without transitions to others; that every race exhibits within itself variations of colour, of hair, of feature, and of form, to such a degree as to bridge over to a large extent the gap that separates it from other races. It is asserted that no race is homogeneous; that there is a tendency to vary; that climate, food, and habits produce and render permanent physical peculiarities, which, though slight in the limited periods allowed to our observation, would, in the long ages during which the human race has existed, have sufficed to produce all the differences that now appear. It is further asserted that the advocates of the opposite theory do not agree among themselves; that some would make three, some five, some fifty or a hundred and fifty species of man; some would have had each species created in pairs, while others require nations to have at once sprung into existence, and that there is no stability or consistency in any doctrine but that of one primitive stock.

The advocates of the original diversity of man, on the other hand, have much to say for themselves. They argue that proofs of change in man have never been brought forward except to the most trifling amount, while evidence of his permanence meets us everywhere. The Portuguese and Spaniards, settled for two or three centuries in South America, retain their chief physical, mental, and moral characteristics; the Dutch boers at the Cape, and the descendants of the early Dutch settlers in the Moluccas, have not lost the features or the colour of the Germanic races; the Jews, scattered over the world in the most diverse climates, retain the same characteristic lineaments everywhere; the Egyptian sculptures and paintings show us that, for at least 4000 or 5000 years, the strongly contrasted features of the Negro and the Semitic races have remained altogether unchanged; while more recent discoveries prove that, in the case at least of the American aborigines, the mound-builders of the Mississippi valley, and the dwellers on Brazilian mountains, had still in the very infancy of the human race the same characteristic type of cranial formation that now distinguishes them.

If we endeavour to decide impartially on the merits of this difficult controversy, judging solely by the evidence that each party has brought forward, it certainly seems that the best of the argument is on the side of those who maintain the primitive diversity of man. Their opponents have not been able to refute the permanence of existing races as far back as we can trace them, and have failed to show, in a single case, that at any former epoch the well marked varieties of mankind approximated more closely than they do at the present day. At the same time this is but negative evidence. A condition of immobility for four or five thousand years, does not preclude an advance at an earlier epoch, and – if we can show that there are causes in nature which would check any further physical change when certain conditions were fulfilled – does not even render such an advance improbable, if there are any general arguments to be adduced in its favour. Such a cause, I believe, does exist, and I shall now endeavour to point out its nature and its mode of operation.

In order to make my argument intelligible, it is necessary for me to explain very briefly the theory of "Natural Selection" promulgated by Mr. Darwin, and the power which it possesses of modifying the forms of animals and plants. The grand feature in the multiplication of organic life is that of close general resemblance, combined with more or less individual variation. The child resembles its parents or ancestors more or less closely in all its peculiarities, deformities, or beauties; it resembles them in general more than it does any other individuals; yet children of the same parents are not all alike, and it often happens that they differ very considerably from their parents and from each other. This is equally true of man, of all animals, and of all plants. Moreover, it is found that individuals do not differ from their parents in certain particulars only, while in all others they are exact duplicates of them. They differ from them and from each other in every particular: in form, in size, in colour, in the structure of internal as well as of external organs; in those subtle peculiarities which produce differences of constitution, as well as in those still more subtle ones which lead to modifications of mind and character. In other words, in every possible way, in every organ and in every function, individuals of the same stock vary.

Now, health, strength, and long life are the results of a harmony between the individual and the universe that surrounds it. Let us suppose that at any given moment this harmony is perfect. A certain animal is exactly fitted to secure its prey, to escape from its enemies, to resist the inclemencies of the seasons, and to rear a numerous and healthy offspring. But a change now takes place. A series of cold winters, for instance, come on, making food scarce, and bringing an immigration of some other animals to compete with the former inhabitants of the district. The new immigrant is swift of foot, and surpasses its rivals in the pursuit of game; the winter nights are colder, and require a thicker fur as a protection, and more nourishing food to keep up the heat of the system. Our supposed perfect animal is no longer in harmony with its universe; it is in danger of dying of cold or of starvation. But the animal varies in its offspring. Some of these are swifter than others – they still manage to catch food enough; some

are hardier and more thickly furred – they manage in the cold nights
to keep warm enough; the slow, the weak, and the thinly clad soon
die off. Again and again, in each succeeding generation, the same
thing takes place. By this natural process, which is so inevitable that it
cannot be conceived not to act, those best adapted to live, live; those
least adapted, die. It is sometimes said that we have no direct evidence
of the action of this selecting power in nature. But it seems to me we
have better evidence than even direct observation would be, because
it is more universal, viz., the evidence of necessity. It must be so; for,
as all wild animals increase in a geometrical ratio, while their actual
numbers remain on the average stationary, it follows that as many die
annually as are born. If therefore, we deny natural selection, it can
only be by asserting that in such a case as I have supposed, the strong,
the healthy, the swift, the well clad, the well organised animals in
every respect, have no advantage over, – do not on the average live
longer than the weak, the unhealthy, the slow, the ill-clad, and the
imperfectly organised individuals; and this no sane man has yet been
found hardy enough to assert. But this is not all; for the offspring on
the average resemble their parents, and the selected portion of each
succeeding generation will therefore be stronger, swifter, and more
thickly furred than the last; and if this process goes on for thousands
of generations, our animal will have again become thoroughly in
harmony with the new conditions in which he is placed. But he will
now be a different creature. He will be not only swifter and stronger,
and more furry, he will also probably have changed in colour, in form,
perhaps have acquired a longer tail, or differently shaped ears; for it
is an ascertained fact, that when one part of an animal is modified,
some other parts almost always change as it were in sympathy with it.
Mr. Darwin calls this "*correlation of growth,*" and gives as instances that
hairless dogs have imperfect teeth; blue eyed cats are deaf; small feet
accompany short beaks in pigeons; and other equally interesting cases.

Grant, therefore, the premises: 1st. That peculiarities of every kind
are more or less hereditary. 2nd. That the offspring of every animal
vary more or less in all parts of their organisation. 3rd. That the
universe in which these animals live, is not absolutely invariable; –
none of which propositions can be denied; and then consider that the
animals in any country (those at least which are not dying out) must
at each successive period be brought into harmony with the surrounding conditions; and we have all the elements for a change of form and
structure in the animals, keeping exact pace with changes of whatever
nature in the surrounding universe. Such changes must be slow, for

the changes in the universe are very slow; but just as these slow changes become important, when we look at results after long periods of action, as we do when we perceive the alterations of the earth's surface during geological epochs; so the parallel changes in animal form become more and more striking according as the time they have been going on is great, as we see when we compare our living animals with those which we disentomb from each successively older geological formation.

This is briefly the theory of "natural selection," which explains the changes in the organic world as being parallel with, and in part dependent on those in the inorganic. What we now have to inquire is, – Can this theory be applied in any way to the question of the origin of the races of man? Or is there anything in human nature that takes him out of the category of those organic existences, over whose successive mutations it has had such powerful sway?

In order to answer these questions, we must consider why it is that "natural selection" acts so powerfully upon animals, and we shall, I believe, find that its effect depends mainly upon their self-dependence and individual isolation. A slight injury, a temporary illness, will often end in death, because it leaves the individual powerless against its enemies. If a herbivorous animal is a little sick and has not fed well for a day or two, and the herd is then pursued by a beast of prey, our poor invalid inevitably falls a victim. So in a carnivorous animal the least deficiency of vigour prevents its capturing food, and it soon dies of starvation. There is, as a general rule, no mutual assistance between adults, which enables them to tide over a period of sickness. Neither is there any division of labour; each must fulfil *all* the conditions of its existence, and, therefore, "natural selection" keeps all up to a pretty uniform standard.

But in man, as we now behold him, this is different. He is social and sympathetic. In the rudest tribes the sick are assisted at least with food; less robust health and vigour than the average does not entail death. Neither does the want of perfect limbs or other organs produce the same effects as among animals. Some division of labour takes place; the swiftest hunt, the less active fish, or gather fruits; food is to some extent exchanged or divided. The action of natural selection is therefore checked; the weaker, the dwarfish, those of less active limbs, or less piercing eyesight, do not suffer the extreme penalty which falls upon animals so defective.

In proportion as these physical characteristics become of less importance, mental and moral qualities will have increasing influence on the well-being of the race. Capacity for acting in concert, for protection and for the acquisition of food and shelter; sympathy, which leads all in turn to assist each other; the sense of right, which checks depredations upon our fellows; the decrease of the combative and destructive propensities; self-restraint in present appetites; and that intelligent foresight which prepares for the future, are all qualities that from their earliest appearance must have been for the benefit of each community, and would, therefore, have become the subjects of "natural selection." For it is evident that such qualities would be for the well-being of man; would guard him against external enemies, against internal dissensions, and against the effects of inclement seasons and impending famine, more surely than could any merely physical modification. Tribes in which such mental and moral qualities were predominant, would therefore have an advantage in the struggle for existence over other tribes in which they were less developed, would live and maintain their numbers, while the others would decrease and finally succumb.

Again, when any slow changes of physical geography, or of climate, make it necessary for an animal to alter its food, its clothing, or its weapons, it can only do so by a corresponding change in its own bodily structure and internal organisation. If a larger or more power-ful beast is to be captured and devoured, as when a carnivorous animal which has hitherto preyed on sheep is obliged from their decreasing numbers to attack buffaloes, it is only the strongest who can hold, – those with most powerful claws, and formidable canine teeth, that can struggle with and overcome such an animal. Natural selection immediately comes into play, and by its action these organs gradually become adapted to their new requirements. But man, under similar circumstances, does not require longer nails or teeth, greater bodily strength or swiftness. He makes sharper spears, or a better bow, or he constructs a cunning pitfall, or combines in a hunting party to circumvent his new prey. The capacities which enable him to do this are what he requires to be strengthened, and these will, therefore, be gradually modified by "natural selection," while the form and struc-ture of his body will remain unchanged. So when a glacial epoch comes on, some animals must acquire warmer fur, or a covering of fat, or else die of cold. Those best clothed by nature are, therefore, preserved by natural selection. Man, under the same circumstances, will make himself warmer clothing, and build better houses; and the necessity of doing this will react upon his mental organisation and social condition – will advance them while his natural body remains naked as before.

When the accustomed food of some animal becomes scarce or totally fails, it can only exist by becoming adapted to a new kind of food, a food perhaps less nourishing and less digestible. "Natural selection" will now act upon the stomach and intestines, and all their individual variations will be taken advantage of to modify the race into harmony with its new food. In many cases, however, it is probable that this cannot be done. The internal organs may not vary quick enough, and then the animal will decrease in numbers, and finally become extinct. But man guards himself from such accidents by superintending and guiding the operations of nature. He plants the seed of his most agreeable food, and thus procures a supply independent of the accidents of varying seasons or natural extinction. He domesticates animals which serve him either to capture food or for food itself, and thus changes of any great extent in his teeth or digestive organs are rendered unnecessary. Man, too, has everywhere the use of fire, and by its means can render palatable a variety of animal and vegetable substances, which he could hardly otherwise make use of, and thus obtains for himself a supply of food far more varied and abundant than that which any animal can command.

Thus man, by the mere capacity of clothing himself, and making weapons and tools, has taken away from nature that power of changing the external form and structure which she exercises over all other animals. As the competing races by which they are surrounded, the climate, the vegetation, or the animals which serve them for food, are slowly changing, they must undergo a corresponding change in their structure, habits, and constitution, to keep them in harmony with the new conditions – to enable them to live and maintain their numbers. But man does this by means of his intellect alone; which enables him with an unchanged body still to keep in harmony with the changing universe.

From the time, therefore, when the social and sympathetic feelings came into active operation, and the intellectual and moral faculties became fairly developed, man would cease to be influenced by "natural selection" in his physical form and structure; as an animal he would remain almost stationary; the changes of the surrounding universe would cease to have upon him that powerful modifying effect which it exercises over other parts of the organic world. But from the moment that his body became stationary, his mind would become subject to those very influences from which his body had escaped; every slight variation in his mental and moral nature which should enable him better to guard against adverse circumstances, and combine for mutual comfort and protection, would be preserved and accumulated; the better and higher specimens of our race would therefore increase and spread, the lower and more brutal would give way and successively die out, and that rapid advancement of mental organisation would occur, which has raised the very lowest races of man so far above the brutes (although differing so little from some of them in physical structure), and, in conjunction with scarcely perceptible modifications of form, has developed the wonderful intellect of the Germanic races.

But from the time when this mental and moral advance commenced, and man's physical character became fixed and immutable, a new series of causes would come into action, and take part in his mental growth. The diverse aspects of nature would now make themselves felt, and profoundly influence the character of the primitive man.

When the power that had hitherto modified the body, transferred its action to the mind, then races would advance and become improved merely by the harsh discipline of a sterile soil and inclement seasons. Under their influence, a hardier, a more provident, and a more social race would be developed, than in those regions where the earth produces a perennial supply of vegetable food, and where neither foresight nor ingenuity are required to prepare for the rigours of winter. And is it not the fact that in all ages, and in every quarter of the globe, the inhabitants of temperate have been superior to those of tropical countries? All the great invasions and displacements of races have been from North to South, rather than the reverse; and we have no record of there ever having existed, any more than there exists to-day, a solitary instance of an indigenous inter-tropical civilisation. The Mexican civilisation and government came from the North, and, as well as the Peruvian, was established, not in the rich tropical plains, but on the lofty and sterile plateaux of the Andes. The religion and civilisation of Ceylon were introduced from North India; the successive conquerors of the Indian peninsula came from the Northwest, and it was the bold and adventurous tribes of the North that overran and infused new life into Southern Europe.

It is the same great law of "*the preservation of favoured races in the struggle for life*," which leads to the inevitable extinction of all those low and mentally undeveloped populations with which Europeans come in contact. The red Indian in North America, and in Brazil; the Tasmanian, Australian and New Zealander in the southern hemisphere, die out, not from any one special cause, but from the inevitable effects of an unequal mental and physical struggle. The intellectual and moral, as well as the physical qualities of the European are superior; the same powers and capacities which have made him rise in a few centuries from the condition of the wandering savage with a scanty and stationary population to his present state of culture and advancement, with a greater average longevity, a greater average strength, and a capacity of more rapid increase, – enable him when in contact with the savage man, to conquer in the struggle for existence, and to increase at his expense, just as the more favourable increase at the expense of the less favourable varieties in the animal and vegetable kingdoms, just as the weeds of Europe overrun North America and Australia, extinguishing native productions by the inherent vigour of their organisation, and by their greater capacity for existence and multiplication.

If these views are correct; if in proportion as man's social, moral and intellectual faculties became developed, his physical structure would cease to be affected by the operation of "natural selection," we have a most important clue to the origin of races. For it will follow, that those striking and constant peculiarities which mark the great divisions of mankind, could not have been produced and rendered permanent after the action of this power had become transferred from physical to mental variations. They must, therefore, have existed since the very infancy of the race; they must have originated at a period when man was gregarious, but scarcely social, with a mind perceptive but not reflective, ere any sense of *right* or feelings of *sympathy* had been developed in him.

By a powerful effort of the imagination, it is just possible to perceive him at that early epoch existing as a single homogeneous race without the faculty of speech, and probably inhabiting some tropical region. He would be still subject, like the rest of the organic world, to the action of "natural selection," which would retain his physical form and constitution in harmony with the surrounding universe. He must have been even then a dominant race, spreading widely over the warmer regions of the earth as it then existed, and, in agreement with what we see in the case of other dominant species, gradually becoming modified in accordance with local conditions. As he ranged farther from his original home, and became exposed to greater extremes of climate, to greater changes of food, and had to contend with new

enemies, organic and inorganic, useful variations in his constitution
would be selected and rendered permanent, and would, on the
principle of "correlation of growth," be accompanied by corresponding external physical changes. Thus arose those striking characteristics
and special modifications which still distinguish the chief races of
mankind. The red, black, yellow, or blushing white skin; the straight,
the curly, the woolly hair; the scanty or abundant beard; the straight
or oblique eyes; the various forms of the pelvis, the cranium, and
other parts of the skeleton.

But while these changes had been going on, his mental development had correspondingly advanced, and had now reached that
condition in which it began powerfully to influence his whole existence, and would therefore, become subject to the irresistible action
of "natural selection." This action would rapidly give the ascendancy
to mind: speech would probably now be first developed, leading to a
still further advance of the mental faculties, and from that moment
man as regards his physical form would remain almost stationary. The
art of making weapons, division of labour, anticipation of the future,
restraint of the appetites, moral, social and sympathetic feelings,
would now have a preponderating influence on his well being, and
would therefore be that part of his nature on which "natural selection"
would most powerfully act; and we should thus have explained that
wonderful persistence of mere physical characteristics, which is the
stumbling-block of those who advocate the unity of mankind.

We are now, therefore, enabled to harmonise the conflicting views
of anthropologists on this subject. Man may have been, indeed I
believe must have been, once a homogeneous race; but it was at a
period of which we have as yet discovered no remains, at a period so
remote in his history, that he had not yet acquired that wonderfully
developed brain, the organ of the mind, which now, even in his lowest
examples, raises him far above the highest brutes; – at a period when
he had the form but hardly the nature of man, when he neither
possessed human speech, nor those sympathetic and moral feelings
which in a greater or less degree everywhere now distinguish the race.
Just in proportion as these truly human faculties became developed
in him would his physical features become fixed and permanent,
because the latter would be of less importance to his well being; he
would be kept in harmony with the slowly changing universe around
him, by an advance in mind, rather than by a change in body. If,
therefore, we are of opinion that he was not really man till these
higher faculties were developed, we may fairly assert that there were

many originally distinct races of men; while, if we think that a being like us in form and structure, but with mental faculties scarcely raised above the brute, must still be considered to have been human, we are fully entitled to maintain the common origin of all mankind.

These considerations, it will be seen, enable us to place the origin of man at a much more remote geological epoch than has yet been thought possible. He may even have lived in the Eocene or Miocene[4] period, when not a single mammal possessed the same form as any existing species. For, in the long series of ages during which the forms of these primeval mammals were being slowly specialised into those now inhabiting the earth, the power which acted to modify them would only affect the mental organisation of man. His brain alone would have increased in size and complexity and his cranium have undergone corresponding changes of form, while the whole structure of lower animals was being changed. This will enable us to understand how the fossil crania of Denise and Engis[5] agree so closely with existing forms, although they undoubtedly existed in company with large mammalia now extinct. The Neanderthal skull may be a specimen of one of the lowest races then existing, just as the Australians are the lowest of our modern epoch. We have no reason to suppose that mind and brain and skull-modification, could go on quicker than that of the other parts of the organisation, and we must, therefore, look back very far in the past to find man in that early condition in which his mind was not sufficiently developed to remove his body from the modifying influence of external conditions, and the cumulative action of "natural selection." I believe, therefore, that there is no à priori reason against our finding the remains of man or his works, in the middle or later tertiary deposits. The absence of all such remains in the European beds of this age has little weight, because as we go further back in time, it is natural to suppose that man's distribution over the surface of the earth was less universal than at present. Besides, Europe was in a great measure submerged during the tertiary epoch, and though its scattered islands may have been uninhabited by man, it by no means follows that he did not at the same time exist in warm or tropical continents. If geologists can point out to us the most extensive land in the warmer regions of the earth, which has not been submerged since eocene or miocene times, it is there that we may expect to find some traces of the very early progenitors of man. It is there that we may trace back the gradually decreasing brain of former races, till we come to a time when the body also begins materially to differ. Then we shall have reached the

starting point of the human family. Before that period, he had not
mind enough to preserve his body from change, and would, therefore,
have been subject to the same comparatively rapid modifications of
form as the other mammals.

If the views I have here endeavoured to sustain have any foundation,
they give us a new argument for placing man apart, as not only the
head and culminating point of the grand series of organic nature, but
as in some degree a new and distinct order of being. From those
infinitely remote ages, when the first rudiments of organic life
appeared upon the earth, every plant, and every animal has been
subject to one great law of physical change. As the earth has gone
through its grand cycles of geological, climatal and organic progress,
every form of life has been subject to its irresistible action, and has
been continually, but imperceptibly moulded into such new shapes as
would preserve their harmony with the ever changing universe. No
living thing could escape this law of its being; none could remain
unchanged and live, amid the universal change around it.

At length, however, there came into existence a being in whom that
subtle force we term *mind*, became of greater importance than his
mere bodily structure. Though with a naked and unprotected body,
this gave him clothing against the varying inclemencies of the seasons.
Though unable to compete with the deer in swiftness, or with the wild
bull in strength, *this* gave him weapons with which to capture or
overcome both. Though less capable than most other animals of living
on the herbs and the fruits that unaided nature supplies, this wonderful faculty taught him to govern and direct nature to his own benefit,
and make her produce food for him when and where he pleased.
From the moment when the first skin was used as a covering, when
the first rude spear was formed to assist in the chase, the first seed
sown or shoot planted, a grand revolution was effected in nature, a
revolution which in all the previous ages of the earth's history had
had no parallel, for a being had arisen who was no longer necessarily
subject to change with the changing universe – a being who was in
some degree superior to nature, inasmuch, as he knew how to control
and regulate her action, and could keep himself in harmony with her,
not by a change in body, but by an advance of mind.

Here, then, we see the true grandeur and dignity of man. On this
view of his special attributes, we may admit that even those who claim
for him a position as an order, a class, or a sub-kingdom by himself,
have some reason on their side. He is, indeed, a being apart, since he
is not influenced by the great laws which irresistibly modify all other

organic beings. Nay more; this victory which he has gained for himself gives him a directing influence over other existences. Man has not only escaped "natural selection" himself, but he actually is able to take away some of that power from nature which, before his appearance, she universally exercised. We can anticipate the time when the earth will produce only cultivated plants and domestic animals; when man's selection shall have supplanted "natural selection"; and when the ocean will be the only domain in which that power can be exerted, which for countless cycles of ages ruled supreme over all the earth.

Briefly to recapitulate the argument; – in two distinct ways has man escaped the influence of those laws which have produced unceasing change in the animal world. By his superior intellect he is enabled to provide himself with clothing and weapons, and by cultivating the soil to obtain a constant supply of congenial food. This renders it unnecessary for his body, like those of the lower animals, to be modified in accordance with changing conditions – to gain a warmer natural covering, to acquire more powerful teeth or claws, or to become adapted to obtain and digest new kinds of food, as circumstances may require. By his superior sympathetic and moral feelings, he becomes fitted for the social state; he ceases to plunder the weak and helpless of his tribe; he shares the game which he has caught with less active or less fortunate hunters, or exchanges it for weapons which even the sick or the deformed can fashion; he saves the sick and wounded from death; and thus the power which leads to the rigid destruction of all animals who cannot in every respect help themselves, is prevented from acting on him.

This power is "natural selection"; and, as by no other means can it be shewn that individual variations can ever become accumulated and rendered permanent so as to form well-marked races, it follows that the differences we now behold in mankind must have been produced before he became possessed of a human intellect or human sympathies. This view also renders possible, or even requires, the existence of man at a comparatively remote geological epoch. For, during the long periods in which other animals have been undergoing modification in their whole structure to such an amount as to constitute distinct genera and families, man's *body* will have remained generically, or even specifically, the same, while his *head* and *brain* alone will have undergone modification equal to theirs. We can thus understand how it is that, judging from the head and brain, Professor [Richard] Owen places man in a distinct sub-class of mammalia, while, as regards the rest of his body, there is the closest anatomical resemblance to

that of the anthropoid apes, "every tooth, every bone, strictly homolo-
gous – which makes the determination of the difference between
Homo and *Pithecus* [ape] the anatomist's difficulty."[6] The present
theory fully recognises and accounts for these facts; and we may
perhaps claim as corroborative of its truth, that it neither requires us
to depreciate the intellectual chasm which separates man from the
apes, nor refuses full recognition of the striking resemblances to them
which exist in other parts of its structure.

In concluding this brief sketch of a great subject, I would point out
its bearing upon the future of the human race. If my conclusions are
just, it must inevitably follow that the higher – the more intellectual
and moral – must displace the lower and more degraded races; and
the power of "natural selection," still acting on his mental organisation, must ever lead to the more perfect adaptation of man's higher
faculties to the conditions of surrounding nature, and to the exigen-
cies of the social state. While his external form will probably ever
remain unchanged, except in the development of that perfect beauty
which results from a healthy and well organised body, refined and
ennobled by the highest intellectual faculties and sympathetic
emotions, his mental constitution may continue to advance and
improve till the world is again inhabited by a single homogeneous
race, no individual of which will be inferior to the noblest specimens
of existing humanity. Each one will then work out his own happiness
in relation to that of his fellows; perfect freedom of action will be
maintained, since the well balanced moral faculties will never permit
any one to transgress on the equal freedom of others; restrictive laws
will not be wanted, for each man will be guided by the best of laws, a
thorough appreciation of the rights, and a perfect sympathy with the
feelings, of all about him; compulsory government will have died away
as unnecessary (for every man will know how to govern himself), and
will be replaced by voluntary associations for all beneficial public
purposes; the passions and animal propensities will be restrained
within those limits which most conduce to happiness; and mankind
will have at length discovered that it was only required of them to
develop the capacities of their higher nature, in order to convert this
earth, which had so long been the theatre of their unbridled passions,
and the scene of unimaginable misery, into as bright a paradise as
ever haunted the dreams of seer or poet.[7] [1864] ✑

The uneasy truce between Wallace and Darwin was not to last. In
drawing attention to man's being "above biology", Wallace, as

Smith has pointed out,[8] had already intimated that the evolution of man might constitute in his view a special case. To what extent Wallace changed his mind on the subject over the years following the publication of the "Origin of Human Races" paper is controversial. Certainly it was during this period that he became interested in spiritualism, and this may well have amplified any incipient non-Darwinian ideas he had on human evolution.

However, other more rational factors may have played a role. In an era when human races were typically regarded as steps *en route* to evolution's crowning achievement, Caucasians, Wallace's travels had made him intimately familiar with the "lowest" representatives of humanity. As we have seen, even the most well travelled of the leading biologists of the time had not had anywhere near the same exposure to "savage" peoples, except perhaps for Bates. An authority on the orang-utan, Wallace was also the only one among the prominent evolutionists of his day to be intimately familiar with the "highest" representatives of the non-human world, the Great Apes. With his first-hand knowledge of the "lowest" humans and the "highest" animals, he uniquely could conceptualize the gulf between humans and other animals. Maybe his recognition of the enormity of that abyss contributed to his refusal to leave the divine out of human evolution – to have bridged the ape–human gap was, for him, asking too much of natural selection. Racism, on the other hand, made the consistent application of natural selection easier because it insinuated that the "savage races" are anyway close to apes, implicitly narrowing the ape–human divide.

In 1869, Wallace published a review of the latest edition of Lyell's *Principles of Geology* in which, for the first time, he explicitly denied the role of natural selection in human evolution: "While admitting to the full extent the agency of the same great laws of organic development in the origin of the human race as in the origin of all organized beings, there yet seems to be evidence of a Power which has guided the action of those laws in definite directions and for special ends."[9] [1869]

Before he had even seen the review, Darwin wrote with reference to their joint production, natural selection, "I hope you have not murdered too completely your own and my child."[10] Within a year Wallace had produced a full statement of his views on the subject; the "child", as far as humans were concerned, was for Wallace dead and buried. This "Limits of Natural Selection" paper is another of Wallace's masterpieces. Arguing from the functionalist stance that had served him so well in his studies of adaptation in animals, he determines that "savages" are mentally over-endowed: essentially they have the same brain capacity as "civilized" humans, and yet have no need for the "advanced" capabilities of a large brain. Natural selection cannot produce a trait that is surplus to requirements; therefore, Wallace argues, another factor must be involved. In advancing this functionalist argument, he was ignoring his own earlier caveat against simple-minded adaptive reasoning (see pp. 49–50).

THE LIMITS OF NATURAL SELECTION AS APPLIED TO MAN

Throughout this volume [CTNS] I have endeavoured to show, that the known laws of variation, multiplication, and heredity, resulting in a "struggle for existence" and the "survival of the fittest," have probably sufficed to produce all the varieties of structure, all the wonderful adaptations, all the beauty of form and of colour, that we see in the animal and vegetable kingdoms. To the best of my ability I have answered the most obvious and the most often repeated objections to this theory, and have, I hope, added to its general strength, by showing how colour – one of the strongholds of the advocates of special creation – may be, in almost all its modifications, accounted for by the combined influence of sexual selection and the need of protection. I have also endeavoured to show, how the same power which has modified animals has acted on man; and have, I believe, proved that, as soon as the human intellect became developed above a certain low stage, man's body would cease to be materially affected by natural selection, because the development of his mental faculties would render important modifications of its form and structure unnecessary. It will, therefore, probably excite some surprise among my readers, to find that I do not consider that all nature can be explained on the principles of which I am so ardent an advocate; and that I am now myself going to state objections, and to place limits, to the power of "natural selection." I believe, however, that there are such limits; and that just as surely as we can trace the action of natural laws in the development of organic forms, and can clearly conceive that fuller knowledge would enable us to follow step by step the whole process of that development, so surely can we trace the action of some unknown higher law, beyond and independent of all those laws of which we have any knowledge. We can trace this action more or less distinctly in many phenomena, the two most important of which are – the origin of sensation or consciousness, and the development of man from the lower animals. I shall first consider the latter difficulty as more immediately connected with the subjects discussed in this volume.

WHAT NATURAL SELECTION CAN NOT DO

In considering the question of the development of man by known natural laws, we must ever bear in mind the first principle of "natural selection," no less than of the general theory of evolution, that all changes of form or structure, all increase in the size of an organ or in its complexity, all greater specialization or physiological division of labour, can only be brought about, in as much as it is for the good of being so modified. Mr. Darwin himself has taken care to impress upon us, that "natural selection" has no power to produce absolute perfection but only relative perfection, no power to advance any being much beyond his fellow beings, but only just so much beyond them as to enable it to survive them in the struggle for existence. Still less has it any power to produce modifications which are in any degree injurious to its possessor, and Mr. Darwin frequently uses the strong expression, that a single case of this kind would be fatal to his theory. If, therefore, we find in man any characters, which all the evidence we can obtain goes to show would have been actually injurious to him on their first appearance, they could not possibly have been produced by natural selection. Neither could any specially developed organ have been so produced if it had been merely useless to him, or if its use were not proportionate to its degree of development. Such cases as these would prove, that some other law, or some other power, than "natural selection" had been at work. But if, further, we could see that these very modifications, though hurtful or useless at the time when they first appeared, became in the highest degree useful at a much later period, and are now essential to the full moral and intellectual development of human nature, we should then infer the action of mind, foreseeing the future and preparing for it, just as surely as we do, when we see the breeder set himself to work with the determination to produce a definite improvement in some cultivated plant or domestic animal. I would further remark that this enquiry is as thoroughly scientific and legitimate as that into the origin of species itself. It is an attempt to solve the inverse problem, to deduce the existence of a new power of a definite character, in order to account for facts which according to the theory of natural selection ought not to happen. Such problems are well known to science, and the search after their solution has often led to the most brilliant results. In the case of man, there are facts of the nature above alluded to, and in calling attention to them, and in inferring a cause for them, I believe that I am as strictly within the bounds of scientific investigation as I have been in any other portion of my work.

THE BRAIN OF THE SAVAGE SHOWN TO BE LARGER THAN HE NEEDS IT TO BE

Size of Brain an important Element of Mental Power. – The brain is universally admitted to be the organ of the mind; and it is almost as universally admitted, that size of the brain is one of the most important of the elements which determine mental power or capacity. There seems to be no doubt that brains differ considerably in quality, as indicated by greater or less complexity of the convolutions, quantity of grey matter, and perhaps unknown peculiarities of organization; but this difference of quality seems merely to increase or diminish the influence of quantity, not to neutralize it. Thus, all the most eminent modern writers see an intimate connection between the diminished size of the brain in the lower races of mankind, and their intellectual inferiority. The collections of Dr. J. B. Davis and Dr. Morton[11] give the following as the average internal capacity of the cranium in the chief races: – Teutonic family, 94 cubic inches; Esquimaux, 91 cubic inches; Negroes, 85 cubic inches; Australians and Tasmanians, 82 cubic inches; Bushmen, 77 cubic inches. These last numbers, however, are deduced from comparatively few specimens, and may be below the average, just as a small number of Finns and Cossacks give 98 cubic inches, or considerably more than that of the German races. It is evident, therefore, that the absolute bulk of the brain is not necessarily much less in savage than in civilised man, for Esquimaux skulls are known with a capacity of 113 inches, or hardly less than the largest among Europeans. But what is still more extraordinary, the few remains yet known of pre-historic man do not indicate any material diminution in the size of the brain case. A Swiss skull of the stone age, found in the lake dwelling of Meilen,[12] corresponded exactly to that of a Swiss youth of the present day. The celebrated Neanderthal skull had a larger circumference than the average, and its

capacity, indicating actual mass of brain, is estimated to have been
not less than 75 cubic inches, or nearly the average of existing Australian crania. The Engis skull, perhaps the oldest known, and
which, according to Sir John Lubbock, "there seems no doubt was
really contemporary with the mammoth and the cave bear," is yet,
according to Professor Huxley, "a fair average skull, which might
have belonged to a philosopher, or might have contained the
thoughtless brains of a savage."[13] Of the cave men of Les Eyzies,[14]
who were undoubtedly contemporary with the reindeer in the
South of France, Professor Paul Broca[15] says (in a paper read
before the Congress of Pre-historic Archæology in 1868) – "The
great capacity of the brain, the development of the frontal region,
the fine elliptical form of the anterior part of the profile of the
skull, are incontestible characteristics of superiority, such as we are
accustomed to meet with in civilised races"; yet the great breadth
of the face, the enormous development of the ascending ramus of
the lower jaw, the extent and roughness of the surfaces for the
attachment of the muscles, especially of the masticators, and the
extraordinary development of the ridge of the femur, indicate
enormous muscular power, and the habits of a savage and brutal
race.

These facts might almost make us doubt whether the size of the
brain is in any direct way an index of mental power, had we not
the most conclusive evidence that it is so, in the fact that, whenever an adult male European has a skull less than nineteen inches
in circumference, or has less than sixty-five cubic inches of brain,
he is invariably idiotic. When we join with this the equally undisputed fact, that great men – those who combine acute perception
with great reflective power, strong passions, and general energy of
character, such as Napoleon, Cuvier,[16] and O'Connell,[17] have always
heads far above the average size, we must feel satisfied that volume
of brain is one, and perhaps the most important, measure of intellect; and this being the case, we cannot fail to be struck with the
apparent anomaly, that many of the lowest savages should have as
much brains as average Europeans. The idea is suggested of a sur-
plusage of power; of an instrument beyond the needs of its
possessor.

Comparison of the Brains of Man and of Anthropoid Apes. – In order to
discover if there is any foundation for this notion, let us compare
the brain of man with that of animals. The adult male Orang-utan

is quite as bulky as a small sized man, while the Gorilla is considerably above the average size of man, as estimated by bulk and
weight; yet the former has a brain of only 28 cubic inches, the
latter, one of 30, or, in the largest specimen yet known, of 34.5
cubic inches. We have seen that the average cranial capacity of the
lowest savages is probably not less than *five-sixths* of that of the
highest civilised races, while the brain of the anthropoid apes
scarcely amounts to *one-third* of that of man, in both cases taking
the average; or the proportions may be more clearly represented
by the following figures – anthropoid apes, 10; savages, 26; civilised
man, 32. But do these figures at all approximately represent the
relative intellect of the three groups? Is the savage really no further
removed from the philosopher, and so much removed from the
ape, as these figures would indicate? In considering this question,
we must not forget that the heads of savages vary in size, almost as
much as those of civilised Europeans. Thus, while the largest Teutonic skull in Dr. Davis' collection is 112.4 cubic inches, there is an
Araucanian[18] of 115.5, an Esquimaux of 113.1, a Marquesan[19] of
110.6, a Negro of 105.8, and even an Australian of 104.5 cubic
inches. We may, therefore, fairly compare the savage with the highest European on the one side, and with the Orang, Chimpanzee,
or Gorilla, on the other, and see whether there is any relative proportion between brain and intellect.

Range of intellectual power in Man. – First, let us consider what this
wonderful instrument, the brain, is capable of in its higher developments. In Mr. Galton's interesting work on "Hereditary Genius,"
[1869] he remarks on the enormous difference between the intellectual power and grasp of the well-trained mathematician or man
of science, and the average Englishman. The number of marks
obtained by high wranglers,[20] is often more than thirty times as
great as that of the men at the bottom of the honour list, who are
still of fair mathematical ability; and it is the opinion of skilled
examiners, that even this does not represent the full difference of
intellectual power. If, now, we descend to those savage tribes who
only count to three or five, and who find it impossible to comprehend the addition of two and three without having the objects
actually before them, we feel that the chasm between them and the
good mathematician is so vast, that a thousand to one will probably
not fully express it. Yet we know that the mass of brain might be
nearly the same in both, or might not differ in a greater propor-

tion than as 5 to 6; whence we may fairly infer that the savage possesses a brain capable, if cultivated and developed, of performing work of a kind and degree far beyond what he ever requires it to do.

Again, let us consider the power of the higher or even the average civilized man, of forming abstract ideas, and carrying on more or less complex trains of reasoning. Our languages are full of terms to express abstract conceptions. Our business and our pleasures involve the continual foresight of many contingencies. Our law, our government, and our science, continually require us to reason through a variety of complicated phenomena to the expected result. Even our games, such as chess, compel us to exercise all these faculties in a remarkable degree. Compare this with the savage languages, which contain no words for abstract conceptions; the utter want of foresight of the savage man beyond his simplest necessities; his inability to combine, or to compare, or to reason on any general subject that does not immediately appeal to his senses. So, in his moral and æsthetic faculties, the savage has none of those wide sympathies with all nature, those conceptions of the infinite, of the good, of the sublime and beautiful, which are so largely developed in civilized man. Any considerable development of these would, in fact, be useless or even hurtful to him, since they would to some extent interfere with the supremacy of those perceptive and animal faculties on which his very existence often depends, in the severe struggle he has to carry on against nature and his fellow-man. Yet the rudiments of all these powers and feelings undoubtedly exist in him, since one or other of them frequently manifest themselves in exceptional cases, or when some special circumstances call them forth. Some tribes, such as the Santals,[21] are remarkable for as pure a love of truth as the most moral among civilized men. The Hindoo and the Polynesian have a high artistic feeling, the first traces of which are clearly visible in the rude drawings of the palæolithic men who were the contemporaries in France of the Reindeer and the Mammoth. Instances of unselfish love, of true gratitude, and of deep religious feeling, sometimes occur among most savage races.

On the whole, then, we may conclude, that the general moral and intellectual development of the savage, is not less removed from that of civilized man than has been shown to be the case in the one department of mathematics; and from the fact that all the moral and intellectual faculties do occasionally manifest themselves, we may fairly conclude that they are always latent, and that the large brain of the savage man is much beyond his actual requirements in the savage state.

Intellect of Savages and of Animals compared. – Let us now compare the intellectual wants of the savage, and the actual amount of intellect he exhibits, with those of the higher animals. Such races as the Andaman Islanders, the Australians, and the Tasmanians, the Digger Indians of North America,[22] or the natives of Fuegia [Tierra del Fuego], pass their lives so as to require the exercise of few faculties not possessed in an equal degree by many animals. In the mode of capture of game or fish, they by no means surpass the ingenuity or forethought of the jaguar, who drops saliva into the water, and seizes the fish as they come to eat it; or of wolves and jackals, who hunt in packs; or of the fox, who buries his surplus food till he requires it. The sentinels placed by antelopes and by monkeys, and the various modes of building adopted by field mice and beavers, as well as the sleeping place of the orang-utan, and the tree-shelter of some of the African anthropoid apes, may well be compared with the amount of care and forethought bestowed by many savages in similar circumstances. His possession of free and perfect hands, not required for locomotion, enable man to form and use weapons and implements which are beyond the physical powers of brutes; but having done this, he certainly does not exhibit more mind in using them than do many lower animals. What is there in the life of the savage, but the satisfying of the cravings of appetite in the simplest and easiest way? What thoughts, ideas, or actions are there, that raise him many grades above the elephant or the ape? Yet he possesses, as we have seen, a brain vastly superior to theirs in size and complexity; and this brain gives him, in an undeveloped state, faculties which he never requires to use. And if this is true of existing savages, how much more true must it have been of the men whose sole weapons were rudely chipped flints, and some of whom, we may fairly conclude, were lower than any existing race; while the only evidence yet in our possession shows them to have had brains fully as capacious as those of the average of the lower savage races.

We see, then, that whether we compare the savage with the higher developments of man, or with the brutes around him, we are alike driven to the conclusion that in his large and well-developed brain he possesses an organ quite disproportionate to his actual requirements – an organ that seems prepared in advance, only to be fully utilized as he progresses in civilization. A brain slightly larger than that of the gorilla would, according to the evidence before us, fully have sufficed for the limited mental development of the savage; and we must therefore admit, that the large brain he actually possesses could never have been solely developed by any of those laws of evolution, whose essence is, that they lead to a degree of organization exactly proportionate to the wants of each species, never beyond those wants – that no preparation can be made for the future development of the race – that one part of the body can never increase in size or complexity, except in strict co-ordination to the pressing wants of the whole. The brain of prehistoric and of savage man seems to me to prove the existence of some power, distinct from that which has guided the development of the lower animals through their ever-varying forms of being.

THE USE OF THE HAIRY COVERING OF MAMMALIA

Let us now consider another point in man's organization, the bearing of which has been almost entirely overlooked by writers on both sides of this question. One of the most general external characters of the terrestrial mammalia is the hairy covering of the body, which, whenever the skin is flexible, soft, and sensitive, forms a natural protection against the severities of climate, and particularly against rain. That this is its most important function, is well shown by the manner in which the hairs are disposed so as to carry off the water, by being invariably directed downwards from the most elevated parts of the body. Thus, on the under surface the hair is always less plentiful, and, in many cases, the belly is almost bare. The hair lies downwards, on the limbs of all walking mammals, from the shoulder to the toes, but in the orang-utan it is directed from shoulder to the elbow, and again from wrist to elbow, in a reverse direction. This corresponds to the habits of the animal, which, when resting, holds its long arms upwards over its head, or clasping a branch above it, so that the rain would flow down both the arm and fore-arm to the long hair which meets at the elbow. In accordance with this principle, the hair is always longer or more dense along the spine or middle of the back from the nape to the tail, often rising into a crest of hair or bristles on the ridge of the back. This character prevails through the entire series of the mammalia, from the marsupials to the quadrumana, and by this long persistence it must have acquired such a powerful hereditary tendency, that we should expect it to reappear continually even after it had been abolished by ages of the most rigid selection; and we may feel sure that it never could have been completely abolished under the law of natural selection, unless it had become so positively injurious as to lead to the almost invariable extinction of individuals possessing it.

THE CONSTANT ABSENCE OF HAIR FROM CERTAIN PARTS OF MAN'S BODY A REMARKABLE PHENOMENON

In man the hairy covering of the body has almost totally disappeared, and, what is very remarkable, it has disappeared more completely from the back than from any other part of the body. Bearded and beardless races alike have the back smooth, and even when a considerable quantity of hair appears on the limbs and breast, the back, and especially the spinal region, is absolutely free, thus completely reversing the characteristics of all other mammalia. The Ainos [Ainu] of the Kurile Islands and Japan are said to be a hairy race; but Mr. Bickmore,[23] who saw some of them, and described them in a paper read before the Ethnological Society, gives no details as to where the hair was most abundant, merely stating generally, that "their chief peculiarity is their great abundance of hair, not only on the head and face, but over the whole body." This might very well be said of any man who had hairy limbs and breast, unless it was specially stated that his back was hairy, which is not done in this case. The hairy family in Birmah[24] have, indeed, hair on the back rather longer than on the breast, thus reproducing the true mammalian character, but they have still longer hair on the face, forehead, and inside the ears, which is quite abnormal; and the fact that their teeth are all very imperfect, shows that this is a case of monstrosity rather than one of true reversion to the ancestral type of man before he lost his hairy covering.

SAVAGE MAN FEELS THE WANT OF THIS HAIRY COVERING

We must now enquire if we have any evidence to show, or any reason to believe, that a hairy covering to the back would be in any degree hurtful to savage man, or to man in any stage of his progress from his lower animal form; and if it were merely useless, could it have been so entirely and completely removed as not to be continually reappearing in mixed races? Let us look to savage man for some light on these points. One of the most common habits of savages is to use some covering for the back and shoulders, even when they have none on any other part of the body. The early voyagers observed with surprise, that the Tasmanians, both men and women, wore the kangaroo-skin, which was their only covering, not from any feeling of modesty, but over the shoulders to keep the back dry and warm. A cloth over the shoulders was also the national dress of the Maories. The Patagonians wear a cloak or mantle over the shoulders, and the Fuegians often wear a small piece of skin on the back, laced on, and shifted from side to side as the wind blows. The Hottentots[25] also wore a somewhat similar skin over the back, which they never removed, and in which they were buried. Even in the tropics most savages take precautions to keep their backs dry. The natives of Timor use the leaf of a fan palm, carefully stitched up and folded, which they always carry with them, and which, held over the back, forms an admirable protection from the rain. Almost all the Malay races, as well as the Indians of South America, make great palm-leaf hats, four feet or more across, which they use during their canoe voyages to protect their bodies from heavy showers of rain; and they use smaller hats of the same kind when travelling by land.

We find, then, that so far from there being any reason to believe that a hairy covering to the back could have been hurtful or even useless to pre-historic man, the habits of modern savages indicate exactly the opposite view, as they evidently feel the want of it, and are obliged to provide substitutes of various kinds. The perfectly erect posture of man, may be supposed to have something to do with the disappearance of the hair from his body, while it remains on his head; but when walking, exposed to rain and wind, a man naturally stoops forwards, and thus exposes his back; and the undoubted fact, that most savages feel the effects of cold and wet most severely in that part of the body, sufficiently demonstrates that the hair could not have ceased to grow there merely because it was useless, even if it were likely that a character so long persistent in the entire order of mammalia, could have so completely disappeared, under the influence of so weak a selective power as a diminished usefulness.

MAN'S NAKED SKIN COULD NOT HAVE BEEN PRODUCED BY NATURAL SELECTION

It seems to me, then, to be absolutely certain, that "Natural Selection" could not have produced man's hairless body by the accumulation of variations from a hairy ancestor. The evidence all goes to show that such variations could not have been useful, but must, on the contrary, have been to some extent hurtful. If even, owing to an unknown correlation with other hurtful qualities, it had been abolished in the ancestral tropical man, we cannot conceive that, as man spread into colder climates, it should not have returned under the powerful influence of reversion to such a long persistent ancestral type. But the very foundation of such a supposition as this is untenable; for we cannot suppose that a character which, like hairiness, exists throughout the whole of the mammalia, can have become, in one form only, so constantly correlated with an injurious character, as to lead to its permanent suppression – a suppression so complete and effectual that it never, or scarcely ever, reappears in mongrels of the most widely different races of man.

Two characters could hardly be wider apart, than the size and development of man's brain, and the distribution of hair upon the surface of his body; yet they both lead us to the same conclusion – that some other power than Natural Selection has been engaged in his production.

FEET AND HANDS OF MAN, CONSIDERED AS DIFFICULTIES ON THE THEORY OF NATURAL SELECTION

There are a few other physical characteristics of man, that may just be mentioned as offering similar difficulties, though I do not attach the same importance to them as to those I have already dwelt on. The specialization and perfection of the hands and feet of man seems difficult to account for. Throughout the whole of the quadrumana the foot is prehensile; and a very rigid selection must therefore have been needed to bring about that arrangement of the bones and muscles, which has converted the thumb into a great toe, so completely, that the power of opposability is totally lost in every race, whatever some travellers may vaguely assert to the contrary. It is difficult to see why the prehensile power should have been taken away. It must certainly have been useful in climbing, and the case of the baboons shows that it is quite compatible with terrestrial locomotion. It may not be compatible with perfectly easy erect locomotion; but, then, how can we conceive that early man, *as an animal*, gained anything by purely erect locomotion? Again, the hand of man contains latent capacities and powers which are unused by savages, and must have been even less used by palæolithic man and his still ruder predecessors. It has all the appearance of an organ prepared for the use of civilized man, and one which was required to render civilization possible. Apes make little use of their separate fingers and opposable thumbs. They grasp objects rudely and clumsily, and look as if a much less specialized extremity would have served their purpose as well. I do not lay much stress on this, but, if it be proved that some intelligent power has guided or determined the development of man, then we may see indications of that power, in facts which, by themselves, would not serve to prove its existence.

The Voice of Man. – The same remark will apply to another peculiarly human character, the wonderful power, range, flexibility, and sweetness, of the musical sounds producible by the human larynx, especially in the female sex. The habits of savages give no indication of how this faculty could have been developed by natural selection; because it is never required or used by them. The singing of savages is more or less monotonous howling, and the females seldom sing at all. Savages certainly never choose their wives for fine voices, but for rude health, and strength, and physical beauty. Sexual selection could not therefore have developed this wonderful power, which only comes into play among civilized people. It seems as if the organ had been prepared in anticipation of the future progress of man, since it contains latent capacities which are useless to him in his earlier condition. The delicate correlations of structure that give it such marvellous powers, could not therefore have been acquired by means of natural selection.

THE ORIGIN OF SOME OF MAN'S MENTAL FACULTIES, BY THE PRESERVATION OF USEFUL VARIATIONS, NOT POSSIBLE

Turning to the mind of man, we meet with many difficulties in attempting to understand, how those mental faculties, which are especially human, could have been acquired by the preservation of useful variations. At first sight, it would seem that such feelings as those of abstract justice and benevolence could never have been so acquired, because they are incompatible with the law of the strongest, which is the essence of natural selection. But this is, I think, an erroneous view, because we must look, not to individuals but to societies; and justice and benevolence, exercised towards members of the same tribe, would certainly tend to strengthen that tribe, and give it a superiority over another in which the right of the strongest prevailed, and where consequently the weak and the sickly were left to perish, and the few strong ruthlessly destroyed the many who were weaker.

But there is another class of human faculties that do not regard our fellow men, and which cannot, therefore, be thus accounted for. Such are the capacity to form ideal conceptions of space and time, of eternity and infinity – the capacity for intense artistic feelings of pleasure, in form, colour, and composition – and for those abstract notions of form and number which render geometry and arithmetic possible. How were all or any of these faculties first developed, when they could have been of no possible use to man in his early stages of barbarism? How could "natural selection," or survival of the fittest in the struggle for existence, at all favour the development of mental powers so entirely removed from the material necessities of savage men, and which even now, with our comparatively high civilization, are, in their farthest developments, in advance of the age, and appear to have relation rather to the future of the race than to its actual status?

DIFFICULTY AS TO THE ORIGIN OF THE MORAL SENSE

Exactly the same difficulty arises, when we endeavour to account for the development of the moral sense or conscience in savage man; for although the *practice* of benevolence, honesty, or truth, may have been useful to the tribe possessing these virtues, that does not at all account for the peculiar *sanctity*, attached to actions which each tribe considers right and moral, as contrasted with the very different feelings with which they regard what is merely *useful*. The utilitarian hypothesis (which is the theory of natural selection applied to the mind) seems inadequate to account for the development of the moral sense. This subject has been recently much discussed, and I will here only give one example to illustrate my argument. The utilitarian sanction for truthfulness is by no means very powerful or universal. Few laws enforce it. No very severe reprobation follows untruthfulness. In all ages and countries, falsehood has been thought allowable in love, and laudable in war; while, at the present day, it is held to be venial by the majority of mankind, in trade, commerce, and speculation. A certain amount of untruthfulness is a necessary part of politeness in the east and west alike, while even severe moralists have held a lie justifiable, to elude an enemy or prevent a crime. Such being the difficulties with which this virtue has had to struggle, with so many exceptions to its practice, with so many instances in which it brought ruin or death to its too ardent devotee, how can we believe that considerations of utility could ever invest it with the mysterious sanctity of the highest virtue, – could ever induce men to value truth for its own sake, and practice it regardless of consequences?

Yet, it is a fact, that such a mystical sense of wrong does attach to untruthfulness, not only among the higher classes of civilized people, but among whole tribes of utter savages. Sir Walter Elliott tells us (in his paper "On the Characteristics of the Population of Central and Southern India," published in the Journal of the Ethnological Society of London, vol. i., p. 107) that the Kurubars and Santals, barbarous hill-tribes of Central India, are noted for veracity. It is a common saying that "a Kurubar *always* speaks the truth;" and Major Jervis says, "the Santals are the most truthful men I ever met with." As a remarkable instance of this quality the following fact is given. A number of prisoners, taken during the Santal insurrection, were allowed to go free on parole, to work at a certain spot for wages. After some time cholera attacked them and they were obliged to leave, but every man of them returned and gave up his earnings to the guard. Two hundred savages with money in their girdles, walked thirty miles back to prison rather than break their word! My own experience among savages has furnished me with similar, although less severely tested, instances; and we cannot avoid asking, how is it, that in these few cases "experiences of utility" have left such an overwhelming impression, while in so many others they have left none? The experiences of savage men as regards the utility of truth, must, in the long run, be pretty nearly equal. How is it, then, that in some cases the result is a sanctity which overrides all considerations of personal advantage, while in others there is hardly a rudiment of such a feeling?

The intuitional theory, which I am now advocating, explains this by the supposition, that there is a feeling – a sense of right and wrong – in our nature, antecedent to and independent of experiences of utility. Where free play is allowed to the relations between man and man, this feeling attaches itself to those acts of universal utility or self-sacrifice, which are the products of our affections and sympathies, and which we term moral; while it may be, and often is, perverted, to give the same sanction to acts of narrow and conventional utility which are really immoral, – as when the Hindoo will tell a lie, but will sooner starve than eat unclean food; and looks upon the marriage of adult females as gross immorality.

The strength of the moral feeling will depend upon individual or racial constitution, and on education and habit; – the acts to which its sanctions are applied, will depend upon how far the simple feelings and affections of our nature, have been modified by custom, by law, or by religion.

It is difficult to conceive that such an intense and mystical feeling of right and wrong, (so intense as to overcome all ideas of personal advantage or utility), could have been developed out of accumulated ancestral experiences of utility; and still more difficult to understand, how feelings developed by one set of utilities, could be transferred to acts of which the utility was partial, imaginary, or altogether absent. But if a moral sense is an essential part of our nature, it is easy to see, that its sanction may often be given to acts which are useless or immoral; just as the natural appetite for drink, is perverted by the drunkard into the means of his destruction.

SUMMARY OF THE ARGUMENT AS TO THE INSUFFICIENCY OF NATURAL SELECTION TO ACCOUNT FOR THE DEVELOPMENT OF MAN

Briefly to resume my argument – I have shown that the brain of the lowest savages, and, as far as we yet know, of the pre-historic races, is little inferior in size to that of the highest types of man, and immensely superior to that of the higher animals; while it is universally admitted that quantity of brain is one of the most important, and probably the most essential, of the elements which determine mental power. Yet the mental requirements of savages, and the faculties actually exercised by them, are very little above those of animals. The higher feelings of pure morality and refined emotion, and the power of abstract reasoning and ideal conception, are useless to them, are rarely if ever manifested, and have no important relations to their habits, wants, desires, or well-being. They possess a mental organ beyond their needs. Natural Selection could only have endowed savage man with a brain a little superior to that of an ape, whereas he actually possesses one very little inferior to that of a philosopher.

The soft, naked, sensitive skin of man, entirely free from that hairy covering which is so universal among other mammalia, cannot be explained on the theory of natural selection. The habits of savages show that they feel the want of this covering, which is most completely absent in man exactly where it is thickest in other animals. We have no reason whatever to believe, that it could have been hurtful, or even useless to primitive man; and, under these circumstances, its complete abolition, shown by its never reverting in mixed breeds, is a demon-

stration of the agency of some other power than the law of the survival
of the fittest, in the development of man from the lower animals.

Other characters show difficulties of a similar kind, though not
perhaps in an equal degree. The structure of the human foot and
hand seem unnecessarily perfect for the needs of savage man, in
whom they are as completely and as humanly developed as in the
highest races. The structure of the human larynx, giving the power of
speech and of producing musical sounds, and especially its extreme
development in the female sex, are shown to be beyond the needs of
savages, and from their known habits, impossible to have been
acquired either by sexual selection, or by survival of the fittest.

The mind of man offers arguments in the same direction, hardly
less strong than those derived from his bodily structure. A number of
his mental faculties have no relation to his fellow men, or to his
material progress. The power of conceiving eternity and infinity, and
all those purely abstract notions of form, number, and harmony,
which play so large a part in the life of civilised races, are entirely
outside of the world of thought of the savage, and have no influence
on his individual existence or on that of his tribe. They could not,
therefore, have been developed by any preservation of useful forms of
thought; yet we find occasional traces of them amidst a low civilization,
and at a time when they could have had no practical effect on the
success of the individual, the family, or the race; and the development
of a moral sense or conscience by similar means is equally
inconceivable.

But, on the other hand, we find that every one of these characteristics is necessary for the full development of human nature. The
rapid progress of civilization under favourable conditions, would not
be possible, were not the organ of the mind of man prepared in
advance, fully developed as regards size, structure, and proportions,
and only needing a few generations of use and habit to co-ordinate its
complex functions. The naked and sensitive skin, by necessitating
clothing and houses, would lead to the more rapid development of
man's inventive and constructive faculties; and, by leading to a more
refined feeling of personal modesty, may have influenced, to a considerable extent, his moral nature. The erect form of man, by freeing
the hands from all locomotive uses, has been necessary for his
intellectual advancement; and the extreme perfection of his hands,
has alone rendered possible that excellence in all the arts of civilization which raises him so far above the savage, and is perhaps but the
forerunner of a higher intellectual and moral advancement. The

perfection of his vocal organs has first led to the formation of articulate speech, and then to the development of those exquisitely toned sounds, which are only appreciated by the higher races, and which are probably destined for more elevated uses and more refined enjoyment, in a higher condition than we have yet attained to. So, those faculties which enable us to transcend time and space, and to realize the wonderful conceptions of mathematics and philosophy, or which give us an intense yearning for abstract truth, (all of which were occasionally manifested at such an early period of human history as to be far in advance of any of the few practical applications which have since grown out of them), are evidently essential to the perfect development of man as a spiritual being, but are utterly inconceivable as having been produced through the action of a law which looks only, and can look only, to the immediate material welfare of the individual or the race.

The inference I would draw from this class of phenomena is, that a superior intelligence has guided the development of man in a definite direction, and for a special purpose, just as man guides the development of many animal and vegetable forms. The laws of evolution alone would, perhaps, never have produced a grain so well adapted to man's use as wheat and maize; such fruits as the seedless banana and bread-fruit; or such animals as the Guernsey milch cow, or the London dray-horse. Yet these so closely resemble the unaided productions of nature, that we may well imagine a being who had mastered the laws of development of organic forms through past ages, refusing to believe that any new power had been concerned in their production, and scornfully rejecting the theory (as my theory will be rejected by many who agree with me on other points), that in these few cases a controlling intelligence had directed the action of the laws of variation, multiplication, and survival, for his own purposes. We know, however, that this has been done; and we must therefore admit the possibility that, if we are not the highest intelligences in the universe, some higher intelligence may have directed the process by which the human race was developed, by means of more subtle agencies than we are acquainted with. At the same time I must confess, that this theory has the advantage of requiring the intervention of some distinct individual intelligence, to aid in the production of what we can hardly avoid considering as the ultimate aim and outcome of all organized existence – intellectual, ever-advancing, spiritual man. It therefore implies, that the great laws which govern the material universe were insufficient for his production, unless we consider (as we may fairly do) that the controlling action of such higher intelligences is a necessary part of those laws, just as the action of all surrounding organisms is one of the agencies in organic development. But even if my particular view should not be the true one, the difficulties I have put forward remain, and I think prove, that some more general and more fundamental law underlies that of "natural selection." The law of "unconscious intelligence" pervading all organic nature, put forth by Dr. Laycock[26] and adopted by Mr. Murphy, is such a law; but to my mind it has the double disadvantage of being both unintelligible and incapable of any kind of proof. It is more probable, that the true law lies too deep for us to discover it; but there seems to me, to be ample indications that such a law does exist, and is probably connected with the absolute origin of life and organization.[27] [1870]

Wallace did not confine his writing on human evolution to grand issues of teleology. His interest in linguistics, for example, prompted him to write on the nuts and bolts of the evolution of language. Despite insisting in typical self-deprecatory fashion, that "[a]nother and more serious defect [of mine] is in verbal memory, which, combined with the inability to reproduce vocal sounds, has rendered the acquirement of all foreign languages very difficult and distasteful",[28] Wallace was in fact a conscientious linguist who recorded extensive vocabularies from local groups on his travels. Both the original editions of the Amazon and South-east Asian travel books had extensive appendices containing these word lists. In the *Malay Archipelago*, for example, Wallace gives the word for "louse" (and 116 other words) in 33 South-east Asian languages. He argued that onomatopoeia may have played an important role in the evolution of language.

HOW SPEECH ORIGINATED

Some of the correspondences which have been here pointed out between words and their meanings, will doubtless be held by many to be mere fantastic imaginings. But if we try to picture to ourselves the condition of mankind when first acquiring and developing spoken language, and struggling in every possible way to produce articulate sounds which should carry in themselves, both to the speaker and the hearer, some expression of the things, motions, or actions represented, it will seem quite natural that they should utilize everything

connected with the act of speaking which could in any way further
that object. We are apt to forget that, though speech is now acquired
by children solely by imitation, and must be to them almost wholly
conventional, this was not its original character. Speech was formed
and evolved, not by children, but by men and women who felt the
need of a mode of communication other than by gesture only.
Gesture-language and word-language doubtless arose together, and
for a long time were used in conjunction and supplemented each
other. It is admitted that gesture-language is never purely conventional, but is based either on direct imitation or on some kind of
analogy or suggestion; and it is therefore almost certain that word-
language, arising at the same time, would be developed in the same
way, and would never originate in purely conventional terms. Gesture
would at first be exclusively used to describe motion, action, and
passion; speech to represent the infinite variety of sounds in nature,
and, with some modification, the creatures or objects that produced
the sounds. But there are many disadvantages in the use of gesture as
compared with speech. It required always a considerable muscular
effort; the hands and limbs must be free; an erect, or partially erect,
posture needed; there must be sufficient light – and, lastly, the
communicators must be in such a position as to see each other. As
articulate speech is free from all these disadvantages, there would be
a constant endeavour to render it capable of replacing gesture; and
the most obvious way of doing this would be to transfer gesture from
the limbs to the mouth itself, and to utilize so much of the corresponding motions as were possible to the lips, tongue and breath.
These mouth-gestures, as we have seen, necessarily lead to distinct
classes of sounds; and thus there arose from the very beginnings of
articulate speech, the use of characteristic sounds to express certain
groups of motions, actions, and sensations which we are still able to
detect even in our highly-developed language, and the more important of which I have here attempted to define and illustrate.

It may be well to give an example of how definite words may have
arisen by such a process. Each of the words – air, wind, breeze, blow,
blast, breathe – has to us a definite meaning, and a form which seems
often to have nothing in common with the rest. Yet they possess the
common character that the essential part of each is a breathing, more
or less pronounced and modulated; and at first they were probably all
alike expressed by a strong and audible breathing or blowing. For
convenience and to save exertion, this would soon be modified into
an articulate sound or word which would enable the act of blowing to

be easily recognized. Then, as time went on and the need arose, some
one or other of the different ideas comprised in the word would be
separated, and this would be most effectually done by the use of
different consonants with the same fundamental form of *breathing* or
blowing, and the distinction caused by the *r* and *l* in these two words
well illustrates the principle. Thus, every such class of expressive words
would have a natural basis, while the detailed modifications to differ-
entiate the various ideas included in it might be to a considerable
extent conventional.[29] [1895] ﺰ

Human Improvement

Dying shortly before World War I arrived to usher in the cynicism of the current era, Wallace belonged to a more innocent time. He was inclined to sanguine views about the improvement of society, and subscribed, albeit with reservations, to the utopian visions of the factory reformer Robert Owen and socialist writer Edward Bellamy.[1] Underpinning such thinking was a notion of *progress*: things had become better, and should continue to do so. Needless to say, Wallace's version of evolutionary change dovetailed comfortably with this worldview. Herbert Spencer, die-hard champion of what he called "universal progress", was positively evangelical on the subject: "Whether it be in the development of the Earth, in the development of Life upon its surface, in the development of Society, of Government, of Manufactures, of Commerce, of Language, Literature, Science, Art, this same evolution of the simple into the complex, through successive differentiations, holds throughout."[2] The question, though, remained: *how* was this improvement going to come about?

Two biology-based doctrines for the improvement of society emerged at the end of the nineteenth century: social Darwinism and eugenics. Social Darwinism, whereby an economic "survival of the fittest" would ensure the elimination of those unfitted to Western European capitalism, is somewhat unfairly attributed to Spencer. He did indeed hold that the welfare state was effectively a way of weakening society, but not strictly for social Darwinian reasons. As a Lamarckian, believing in the inheritance of acquired characters, Spencer asserted that the only way for those in society's basement to improve their lot was through striving. Lamarckian inheritance rewards effort – the giraffe's offspring have longer necks because it had craned to reach the highest foliage – and Spencer maintained that the welfare state, by eliminating the incentive to strive, would be condemning the basement's occupants to perpetual basement occupation. Spencer was accordingly a staunch defender of Lamarckism. Wallace wrote in a letter that he had heard "that H. Spencer is dreadfully disturbed on the question. He fears that acquired characters may not be inherited, in which case the foundation of his whole philosophy is undermined!"[3]

The contrast between a present-day conception of Spencer as the villainous social Darwinian and the views of his contemporaries is interestingly illustrated by Wallace. The present-day conception would suggest that no Victorian could have been more opposed to Spencer's views than Wallace, the mystic, socialist, and humanitarian. In fact, despite their many disagreements, Wallace admired Spencer: "I may remark that, although I differ greatly from him on certain important matters, both of natural and social science, and have never hesitated to state my reasons for those differences with whatever force of fact and argument I could bring to bear upon them, I yet look upon these as but spots on the sun of his great intellectual powers, and feel it to be an honour to have been his contemporary, and, to a limited extent, his friend and coadjutor."[4] The extent of Wallace's admiration can be gauged from his choice of name for his first-born son, Herbert Spencer Wallace.[5] (Wallace's tribute to Darwin, on the other hand, was the title-page dedication of *The Malay Archipelago*.)

"Eugenics" is a term coined by a first cousin of Darwin, Francis Galton. In 1869 he published *Hereditary Genius*, a study of the genetic transmission of talents, whose conclusions – that talents do indeed run in families, and must therefore be heritable – prompted him to think in terms of enriching society for these traits. That his analysis was fundamentally flawed – environmental factors like education probably play a more important role in determining an individual's abilities than genetic ones, and, of course, educated parents are more likely to invest time and money in the education of their children – did not prevent him from borrowing from agricultural breeders their basic technique for improving stock. This was the artificial selection process so famously highlighted by Darwin as analogous to natural selection: in order to improve the milk yield of his herd, a farmer only breeds from those individuals with the highest yield. Galton advocated the selective breeding of humans, eugenics. This became a popular doctrine as the Darwin–Wallace theory caught on. Natural selection was apparently no longer operating in Victorian Britain. Indeed it seemed that the poorest and most depraved were the ones with the largest families, while those middle-class pillars of respectability – the ones with the "good genes" – were having only small families. The bad genes were out-reproducing the good ones, and a "eugenic crisis" was in the offing. Wallace reported that Darwin himself was concerned:

In one of my latest conversations with Darwin he expressed himself very gloomily on the future of humanity, on the ground that in our modern civilisation natural selection had no play, and the fittest did not survive. Those who succeed in the race for wealth are by no means the best or the most intelligent, and it is notorious that our population is more largely renewed in each generation from the lower than from the middle and upper classes.[6] [1890]

Eugenics could be positive – encouraging the right kind of people to have children – or negative – discouraging the wrong kind of people from having children. During the first part of the twentieth century it was generally embraced, both by biologists and by the public. In view of its ghastly apotheosis at the hands of the Nazis, for whom the good genes belonged to Aryans, and the bad to Jews and Gypsies, it is difficult for us to conceive of a time when the very word "eugenics" seemed to promise genuine science-based social improvement. In fact, eugenics was enthusiastically endorsed by many liberals, including Fabian socialists like George Bernard Shaw, H. G. Wells and Sidney and Beatrice Webb.[7] Wallace himself, however, was not impressed.

Segregation of the unfit, indeed! It is a mere excuse for establishing a medical tyranny. And we have enough of this kind of tyranny already. Even now, the lunacy laws give dangerous powers to the medical fraternity. At the present moment, there are some perfectly sane people incarcerated in lunatic asylums simply for believing in spiritualism. The world does not want the eugenist to set it straight. Give the people good conditions, improve their environment, and all will tend towards the highest type. Eugenics is simply the meddlesome interference of an arrogant, scientific priestcraft.[8] [1912]

Wallace's utopian vision did in fact contain elements of selection, but not selection directed by a "scientific priestcraft". Rather, his system was based on free choice of husbands by women. In essence, he argued that women married for the wrong reasons, typically forced into marriage by economic considerations. In an ideal socialist world in which those economic concerns were of no matter, a woman would be able to choose freely, and she would, of course, choose from among the "better" men. Thus Wallace aspired to achieve the effects of positive eugenics, but via the mediation of female choice. Darwin, had he been alive (Wallace first published these ideas in 1890, eight years after Darwin's death), could have been forgiven for thinking that Wallace was being deliberately contrary on the subject of humans. First he had disavowed a role for natural selection in human evolution, and now he was advocating female choice-mediated sexual selection for future improvement. Wallace and Darwin disagreed strongly on the importance of female choice in the natural world; Darwin thought it was important, and Wallace thought it was not.[9] It is ironic, therefore, that Wallace chose female choice as the foundation of his vision for the future. Nevertheless, it is an interesting idea and one that highlights Wallace's proto-feminism. He gave a brief summary of it at the end of his life in *Social Environment and Moral Progress* (1913).

EUGENICS, OR RACE IMPROVEMENT THROUGH MARRIAGE

The total cessation of the action of natural selection as a cause of improvement in our race, either physical or mental, led to the proposal of the late Sir F. Galton to establish a new science, which he termed Eugenics. A society has been formed,[10] and much is being written about checking degeneration and elevating the race to a higher level by its means. Sir F. Galton's own proposals were limited to giving prizes or endowments for the marriage of persons of high character, determined by some form of inquiry or examination. This may, perhaps, not do much harm, but it would certainly do very little good. Its range of action would be extremely limited, and so far as it induced any couples to marry each other for the pecuniary reward, it would be absolutely immoral in its nature, and probably result in no perceptible improvement of the race.

But there is great danger in such a process of artificial selection by experts, who would certainly soon adopt methods very different from those of the founder. We have already had proposals made for the "'segregation of the feeble-minded," while the "sterilization of the unfit" and of some classes of criminals is already being discussed. This might soon be extended to the destruction of deformed infants, as was actually proposed by the late Grant Allen;[11] while Mr. Hiram M.

Stanley, in a work on *Our Civilization and the Marriage Problem,* proposed more far reaching measures. He says: "The drunkard, the criminal, the diseased, the morally weak, should never come into society. Not reform, but prevention, should be the cry." And he hints at the methods he would adopt, in the following passages: "In the true golden age, which lies not behind but before us, the privilege of parentage will be esteemed an honor for the comparatively few, and no child will be born who is not only sound in body and mind, but also above the average as to natural ability and moral force." And he concludes: "The most important matter in society, the inherent quality of the members of which it is composed, should be regulated by trained specialists."

Of course, our modern eugenists will disclaim any wish to adopt such measures as are here hinted at, which are in every way dangerous and detestable. But I protest strenuously against any direct interference with the freedom of marriage, which, as I shall show, is not only totally unnecessary, but would be a much greater source of danger to morals and to the well-being of humanity than the mere temporary evils it seeks to cure. I trust that all my readers will oppose any *legislation* on this subject by a chance body of elected persons who are totally unfitted to deal with far less complex problems than this one, and as to which they are sure to bungle disastrously.

It is in the highest degree presumptuous and irrational to attempt to deal by compulsory enactments with the most vital and most sacred of all human relations, regardless of the fact that our present phase of social development is not only extremely imperfect but, as I have already shown, vicious and rotten at the core. How can it be possible to determine by legislation those relations of the sexes which shall be best alike for individuals and for the race, in a society in which a large proportion of our women are forced to work long hours daily for the barest subsistence, with an almost total absence of the rational pleasures of life, for the want of which thousands are driven into wholly uncongenial marriages in order to secure some amount of personal independence or physical well-being?

Let anyone consider, on the one hand, the lives of the wealthy as portrayed in the society newspapers of the day, with their endless round of pleasure and luxury, their almost inconceivable wastefulness and extravagance, indicated by the cost of female dress and the fact of a thousand pounds or more being expended on the flowers for a single entertainment. On the other hand, let him contemplate the awful lives of millions of workers, so miserably paid and with such uncertainty of work that many thousands of the women and young girls are driven on the streets as the only means of breaking the monotony of their unceasing labour and obtaining some taste of the enjoyments of life at whatever cost; and then ask himself if the legislature which cannot remedy this state of things should venture to meddle with the great problems of marriage and the sanctities of family life. Is it not a hideous mockery that the successive governments which for forty years have seen the people they profess to govern so driven to despair by the vile conditions of their existence that in an ever larger and larger proportion they seek death by suicide as their only means of escape – that governments which have done nothing to put an end to this continuous horror of starvation and suicide should be thought capable of remedying some of its more terrible *results*, while leaving its *causes* absolutely untouched?

It is my firm conviction, for reasons I shall give farther on, that, when we have cleansed the Augean stable of our present social organization, and have made such arrangements that *all* shall contribute their share either of physical or mental labour, and that every one shall obtain the full and equal reward for their work, the future progress of the race will be rendered certain by the fuller development of its higher nature acted on by a special form of selection which will then come into play.

When men and women are, for the first time in the course of civilization, alike free to follow their best impulses; when idleness and vicious or hurtful luxury on the one hand, oppressive labour and the dread of starvation on the other, are alike unknown; when all receive the best and broadest education that the state of civilization and knowledge will admit; when the standard of public opinion is set by the wisest and the best among us, and that standard is systematically inculcated on the young; then we shall find that a system of *truly natural* selection will come spontaneously into action which will steadily tend to eliminate the lower, the less developed, or in any way defective types of men, and will thus continuously raise the physical, moral, and intellectual standard of the race. The exact mode in which this selection will operate will now be briefly explained.

FREE-SELECTION IN MARRIAGE

It will be generally admitted that, although many women now remain unmarried from necessity rather than from choice, there are always considerable numbers who feel no strong impulse to marriage, and accept husbands to secure subsistence and a home of their own rather than from personal affection or strong sexual emotion. In a state of society in which all women were economically independent, were all fully occupied with public duties and social or intellectual pleasures, and had nothing to gain by marriage as regards material well-being or social position, it is highly probable that the numbers of the unmarried from choice would increase. It would probably come to be considered a degradation for any woman to marry a man whom she could not love and esteem, and this reason would tend at least to delay marriage till a worthy and sympathetic partner was encountered.

In man, on the other hand, the passion of love is more general and usually stronger; and in such a society as here postulated there would be no way of gratifying this passion but by marriage. Every woman, therefore, would be likely to receive offers, and a powerful selective agency would rest with the female sex. Under the system of education and public opinion here supposed, there can be little doubt how this selection would be exercised. The idle or the utterly selfish would be almost universally rejected; the chronically diseased or the weak in intellect would also usually remain unmarried, at least till an advanced period of life, while those who showed any tendency to insanity or exhibited any congenital deformity would also be rejected by the younger women, because it would be considered an offence against society to be the means of perpetuating any such diseases or imperfections.

We must also take account of a special factor, hitherto almost unnoticed, which would tend to intensify the selection thus exercised. It is a fact well known to statisticians that, although females are in excess in almost all civilized populations, yet this is not due to a law of Nature; for with us, and I believe in all parts of the Continent, more males than females are born to an amount of about 3½ to 4 per cent. But between the ages of five and thirty-five there were, in 1910, 4.225 deaths of males from accident or violence and only 1.300 of females, showing an excess of male deaths of 2.925 in one year; and for many years the numbers of this class of deaths have not varied much, the excess of preventable deaths of males at those ages being very nearly 3,000 annually. This excess is no doubt due to boys and young men being more exposed, both in play and work, to various kinds of accidents than are women, and this brings about the constant excess of females in what may be termed normal civilized populations.

In 1901 it was about a million; while fifty years earlier, when the population was about half, it was only 359,000, or considerably less than half the present proportion. This is what we should expect from the constant increase of accidents and of emigration, the effects of both of which fall most upon males.

It appears, therefore, that the larger number of women in our population today is not a natural phenomenon, but is almost wholly the result of our own man-made social environment. When the lives of *all* our citizens are accounted of equal value to the community, irrespective of class or of wealth, a much smaller number will be allowed to suffer from such preventable causes; while, as our colonies fill up with a normal population, and the enormous areas of uncultivated or half cultivated land at home are thrown open to our own people on the most favourable terms, the great tide of emigration will be diminished and will then cease to affect the proportion of the sexes. The result of these various causes, now all tending to increase the numbers of the female population, will, in a rational and just system of society, of which we may hope soon to see the commencement, act in a contrary direction, and will in a few generations bring the sexes first to an equality, and later on to a majority of males.

There are some, no doubt, who will object that, even when women have a free choice, owing to improved economic conditions, they will not choose wisely so as to advance the race. But no one has the right to make such a statement without adducing very strong evidence in support of it. We have for generations degraded women in every possible way; but we now know that such degradation is not hereditary, and therefore not permanent. The great philosopher and seer, Swedenborg,[12] declared that whereas men loved justice, wisdom, and power for their own sakes, women loved them as seen in the characters of men. It is generally admitted that there is truth in this observation; but there is surely still more truth in the converse, that they do not admire those men who are palpably unjust, stupid, or weak, and still less those who are distorted, diseased, or grossly vicious, though under present conditions they are often driven to marry them. It may be taken as certain, therefore, that when women are economically and socially free to choose, numbers of the worst men among all classes who now readily obtain wives *will be almost universally rejected.*

Now, this mode of improvement by elimination of the less desirable has many advantages over that of securing early marriages of the more admired; for what we most require is to improve the *average* of our population by rejecting its lower types rather than by raising the advanced types a little higher. Great and good men are always produced in sufficient numbers and have always been so produced in every phase of civilization. We do not need more of these so much as we want a diminution of the weaker and less advanced types. This weeding-out process has been the method of *natural selection,* by which the whole of the glorious vegetable and animal kingdoms have been developed and advanced. The survival of the fittest is really the extinction of the unfit; and it is the one brilliant ray of hope for humanity that, just as we advance in the reform of our present cruel and disastrous social system, we shall set free a power of selection in marriage that will steadily and certainly improve the character, as well as the strength and the beauty, of our race.[13] [1913] ❧

Spiritualism and Metaphysics

Wallace attended his first seance in 1865, three years after returning from South-east Asia, and had published a pamphlet on "The
Scientific Aspect of the Supernatural" within a year. He remained
a convinced spiritualist for the rest of his life. The extent to which
Wallace's metaphysical interests affected his science remains controversial, but Martin Fichman[1] has shown convincingly that
Wallace's spiritualism evolved gradually into a nebulous form of
theism which became a major component of his scientific worldview. This is Wallace at his most enigmatic: on one hand we have
the co-champion of a revolutionary materialistic interpretation of
the natural world, evolution by natural selection, and on the
other we have an avowed anti-materialist. G. K. Chesterton,
writing in 1904 when Wallace was eighty-one, deemed Wallace
the second "greatest man of our time" (the first was Walt Whit-
man) because simultaneously "he has been the leader of a
revolution and the leader of a counter revolution."[2]

"Strange doings": Conversion

Three early strands combined in Wallace to create a fully fledged theist: mesmerism (hypnotic trance induction),[1] phrenology,[2] and spiritualism. There might at first sight appear to be no obvious connection between them, but all three were scientifically marginal: mesmerism was typically demonstrated by travelling showmen, reducing it to little more than a form of vaudeville; phrenology was popular during the first half of the nineteenth century but became progressively eclipsed by anatomical discoveries; spiritualism, despite its faddish embrace by the Victorians, was forever – and often unsuccessfully – fending off charges of fraud. Perhaps Wallace's affinity for the underdog – so superbly articulated in his socialism – made him sympathetic to these pursuits. Also it has to be said that Wallace retained throughout his life a quality of innocence that in some circumstances translated into gullibility.

Despite being raised in a conventionally religious household, Wallace was quick to disavow organized religion.

❧ I have already shown that my early home training was in a thoroughly religious but by no means rigid family, where, however, no religious doubts were ever expressed, and where the word "atheist" was used with bated breath as pertaining to a being too debased almost for human society. The only regular teaching I received was to say or hear a formal prayer before going to bed, hearing grace before and after dinner, and learning a collect every Sunday morning, the latter certainly one of the most stupid ways of inculcating religion ever conceived. On Sunday evenings, if we did not go to church or chapel, my father would read some old sermon, and when we did go we were asked on our return what was the text. The only books allowed to be read on Sundays were the "Pilgrim's Progress" or "Paradise Lost," or some religious tracts or moral tales, or the more interesting parts of the Bible were read by my mother, or we read ourselves about Esther and Mordecai[3] or Bel and the Dragon,[4] which were as good as any story book. But all this made little impression upon me, as it never dealt sufficiently with the mystery, the greatness, the ideal and emotional aspects of religion, which only appealed to me occasionally in some of the grander psalms and hymns, or through the words of some preacher more impassioned than usual.

As might have been expected, therefore, what little religious belief I had very quickly vanished under the influence of philosophical or scientific scepticism. This came first upon me when I spent a month or two in London with my brother John... and during the seven years I lived with my brother William, though the subject of religion was not often mentioned, there was a pervading spirit of scepticism, or free-thought as it was then called, which strengthened and confirmed my doubts as to the truth or value of all ordinary religious teaching.[5] [1905] ও

Hellfire and brimstone did not sit well with Wallace.

ও After the most terrible description had been given of the unimaginable torments of hell-fire, we were told to suppose that the whole earth was a mass of fine sand, and that at the end of a thousand years one single grain of this sand flew away into space. Then – we were told – let us try to imagine the slow procession of the ages, while grain by grain the earth diminished, but still remained apparently as large as ever, – and still the torments went on. Then let us carry on the imagination through thousands of millions of millions of ages, till at last the globe could be seen to be a little smaller – and then on and on, and on for other and yet other myriads of ages, till after periods which to finite beings would seem almost infinite the last grain flew away, and the whole material of the globe was dissipated in space. And then, asked the preacher, is the sinner any nearer the end of his punishment? No! for his punishment is to be infinite, and after thousands of such globes had been in the same way dissipated, his torments are still to go on and on for ever! I myself had heard such horrible sermons as these in one of the churches in Hertford, and a lady we knew well had been so affected by them that she had tried to commit suicide. I therefore thoroughly agreed with Mr. Dale Owen's[6] conclusion, that the orthodox religion of the day was degrading and hideous, and that the only true and wholly beneficial religion was that which inculcated the service of humanity, and whose only dogma was the brotherhood of man. Thus was laid the foundation of my religious scepticism.[7] [1905] ও

Perhaps the most candid account of Wallace's early views on religion is given in a "PS" of a letter written during his travels from Timor to his brother-in-law Thomas Sims. Sims is admonished, "This for yourself; show the *letter only* to my mother."

In my early youth I heard, as ninety-nine-hundredths of the world do, only the evidence on one side, and became impressed with a veneration for religion which has left some traces even to this day. I have since heard and read much on both sides, and pondered much upon the matter in all its bearings. I spent, as you know, a year and a half in a clergyman's family[8] and heard almost every Tuesday the very best, most earnest and most impressive preacher it has ever been my fortune to meet with, but it produced no effect whatever on my mind. I have since wandered among men of many races and many religions. I have studied man and nature in all its aspects, and I have sought after truth. In my solitude I have pondered much on the incomprehensible subjects of space, eternity, life and death. I think I have fairly heard and fairly weighed the evidence on both sides, and I remain an *utter disbeliever* in almost all that you consider the most sacred truths. I will pass over as utterly contemptible the oft-repeated accusation that sceptics shut out evidence because they will not be governed by the morality of Christianity. You I know will not believe that in my case, and I know its falsehood as a general rule. I only ask, Do you think I can change the self-formed convictions of twenty-five years, and could you think such a change would have anything in it to merit *reward* from *justice*? I am thankful I can see much to admire in all religions. To the mass of mankind religion of some kind is a necessity. But whether there be a God and whatever be His nature; whether we have an immortal soul or not, or whatever may be our state after death, I can have no fear of having to suffer for the study of nature and the search for truth, or believe that those will be better off in a future state who have lived in the belief of doctrines inculcated from childhood, and which are to them rather a matter of blind faith than intelligent conviction.[9] [1861]

It was Wallace's first encounter with mesmerism – and in turn with phrenology – in 1844 that provoked his lifelong interest in these "alternative" worldviews.

It was at Leicester that I was first introduced to a subject which I had at that time never heard of, but which has played an important part in my mental growth – psychical research, as it is now termed. Some time in 1844 Mr. Spencer Hall[10] gave some lectures on mesmerism illustrated by experiments, which I, as well as a few of the older boys, attended. I was greatly interested and astonished at the phenomena exhibited, in some cases with persons who volunteered from the audience; and I was also impressed by the manner of the lecturer, which was not at all that of the showman or the conjurer. At the conclusion of the course he assured us that most persons possessed in some degree the power of mesmerising others, and that by trying with a few of our younger friends or acquaintances, and simply doing what we had seen him do, we should probably succeed. He also showed us how to distinguish between the genuine mesmeric trance and any attempt to imitate it.

In consequence of this statement, one or two of the elder boys tried to mesmerise some of the younger ones, and in a short time succeeded; and they asked me to see their experiments. I found that they could produce the trance state, which had all the appearance of being genuine, and also a cataleptic rigidity of the limbs by passes and by suggestion, both in the trance and afterwards in the normal waking state. This led me to try myself in the privacy of my own room, and I succeeded after one or two attempts in mesmerising three boys from twelve to sixteen years of age, while on others within the same ages I could produce no effect, or an exceedingly slight one. During the trance they seemed in a state of semi-torpor, with apparently no volition. They would remain perfectly quiescent so long as I did not notice them, but would at once answer any questions or do anything I told them. On the two boys with whom I continued to experiment for some time, I could produce catalepsy of any limb or the whole body, and in this state they could do things which they could not, and certainly would not have done in their normal state. For example, on the rigid outstretched arm I would hang an ordinary chair at the wrist, and the boy would hold it there for several minutes, while I sat down and wrote a short letter for instance, without any complaint, or making any remark when 1 took it off. I never left it more than five minutes because I was afraid that some injury might be caused by it. I soon found that this rigidity could be produced in those who had been mesmerised by suggestion only, and in this way often fixed them in any position, notwithstanding their efforts to change it. One experiment was to place a shilling on the table in front of a boy, and then say to him, "Now, you can't touch that shilling." He would at once move his hand towards it, but when halfway it would seem to stick fast, and all his efforts could not bring it nearer, though he was promised the shilling if he could take it.

Every phenomenon of suggestion I had seen at the lecture, and many others, I could produce with this boy. Giving him a glass of water and telling him it was wine or brandy, he would drink it, and soon show all the signs of intoxication, while if I told him his shirt was on fire he would instantly strip himself naked to get it off. I also found that he had community of sensation with myself when in the trance. If I held his hand he tasted whatever I put in my mouth, and the same thing occurred if one or two persons intervened between him and myself; and if another person put substances at random into my mouth, or pinched or pricked me in various parts of the body, however secretly, he instantly felt the same sensation, would describe it, and put his hand to the spot where he felt the pain.

In like manner any sense could be temporarily paralyzed so that a light could be flashed on his eyes or a pistol fired behind his head without his showing the slightest sign of having seen or heard anything. More curious still was the taking away the memory so completely that he could not tell his own name, and would adopt any name that was suggested to him, and perhaps remark how stupid he was to have forgotten it; and this might be repeated several times with different names, all of which he would implicitly accept. Then, on saying to him, "Now you remember your own name again; what is it?" an inimitable look of relief would pass over his countenance, and he would say, "Why, P-, of course," in a way that carried complete conviction.

But perhaps the most interesting group of phenomena to me were those termed phreno-mesmerism. I had read, when with my brother, George Combe's[11] "Constitution of Man," with which I had been greatly interested, and afterwards one of the writer's works on Phrenology, and at the lecture I had seen some of the effects of exciting the phrenological organ by touching the corresponding parts of the patient's head. But as I had no book containing a chart of the organs, I bought a small phrenological bust to help me in determining the positions.

Having my patient in the trance, and standing close to him, with the bust on my table behind him, I touched successively several of the organs, the position of which it was easy to determine. After a few seconds be would change his attitude and the expression of his face in correspondence with the organ excited. In most cases the effect was unmistakable, and superior to that which the most finished actor could give to a character exhibiting the same passion or emotion.

At this very time the excitement caused by painless surgical operations during the mesmeric trance was at its full height, as I have described in my "Wonderful Century" (chapter xxi), and I had read a good deal about these, and also about the supposed excitement of the phrenological organs, and the theory that these latter were caused by mental suggestion from the operator to the patient, or what is now termed telepathy. But as the manifestations often occurred in a different form from what I expected, I felt sure that this theory was not correct. One day I intended to touch a particular organ, and the effect on the patient was quite different from what I expected, and looking at the bust while my finger was still on the boy's head, I found that I was not touching the part I supposed, but an adjacent part, and that the effect exactly corresponded to the organ touched and not to the organ I thought I had touched, completely disproving the theory of suggestion. I then tried several experiments by looking away from the boy's head while I put my finger on it at random, when I always found that the effect produced corresponded to that indicated by the bust. I thus established, to my own satisfaction, the fact that a real effect was produced on the actions and speech of a mesmeric patient by the operator touching various parts of the head; that the effect corresponded with the natural expression of the emotion due to the phrenological organ situated at that part – as combativeness, acquisitiveness, fear, veneration, wonder, tune, and many others; and that it was in no way caused by the will or suggestion of the operator.

As soon as I found that these experiments were successful I informed Mr. Hill,[12] who made no objection to my continuing them, and several times came to see them. He was so much interested that one evening he invited two or three friends who were interested in the subject, and with my best patient I showed most of the phenomena. At the suggestion of one of the visits I told the boy he was a jockey, and was to get on his horse and be sure to win the race. Without another word from me he went through the motions of getting on horseback, of riding at a gallop, and after a minute or two he got excited, spoke to his horse, appeared to use his spurs, shake the reins, then suddenly remain quiet, as if he had passed the winning-post; and the gentleman who had suggested the experiment declared that his whole motions, expressions, and attitudes were those of a jockey riding a race. At that time I myself had never seen a race. The importance of these experiments to me was that they convince me, once for all, that the antecedently incredible may nevertheless be true; and, further, that the accusations of imposture by scientific men should have no weight whatever against the detailed observations and statements of other men, presumably as sane and sensible as their opponents, who had witnessed and tested the phenomena, as I had done myself in the case of some of them. At that time lectures on this subject were frequent, and during the holidays, which I generally spent in London with my brother [John], we took every opportunity of attending these lectures and witnessing as many experiments as possible. Knowing by my own experience that it is quite unnecessary to resort to trickery to produce the phenomena, I was relieved from that haunting idea of imposture which possesses most people who first see them, and which seems to blind most medical and scientific men to such an extent as to render them unable to investigate the subject fairly, or to arrive at any trustworthy conclusions in regard to it.[13] [1905]   ஐ

When in the Amazon, Wallace's younger brother, Herbert, also carried out a "culturally controlled" experiment on mesmerism.

ஐ   I will here only add that my brother Herbert also possessed the power, and that when we were residing together at Manaos, he used to call up little Indian boys out of the street, give them a copper, and by a little gazing and a few passes send them into the trance state, and then produce all the curious phenomena of catalepsy, loss of sensation, etc., which I have already described. This was interesting because it showed that the effects could be produced without any expectation on the part of the patients, and, further, that similar phenomena followed as in Europe, although these boys had certainly no knowledge of such phenomena.[14] [1905]   ஐ

In the days before chemical anaesthesia, hypnosis was sometimes used to anaesthetize patients during medical procedures. Wallace's allegiance to mesmerism, however, wavered in the face of the prospect of a painful bout of dentistry, much to the annoyance of Dr Purland, his dentist.

ஐ   Dr. Purland was ... a powerful and enthusiastic mesmerist, and had given his services to many surgical operations. Just as the opposition of the chiefs of the medical profession was dying away, and they were beginning to acknowledge the great value of the mesmeric sleep in alleviating pain and greatly facilitating serious operations, the discovery of anaesthetics offered a rival, which, though much more dangerous, was more certain and more easily applied in emergencies, and this led to the discontinuance of the use of mesmerism as a remedial agent. This naturally disgusted Dr. Purland, who, with the whole energy of his character, hated chloroform, ether, and nitrous-oxide gas, and would have nothing to do with them in his profession. Besides, he despised any one who could not bear the pain of tooth-drawing, and would turn away any patient who required the gas to be administered. A year or two after the date of his last letter my teeth were in a very bad state, and I had a number of broken stumps which required to be extracted preparatory to having a complete set of artificials. Entirely forgetting his objections, which, in fact, I had hardly believed to be real, after making an appointment I asked him to get a doctor to administer nitrous-oxide, as I could not stand the pain of three or four extractions of stumps of molars in succession. This thoroughly enraged him. He wrote me a most violent letter, saying he could not continue to be the friend of a man who could ask him to do such a thing, and gave me the name of an acquaintance of his who had no such scruples and whose work was thoroughly good. And that was the last communication I ever had from Dr. Purland.[15] [1905] ❧

Wallace himself underwent phrenological analysis twice. "Mr. Edwin Thomas Hicks, who called himself 'Professor of Phrenology'", determined Wallace to be mathematically inclined: "You have a good development of number and order, will therefore be a good calculator, will excell in mathematics. . . ." "Mr. James Quilter Rumball, an M. R. C. S. [Member of the Royal College of Surgeons] and author of some medical works", however, was less convinced of Wallace's mathematical prowess: "If Wit were larger he would be a good Mathematician; but without it, however clear and analytical the mind may be, it wants breadth and depth, and so I do not put down his mathematical talents as first-rate. . . ."[16] That the two surveys of the same head produced rather different results did not seem to bother Wallace. He remained convinced that phrenology constituted a legitimate science, closing his essay on "The Neglect of Phrenology" in *The Wonderful Century* with a rousing, if completely mistaken, call to arms.

❧ In the coming century Phrenology will assuredly gain general acceptance. It will prove itself the true science of mind. Its practical uscs in education, in self-discipline, in the reformatory treatment of criminals, and in the remedial treatment of the insane, will give it one of the highest places in the hierarchy of the sciences; and its persistent neglect and obloquy during the last sixty years, will be referred to as an example of the almost incredible narrowness and prejudice which prevailed among men of science, at the very time they were making such splendid advances in other fields of thought and discovery.[17] [1898] ঌ

Wallace was inconsistent in his application of scientific scepticism. Some phrenologists he saw through immediately:

ঌ He [Baker, a deaf and dumb American living in Brazil] made himself at home in every house in Barra [Manaus], walking in and out as he liked, and asking by signs for whatever he wanted. He was very merry, fond of practical jokes, and of making strange gesticulations. He pretended to be a phrenologist; and on feeling the head of a Portuguese or a Brazilian would always write down on his slate, "Very fond of the ladies"; which on being translated would invariably elicit, "He verdade" (that's very true), and signs of astonishment at his penetration.[18] [1853] ঌ

Similarly, he retold this tale of visitation from beyond the grave from his Amazon travels under the heading "Negro Credulity" and yet was himself the credulous dupe of many a medium's trick.

ঌ He [Señor Calistro, Wallace's host] related to us many anecdotes, of which the following is a specimen, serving to illustrate the credulity of the Negroes. "There was a Negro," said he, "who had a pretty wife, to whom another Negro was rather attentive when he had the chance. One day the husband went out to hunt, and the other party thought it a good opportunity to pay a visit to the lady. The husband, however, returned rather unexpectedly, and the visitor climbed up on the rafters to be out of sight among the old boards and baskets that were stowed away there. The husband put his gun by in a corner, and called to his wife to get his supper, and then sat down in his hammock. Casting his eyes up to the rafters, he saw a leg protruding from among the baskets, and, thinking it something supernatural, crossed himself, and said, 'Lord, deliver us from the legs appearing overhead!' The other, hearing this, attempted to draw up his legs out of sight, but, losing his balance, came down suddenly on the floor in front of the astonished husband, who, half frightened, asked, 'Where do you come from?' 'I have just come from heaven,' said the other, 'and have brought you news of your little daughter Maria.' 'Oh! wife, wife! Come and see a man who has brought us news of our little daughter Maria'; then, turning to the visitor, continued, 'And what was my little daughter doing when you left?' 'Oh! she was sitting at the feet of the Virgin, with a golden crown on her head, and smoking a golden pipe a yard long.' 'And did she not send any message to us?' 'Oh yes, she sent many remembrances, and begged you to send her two pounds of your tobacco from the little rhossa [store], they have not got any half so good up there.' 'Oh! wife, wife! bring two pounds of our tobacco from the little rhossa, for our daughter Maria is in heaven, and she says they have not any half so good up there.' So the tobacco was brought, and the visitor was departing, when he was asked, 'Are there many white men up there?' 'Very few,' he replied; 'they are all down below with the *diabo* [*sic*].' 'I thought so,' the other replied, apparently quite satisfied; 'good-night!'"[19] [1853]   ঙ

Wallace recounted his first encounters with the spirit world in the third edition of *On Miracles and Modern Spiritualism*.

ঙ   During twelve years of tropical wanderings between the year 1848 and 1862, occupied in the study of natural history, I heard occasionally of the strange phenomena said to be occurring in America[20] and Europe under the general names of "table-turning" and "spirit-rapping;" and being aware, from my own knowledge of Mesmerism, that there were mysteries connected with the human mind which modern science ignored because it could not explain, I determined to seize the first opportunity on my return home to examine into these matters. It is true, perhaps, that I ought to state that for twenty-five years I had been an utter sceptic as to the existence of any preter-human or super-human intelligences, and that I never for a moment contemplated the possibility that the marvels related by Spiritualists could be literally true. If I have now changed my opinion, it is simply by the force of evidence. It is from no dread of annihilation that I have gone into this subject; it is from no inordinate longing for eternal existence that I have come to believe in facts which render this highly probable, if they do not actually prove it. At least three times during my travels I have had to face death as imminent or probable within a few hours, and what I felt on those occasions was at

most a gentle melancholy at the thought of quitting this wonderful
and beautiful earth to enter on a sleep which might know no waking.
In a state of ordinary health I did not feel even this. I knew that the
great problem of conscious existence was one beyond man's grasp,
and this fact alone gave some hope that existence might be independent of the organised body. I came to the inquiry, therefore,
utterly unbiassed [*sic*] by hopes or fears, because I knew that my belief
could not affect the reality, and with an ingrained prejudice against
even such a word as "spirit," which I have hardly yet overcome.

It was in the summer of 1865 that I first witnessed any of the
phenomena of what is called Spiritualism, in the house of a friend – a
sceptic, a man of science, and a lawyer, with none but members of his
own family present. Sitting at a good-sized round table, with our hands
placed upon it, after a short time slight movements would commence
– not often "turnings" or "tiltings," but a gentle intermittent movement, like steps, which after a time would bring the table quite across
the room. Slight but distinct tapping sounds were also heard. The
following notes made at the time were intended to describe exactly
what took place: – "July 22nd, 1865. – Sat with my friend, his wife, and
two daughters, at a large loo table [circular card table], by daylight.
In about half-an-hour some faint motions were perceived, and some
faint taps heard. They gradually increased; the taps became very
distinct, and the table moved considerably, obliging us all to shift our
chairs. Then a curious vibratory motion of the table commenced,
almost like the shivering of a living animal. I could feel it up to my
elbows. These phenomena were variously repeated for two hours. On
trying afterwards, we found the table could not be voluntarily moved
in the same manner without a great exertion of force, and we could
discover no possible way of producing the taps when our hands were
upon the table."

On other occasions we tried the experiment of each person in
succession leaving the table, and found that the phenomena continued the same as before, both taps and the table movement. Once I
requested one after another to leave the table; the phenomena
continued, but as the number of sitters diminished with decreasing
vigour, and just after the last person had drawn back leaving me alone
at the table, there were two dull taps or blows, as with a fist on the
pillar or foot of the table, the vibration of which I could feel as well as
hear. No one present but myself could have made these, and I
certainly did not make them. These experiments clearly indicated that
all were concerned in producing the sounds and movements, and that

if there was any wilful deception the whole party were engaged in deceiving me. Another time we sat half-an-hour at the large table, but had no manifestations whatever. We then removed to the small table, where taps immediately commenced and the table moved. After some time we returned to the large table, and after a few minutes the taps and movements took place as at the small one.

The movement of the table was almost always in curves, as if turning on one of the claws, so as to give a progressive motion. This was frequently reversed, and sometimes regularly alternate, so that the table would travel across the room in a zigzag manner. This gives an idea of what took place with more or less regularity during more than a dozen sittings. Now there can be no doubt that the whole of the *movements* of the table could have been produced by any of the persons present if not counteracted by the others, but our experiments showed that this could not *always* be the case, and we have therefore no right to conclude that it was *ever* the case. The taps, on the other hand, we could not make at all. They were of about the quality that would be produced by a long finger-nail tapping underneath the leaf of the table. As all hands were on the table, and my eyes at least always open, I know they were not produced by the hands of any one present. They might possibly have been produced by the feet if properly armed with some small hard point to strike with; but if so, the experiments already related show that *all* must have practised the deception. And the fact that we often sat half an hour in one position without a single sound, and that the phenomena never progressed further than I have related, weighs I think very strongly against the supposition that a family of four highly intelligent and well-educated persons should occupy themselves for so many weary hours in carrying out what would be so poor and unmeaning a deception. The following remark occurs at the end of my notes made at the time: "These experiments have satisfied me that there is an unknown power developed from the bodies of a number of persons placed in connection by sitting round a table with all their hands upon it."[21] [1896] ❧

"To excite to inquiry": Spiritualism and Science

A scientist convinced of the existence of a contactable spirit world, Wallace was inevitably on the defensive among his scientific colleagues. However, he was not alone. Other eminent Victorian scientists, most notably the chemist Sir William Crookes[1] and the physicist Sir Oliver Lodge,[2] were spiritualists.[3] Perhaps naïvely, Wallace printed up copies of his 1866 pamphlet "The Scientific Aspect of the Supernatural" and sent them to various leading intellectual lights of the day. Responses were unenthusiastic, Huxley's particularly so:

"I am neither shocked nor disposed to issue a Commission of Lunacy against you. It may be all true, for anything I know to the contrary, but really I cannot get up any interest in the subject. I never cared for gossip in my life, and disembodied gossip, such as these worthy ghosts supply their friends with, is not more interesting to me than any other. As for investigating the matter, I have half-a-dozen investigations of infinitely greater interest to me to which any spare time I may have will be devoted. I give it up for the same reason I abstain from chess – it's too amusing to be fair work, and too hard work to be amusing."[4] [1866]

Wallace explained that his goal was not to convert people to his beliefs but merely to add spiritual phenomena to the scientific agenda as worthy of serious investigation.

I have reached my present standpoint by a long series of experiences under such varied and peculiar conditions as to render unbelief impossible. As Dr. W. B. Carpenter[5] well remarked many years ago, people can only believe new and extraordinary facts if there is a place for them in their existing "fabric of thought." The majority of people to-day have been brought up in the belief that miracles, ghosts, and the whole series of strange phenomena here described cannot exist; that they are contrary to the laws of nature; that they are the superstitions of a bygone age; and that therefore they are necessarily either impostures or delusions. There is no place in the fabric of their thought into which such facts can be fitted. When I first began this inquiry it was the same with myself. The facts did not fit into my then existing fabric of thought. All my preconceptions, all my knowledge, all my belief in the supremacy of science and of natural law were against the possibility of such phenomena. And even when, one by one, the facts were forced upon me without possibility of escape from them, still, as Sir David Brewster[6] declared after being at first astounded by the phenomena he saw with Mr. Home,[7] "spirit was the last thing I could give in to." Every other possible solution was tried and rejected. Unknown laws of nature were found to be of no avail when there was always an unknown intelligence behind the phenomena – an intelligence that showed a human character and individuality, and an individuality which almost invariably *claimed* to be that of some person who had lived on earth, and who, in many cases, was able to prove his or her identity. Thus, little by little, a place was made in my fabric of thought, first for all such well-attested facts, and then, but more slowly, for the spiritualistic interpretation of them.

Unfortunately, at the present day most inquirers begin at the wrong end. They want to see, and sometimes do see the most wonderful phenomena first, and being utterly unable to accept them as facts denounce them as impostures, as did Tyndall[8] and G. H. Lewes,[9] or declare, as did Huxley, that such phenomena do not interest them. Many people think that when I and others publish accounts of such phenomena, we wish or require our readers to believe them on *our* testimony. But that is not the case. Neither I nor any other well-instructed spiritualist expects anything of the kind. We write not to convince, but to excite to inquiry. We ask our readers not for *belief*, but for doubt of their own infallibility on this question; we ask for inquiry and patient experiment before hastily concluding that we are, all of us, mere dupes and idiots as regards a subject to which we have devoted our best mental faculties and powers of observation for many years.[10] [1905]   ≥∞

Rather than being excited "to inquiry", some scientists were excited to litigation. In 1876 E. Ray Lankester,[11] a student of Huxley, took it upon himself to expose as a fraud a well-known American medium, one "Dr" Henry Slade,[12] and Charles Darwin contributed £10 – a substantial amount at that time – to the costs of the prosecution.[13] Wallace was the defence's star witness, but his characterization of the defendant as an "earnest inquirer after truth in the department of Natural Science" failed to prevent Slade from being convicted. Darwin was delighted; he had no time for the "clever rogues" who preyed upon grieving relatives anxious to contact a loved one.

Wallace nevertheless defended himself capably – indeed, in the opening pages of *On Miracles and Modern Spiritualism*, he did so with *élan*.

❧ I am well aware that my scientific friends are somewhat puzzled to account for what they consider to be my delusion, and believe that it has injuriously affected whatever power I may once have possessed of dealing with the philosophy of Natural History. One of them – Mr. Anton Dohrn[14] – has expressed this plainly. I am informed that, in an article entitled "Englische Kritiker und Anti-Kritiker des Darwinismus," published in 1861, he has put forth the opinion that Spiritualism and Natural Selection are incompatible, and that my divergence from the views of Mr. Darwin arises from my belief in Spiritualism. He also supposes that in accepting the spiritual doctrines I have been to some extent influenced by clerical and religious prejudice. As Mr. Dohrn's views may be those of other scientific friends, I may perhaps be excused for entering into some personal details in reply.

From the age of fourteen I lived with an elder brother [John], of advanced liberal and philosophical opinions, and I soon lost (and have never since regained) all capacity for being affected in my judgements either by clerical influence or religious prejudice. Up to the time when I first became acquainted with the facts of Spiritualism, I was a confirmed philosophical sceptic, rejoicing in the works of Voltaire, Strauss,[15] and Carl Vogt,[16] and an ardent admirer (as I am still) of Herbert Spencer. I was so thorough and confirmed a materialist that I could not at that time find a place in my mind for the conception of spiritual existence, or for any other agencies in the universe than matter and force. Facts, however, are stubborn things. My curiosity was at first excited by some slight but inexplicable phenomena occurring in a friend's family, and my desire for knowledge and love of truth forced me to continue the inquiry. The facts became more and more assured, more and more varied, more and more removed from anything that modern science taught or modern philosophy speculated on. The facts beat me. They compelled me to accept them *as facts* long before I could accept the spiritual explanation of them; there was at that time "no place in my fabric of thought into which it could be fitted." By slow degrees a place was made; but it was made, not by any preconceived or theoretical opinions, but by the continuous action of fact after fact, which could not be got rid of in any other way. So much for Mr. Anton Dohrn's theory of the causes that led me to accept Spiritualism. Let us now consider the statement as to its incompatibility with Natural Selection.

Having, as above indicated, been led, by a strict induction from the facts, to a belief – 1stly, In the existence of a number of preterhuman intelligences of various grades, and 2ndly, That some of these intelligences, although usually invisible and intangible to us, can and do act on matter, and do influence our minds, – I am surely following a strictly logical and scientific course in seeing how far this doctrine will enable us to account for some of those residual phenomena which Natural Selection will not explain. In the 10th chapter of my *Contributions to the Theory of Natural Selection* I have pointed out what I consider to be some of these residual phenomena; and I have suggested that they may be due to the action of some of the various intelligences above referred to. This view was, however, put forward with hesitation, and I myself suggested difficulties in the way of its acceptance; but I maintained, and still maintain, that it is one which is logically tenable, and is in no way inconsistent with a thorough acceptance of the grand doctrine of Evolution through Natural Selection, although implying (as indeed many of the chief supporters of that doctrine admit) that it is not the all-powerful, all-sufficient, and only cause of the development of organic forms.[17] [1875]   ⮞

Wallace persisted in peppering his scientific friends with accounts of his experiences. This one, recounted in a letter to Tyndall, involved the levitation of the substantial Miss Nichol.

⮜ During the last two years I have witnessed a great variety of phenomena, under such varied conditions that each objection as it arose was answered by other phenomena. The further I inquire, and the more I see, the more impossible becomes the theory of imposture or delusion. I know that the facts are real natural phenomena, just as certainly as I know any other curious facts in nature.

Allow me to narrate *one* of the scores of equally remarkable things I have witnessed, and this one, though it certainly happened in the dark, is thereby only rendered more difficult to explain as a trick.

The *place* was the drawing-room of a friend of mine, a brother of one of our best artists. The *witnesses* were his own and his brother's family, one or two of their friends, myself, and Mr. John Smith, banker, of Malton, Yorkshire, introduced by me. The medium was Miss Nichol. We sat round a pillar-table in the middle of the room, exactly under a glass chandelier. Miss Nichol sat opposite me, and my friend, Mr. Smith, sat next her. We all held our neighbour's hands, and Miss Nichol's hands were both held by Mr. Smith, a stranger to all but myself, and who had never met Miss N. before. When comfortably arranged in this manner the lights were put out, one of the party holding a box of matches ready to strike a light when asked.

After a few minutes' conversation, during a period of silence, I heard the following sounds in rapid succession: a slight *rustle,* as of a lady's dress; a little *tap,* such as might be made by setting down a wineglass on the table; and a very slight jingling of the drops of the glass chandelier. An instant after Mr. Smith said, "Miss Nichol is gone." The match-holder struck a light, and on the table (which had no cloth) was Miss Nichol *seated in her chair,* her head just touching the chandelier.

I had witnessed a similar phenomenon before, and was able to observe coolly; and the facts were noted down soon afterwards. Mr. Smith assured me that Miss Nichol simply glided out of his hands. No one else moved or quitted hold of their neighbour's hands. There was not more noise than I have described, and no motion or even tremor of the table, although our hands were upon it.

You know Miss N.'s size and probable weight, and can judge of the force and exertion required to lift her and her chair on to the exact centre of a large pillar-table, as well as the great surplus of force required to do it almost instantaneously and noiselessly, in the dark, and without pressure on the side of the table which would have tilted it up. Will any of the known laws of nature account for this?[18] [1868]

Wallace took his campaign to the letters page of *The Times.*

Having been named by several of your correspondents as one of the scientific men who believe in spiritualism, you will perhaps allow me to state briefly what amount of evidence has forced the belief upon me. I began the investigation about eight years ago, and I esteem it a fortunate thing that at that time the more marvellous phenomena were far less common and less accessible than they are now, because I was led to experiment largely at my own house, and among friends whom I could trust, and was able to establish to my own satisfaction, by means of a great variety of tests, the occurrence of sounds and movements not traceable to any known or conceivable physical cause. Having thus become thoroughly familiar with these undoubtedly genuine phenomena, I was able to compare them with the more powerful manifestations of several public mediums, and to recognize an identity of cause in both by means of a number of minute but highly characteristic resemblances. I was also able, by patient observation, to obtain tests of the reality of some of the more curious phenomena which appeared at the time, and still appear to me, to be conclusive. To go into details as to those experiences would require a volume, but I may, perhaps, be permitted briefly to describe one, from notes kept at the time, because it serves as an example of the complete security against deception which often occurs to the patient observer without seeking for it.

A lady who had seen nothing of the phenomena asked me and my sister to accompany her to a well-known public medium. We went, and had a sitting alone in the bright light of a summer's day. After a number of the usual raps and movements our lady friend asked if the name of the deceased person she was desirous of communicating with could be spelt out. On receiving an answer in the affirmative, the lady pointed successively to the letters of a printed alphabet while I wrote down those at which three affirmative raps occurred. Neither I nor my sister knew the name the lady wished for, nor even the names of any of her deceased relatives; her own name had not been mentioned, and she had never been near the medium before. The following is exactly what happened, except that I alter the surname, which was a very unusual one, having no authority to publish it. The letters I wrote down were of the following kind: – y n r e h n o s p m o h t. After the first three – y n r – had been taken down, my friend said, "This is nonsense, we had better begin again." Just then her pencil was at e, and raps came, when a thought struck me (having read of, but never witnessed a similar occurrence) and I said "Please go on, I think I see what is meant." When the spelling was finished I handed the paper to her, but she could see no meaning in it till I divided it at the first h, and asked her to read each portion backwards, when to her intense astonishment the name "Henry Thompson" came out, that of a deceased son of whom she had wished to hear, correct in every letter.[19] Just about that time I had been hearing ad nauseam of the superhuman acuteness of mediums who detect the letters of the name the deluded visitors expect, notwithstanding all their care to pass the pencil over the letters with perfect regularity. This experience, however (for the substantial accuracy of which as above narrated I vouch), was and is, to my mind, a complete disproof of every explanation yet given of the means by which the names of deceased persons are rapped out. Of course, I do not expect any sceptic, whether scientific or unscientific, to accept such facts, of which I could give many, on my testimony, but neither must they expect me, nor the thousands of intelligent men to whom equally conclusive tests have occurred, to accept their short and easy methods of explaining them.

If I am not occupying too much of your valuable space I should like to make a few remarks on the misconceptions of many scientific men as to the nature of this inquiry, taking the letters of your correspondent Mr. Dircks as an example. In the first place, he seems to think that it is an argument against the facts being genuine that they cannot all be produced and exhibited at will; and another argument against them, that they cannot be explained by any known laws. But neither can catalepsy, the fall of meteoric stones, nor hydrophobia be produced at will; yet these are all facts, and none the less so that the first is sometimes imitated, the second was once denied, and the symptoms of the third are often greatly exaggerated, while none of them are yet brought under the domain of strict science; yet no one would make this an argument for refusing to investigate these subjects. Again, I should not have expected a scientific man to state, as a reason for not examining it, that spiritualism "is opposed to every known natural law, especially the law of gravity," and that it "sets chymistry, human physiology, and mechanics at open defiance"; when the facts simply are that the phenomena, if true, depend upon a cause or causes which can overcome or counteract the action of these several forces, just as some of these forces often counteract or overcome others; and this should surely be a strong inducement to a man of science to investigate the subject.

While not laying any claim myself to the title of "a really scientific man," there are some who deserve that epithet who have not yet been mentioned by your correspondents as at the same time spiritualists. Such I consider the late Dr. Robert Chambers,[20] as well as Dr. Elliotson,[21] Professor William Gregory,[22] of Edinburgh; and Professor Hare,[23] of Philadelphia – all unfortunately deceased; while Dr. Gully,[24] of Malvern, as a scientific physician, and Judge Edmonds,[25] one of the best American lawyers, have had the most ample means of investigation; yet all these not only were convinced of the reality of the most marvellous facts, but also accepted the theory of modern spiritualism as the only one which would embrace and account for the facts. I am also acquainted with a living physiologist of high rank as an original investigator, who is an equally firm believer.

In conclusion I may say that, although I have heard a great many accusations of imposture, I have never detected it myself; and, although a large proportion of the more extraordinary phenomena are such, that, if impostures, they could only be performed by means of ingenious apparatus or machinery, none has ever been discovered. I consider it no exaggeration to say, that the main facts are now as well established and as easily verifiable as any of the more exceptional phenomena of nature which are not yet reduced to law. They have a most important bearing on the interpretation of history, which is full of narratives of similar facts, and on the nature of life and intellect, on which physical science throws a very feeble and uncertain light; and it is my firm and deliberate belief that every branch of philosophy must suffer till they are honestly and seriously investigated, and dealt with as constituting an essential portion of the phenomena of human nature.[26] [1873]   ❧

Wallace resorted, as all good scientific polemicists do, to the plea that his ideas were ahead of their time. This, of course, is a problematic argument: that Galileo was right but initially opposed does not imply that all theories that are opposed are right!

❧ In the history of human progress, we look back in vain for a case parallel to the present one, in which the professed teachers of science have been right. The time-honoured names of Galileo, Harvey,[27] and Jenner,[28] are associated with the record of a blind opposition to new and important truths. [Benjamin] Franklin and Young[29] were laughed and sneered at for discoveries which seemed wild and absurd to their scientific contemporaries. Nearer to our own day, painless operations during mesmeric trance were again and again denounced as imposture; and the various phenomena of mesmerism, as due to collusion and fraud: yet both are now universally acknowledged to be genuine phenomena.[30] Even such a question of pure science as the evidence of the antiquity of man has met with similar treatment till quite recently. Papers by good observers, recording facts since verified, were rejected by our scientific societies, as too absurd for publication; and careful researches now proved to be accurate were ignored, merely because they were opposed to the general belief of geologists.

It appears, then, that men of science are at least consistent in treating the phenomena of Spiritualism with contempt and derision.

They have always done so with new and important discoveries; and, in every case in which the evidence has been even a tenth part of that now accumulated in favour of the phenomena of Spiritualism, they have *always been in the wrong*. It is, nevertheless, a curious psychological fact, that they do not learn by experience to detect a truth when it comes before them, or take any heed of the warnings of their greatest men against preconceived opinions as to what may, or may not, be true. Thus Humboldt declares, that "a presumptuous skepticism, which rejects facts without examination of their truth, is, in some respects, more injurious than an unquestioning incredulity." Sir Humphry Davy[31] warns them, that "one good experiment is of more value than the ingenuity of a brain like Newton's. Facts are more useful when they contradict, than when they support, received theories." And Sir John Herschel[32] says, that "the perfect observer in any department of Nature will have his eyes open for any occurrence, *which, according to received theories, ought not to happen*; for these are the facts which serve as clews to new discoveries." Yet in the present day, when so many things deemed absurd and impossible a few years ago have become every-day occurrences, and in direct opposition to the spirit of the advice of their most eminent teachers, a body of new and most remarkable phenomena is ignored or derided without examination, merely because, *according to received theories, such phenomena ought not to happen*.

The day will assuredly come when this will be quoted as the most striking instance on record of blind prejudice and unreasoning credulity.[33] [1871] ❧

Rather more effectively, Wallace insisted that paranormal phenomena should merely be treated as a hitherto inadequately investigated extension of the natural world.

❧ That intelligent beings may exist around and among us, unperceived during our whole lives, and yet capable under certain conditions of making their presence known by acting on matter, will be inconceivable to some, and will be doubted by many more, but we venture to say that no man acquainted with the latest discoveries and the highest speculations of modern science will deny its possibility. The difficulty which this conception presents will be of quite a different nature from that which obstructs our belief in the possibility of miracle, when defined as a contravention of those great natural laws which the whole tendency of modern science declares to be absolute and immutable. The existence of sentient beings uncognisable by our senses would no more contravene these laws than did the discovery of the true nature of the Protozoa, those structureless gelatinous organisms which exhibit so many of the higher phenomena of animal life without any of that differentiation of parts or specialisation of organs which the necessary functions of animal life seem to require. The existence of such preter-human intelligences, if proved, would only add another and more striking illustration than any we have yet received of how small a portion of the great cosmos our senses give us cognisance.[34] [1875] ε∾

Wallace assumed that new laws of nature would be uncovered that would explain these phenomena.

ε∾ The declaration so often made or implied, that facts witnessed thousands of times by honest and intelligent men, and thousands of times carefully examined to detect fraud or delusion which has never been discovered, can not exist, because they imply a subversion of the laws of Nature, is a most weak and illogical objection, since all we know of the laws of Nature is derived from the observation of facts. No fact can possibly subvert the laws of Nature; and to declare that it does so is to declare that we have exhausted Nature, and know all her laws.[35] [1871] ε∾

For this to be logically tenable, he had first to dismiss contemporary views of what constituted a "miracle".

ε∾ It is now generally admitted, that those opinions and beliefs in which men have been educated generation after generation, and which have thus come to form part of their mental nature, are especially liable to be erroneous, because they keep alive and perpetuate the ideas and prejudices of a bygone and less enlightened age. It is therefore in the interest of truth, that every doctrine or belief, however well established or sacred they may appear to be, should at certain intervals be challenged to arm themselves with such facts and reasonings as they possess, to meet their opponents in the open field of controversy, and do battle for their right to live. Nor can any exemption be claimed in favour of those beliefs which are the product of modern civilisation, and which have for several generations been unquestioned by the great mass of the educated community; for the prejudice in their favour will be proportionately great, and, as was the case with the doctrines of Aristotle, and the dogmas of the schoolmen, they may live on by mere weight of authority and force of habit, long after they have been shown to be opposed alike to fact and to reason. There have been times when popular beliefs were defended by the terrors of the law, and when the sceptic could only attack them at the peril of his life. Now we all admit that truth can take care of itself, and that only error needs protection. But there is another mode of defence which equally implies a claim to certain and absolute truth, and which is therefore equally unworthy and unphilosophical – that of ridicule, misrepresentation, or a contemptuous refusal to discuss the question at all. This method is used among us even now, for there is one belief, or rather disbelief, whose advocates claim more than papal infallibility, by refusing to examine the evidence brought against it, and by alleging general arguments which have been in use for two centuries to prove that it cannot be erroneous. The belief to which I allude is, that all alleged miracles are false; that what is commonly understood by the term *supernatural* does not exist, or if it does, is incapable of proof by any amount of human testimony; that all the phenomena we can have cognizance of depend on ascertainable physical laws, and that no other intelligent beings than man and the inferior animals can or do act upon our material world. These views have been now held almost unquestioned for many generations; they are inculcated as an essential part of a liberal education; they are popular, and are held to be one of the indications of our intellectual advancement; and they have become so much a part of our mental nature, that all facts and arguments brought against them are either ignored as unworthy of serious consideration, or listened to with undisguised contempt. Now this frame of mind is certainly not one favourable to the discovery of truth, and strikingly resembles that by which, in former ages, systems of error have been fostered and maintained. The time has therefore come when it must be called upon to justify itself.

This is the more necessary, because the doctrine, whether true or false, actually rests upon a most unsafe and rotten foundation. I propose to show you that the best arguments hitherto relied upon to prove it are, one and all, fallacious – and prove nothing of the kind. But a theory or belief may be supported by very bad arguments, and yet be true; while it may be supported by some good arguments, and yet be false. But there was never a true theory which had no good arguments to support it. If therefore all the arguments hitherto used against miracles in general can be shown to be bad, it will behove sceptics to discover good ones; and if they cannot do so, the evidence in favour of miracles must be fairly met and judged on its own merits, not ruled out of court as it is now.

It will be perceived therefore, that my present purpose is to clear the ground for the discussion of the great question of the so-called supernatural. I shall not attempt to bring arguments either for or against the main proposition, but shall confine myself to an examination of the allegations and the reasonings which have been supposed to settle the whole question on general grounds.

DAVID HUME AND HIS FALSE DEFINITION OF A MIRACLE

One of the most remarkable works of the great Scotch philosopher, David Hume, is *An Inquiry Concerning Human Understanding*, and the tenth chapter of this work is "On Miracles," in which occur the arguments which are so often quoted to show that no evidence can prove a miracle. Hume himself had a very high opinion of this part of his work, for he says at the beginning of the chapter: – "I flatter myself that I have discovered an argument which, if just, will with the wise and learned be an everlasting check to all kinds of superstitious delusion, and consequently will be useful as long as the world endures; for so long, I presume, will the accounts of miracles and prodigies be found in all history, sacred and profane."

After a few general observations on the nature of evidence, and the value of human testimony in different cases, he proceeds to define what he means by miracle. And here, at the very beginning of the subject, we find that we have to take objection to Hume's definition of a miracle, which exhibits unfounded assumptions and false premises. He gives two definitions in different parts of his essay. The first is – "A miracle is a violation of the laws of nature." The second is – "A miracle is a transgression of a law of nature, by a particular volition of the Deity, or by the interposition of some invisible agent."[36] Now both these definitions are bad or imperfect. The first assumes that we know all the laws of nature – that the particular effect could not be produced by some unknown law of nature overcoming the law we do know; it assumes also, that if an invisible intelligent being held an apple suspended in the air, that act would violate the law of gravity. The second is not precise; it should be "some invisible *intelligent* agent," otherwise the action of galvanism or electricity, when these agents were first discovered, and before they were ascertained to form part of the order of nature, would answer accurately to this definition of a miracle. The words "violation" and "transgression" are both improperly used, and really beg the question by the definition. How does Hume know that any particular miracle is a violation of a law of nature? He assumes this without a shadow of proof, and on these words, as we shall see, rests his whole argument.

THE TRUE DEFINITION OF A MIRACLE

Before proceeding any further, it is necessary for us to consider what is the true definition of a miracle, or what is most commonly meant by that word. A miracle, as distinguished from a new and unheard-of natural phenomenon, supposes an intelligent superhuman agent either visible or invisible – it is not necessary that what is done should be beyond the power of man to do. The simplest action, if performed independently of human or visible agency, such as a tea-cup lifted in the air at request, as by an invisible hand and without assignable cause, would be universally admitted to be a miracle, as much so as the lifting of a house into the air, the instantaneous healing of a wound, or the instantaneous production of an elaborate drawing. My definition of a miracle therefore is as follows – "Any act or event implying the existence and agency of superhuman intelligences," considering the human soul or spirit, if manifested out of the body, as one of these superhuman intelligences. This definition is more complete than that of Hume, and defines more accurately the essence of that which is commonly termed a miracle.

HUME'S FIRST ARGUMENT A RADICAL FALLACY

We now have to consider Hume's arguments. The first is as follows:

"A miracle is a *violation of the laws of nature*; and as a firm and *unalterable experience* has established these laws, the proof against a miracle, from the very nature of the fact, is as entire as any argument from experience can possibly be imagined. Why is it more than probable that all men must die; that lead cannot *of itself remain suspended in the air*; that fire consumes wood, and is extinguished by water; unless it be, that these events are found agreeable to the laws of nature, and there is required a *violation of these laws*, or, in other words a *miracle*, to prevent them? Nothing is esteemed a miracle, if it ever happened in the *common* course of nature. It is no miracle that a man seemingly in good health should die on a sudden;

because such a kind of death, though more unusual than any other, has yet been frequently observed to happen. But it is a miracle that a dead man should come to life; because *that has never been observed in any age or country.* There must, therefore, be an uniform experience against every miraculous event, otherwise the event would not merit that appellation. And as an *uniform* experience amounts to a *proof,* there is here a direct and full proof, from the nature of the fact, against the existence of any miracle; nor can such a proof be destroyed, or the miracle rendered credible, but by an opposite proof, which is superior."[37]

This argument is radically fallacious, because if it were sound, no perfectly new fact could ever be proved, since the first and each succeeding witness would be assumed to have universal experience against him. Such a simple fact as the existence of flying fish could never be proved, if Hume's argument is a good one; for the first man who saw and described one, would have the universal experience against him that fish do not fly, or make any approach to flying, and his evidence being rejected, the same argument would apply to the second, and to every subsequent witness, and thus no man at the present day who has not seen a flying fish ought to believe that such things exist.

Again – painless operations in a state produced by mere passes of the hand, were, twenty-five years ago, maintained to be contrary to the laws of nature, contrary to all human experience, and therefore incredible. On Hume's principles they were miracles, and no amount of testimony could ever prove them to be real. But miracles do not stand alone, single facts opposed to uniform experience. Reputed miracles abound in all periods of history; every one has a host of others leading up to it; and every one has strictly analogous facts testified to at the present day. The uniform opposing experience therefore on which Hume lays so much stress does not exist. What, for instance, can be a more striking miracle than the levitation or raising of the human body into the air without visible cause, yet this fact has been testified to during a long series of centuries.

A few well-known examples are those of St. Francis d'Assisi, who was often seen by many persons to rise in the air, and the fact is testified to by his secretary, who could only reach his feet. St. Theresa, a nun in a convent in Spain, was often raised into the air in the sight of all the sisterhood. Lord Orrery[38] and Mr. Valentine Greatorex[39] both informed Dr. Henry More[40] and Mr. Glanvil,[41] that at Lord Conway's house at Ragley in Ireland, a gentleman's butler, in their presence and in broad daylight, rose into the air, and floated about the room above their heads. This is related by Glanvil in his *Sadducismus Triumphatus.* A similar fact is narrated by eye-witnesses of Ignatius de Loyola, and Mr. Madden, in his *Life of Savonarola*,[42] after narrating a similar circumstance of that saint, remarks that similar phenomena are related in numerous instances, and that the evidence upon which some of the narratives rest is as reliable as any human testimony can be. Butler,[43] in his *Lives of the Saints,* says that many such facts are related by persons of undoubted veracity, who testify that they themselves were eye-witnesses of them. So we all know that at least fifty persons of high character may be found in London, who will testify that they have seen the same thing happen to Mr. Home. I do not adduce this testimony as proving that the circumstances related really took place; I merely bring it forward now to show how utterly unfounded is Hume's argument, which rests upon universal testimony on the one side, and no testimony on the other.[44] [1870] &

A World Viewed Through the Lens of Spiritualism

For Wallace, spiritualism was much more than a way of communing with the dead. It was a first step into a non-material world; once admitted, he took many more steps, embellishing his views until they became undeniably theistic. Critically, he devised schemes to integrate spiritualism into other components of his worldview. Thus he argued that morality followed necessarily from spiritualist beliefs.

❧ It has been shown and will, I am sure, be admitted by all unprejudiced readers, that we have derived from Spiritualism a conception of a future state and of its connection with our life here very different from, and far superior to, the ordinary religious teaching which formerly prevailed. That teaching has now been partly modified through the influence of Spiritualistic ideas; but by the religious preacher it is taught dogmatically, not as it comes to the Spiritualist with all the force of personal communication with those called dead, but who, again and again, tell us they are far more alive than ever they were here. This Spiritualistic teaching as to another life enforces upon us, that our condition and happiness in the future life depends, by the action of strictly natural law, on our life and conduct here. There is no reward or punishment meted out to us by superior beings; but, just as surely as cleanliness and exercise and wholesome food and air produce health of body, so surely does a moral life here produce health and happiness in the spirit-world. Every well-informed Spiritualist realises that, by every thought and word and deed of his daily earth-life, he is actually and inevitably determining his own happiness or misery in a future life which is continuous with this – that he has the power of creating for himself his own heaven or hell. The Spiritualists alone, therefore, or those who accept with equal confidence the Spiritualistic teachings in this respect, can give fully adequate reasons why they should live a moral life. These reasons are in no way dependent on public opinion, or on any relation to success or happiness here, and are, therefore, calculated to influence conduct

under the most extreme conditions of temptation or secrecy Hence
the only Rationalistic and adequate incentive to morality – the only
full and complete affirmative answer to the question, "Why live a
moral life?" – is that which is based upon the conception of a future
state of existence systematically taught by Modern Spiritualism.[1]
[1895] ঌ

From this starting point, we see how Wallace fitted together
apparently disparate pieces of his personal jigsaw puzzle, socialism
and spiritualism.

ঌ The old doctrine as to the nature of the future life was based
upon the idea of rewards and punishments, which were supposed to
be dependent upon *dogmatic beliefs* and ceremonial *observances.* The
atheist, the agnostic, even the Unitarian, were for centuries held to be
certain of future punishment; and, with the unbaptised infant, the
Sabbath-breaker, and the abstainer from church-going, were alike
condemned to hell-fire. Beliefs and observances were then held to be
of the first importance; disposition, conduct, health, and happiness
were of no account.

The new doctrines – founded almost wholly on the teachings of
Modern Spiritualism, though now widely accepted, even among non-
Spiritualists – are the very reverse of all this. They are based upon the
conception of mental and moral continuity; that there are no imposed
punishments; that dogmatic beliefs are absolutely unimportant, except
so far as they affect our relations with our fellows; and that forms and
ceremonies, and the complex observances of most religions, are
equally unimportant. On the other hand, what are of the most vital
importance are motives, with the actions that result from them, and
everything that develops and exercises the whole mental, moral, and
physical nature, resulting in happy and healthy lives for every human
being. The future life will be simply a continuation of the present,
under new conditions; and its happiness or misery will be dependent
upon how we have developed all that is best in our nature here.

Under the old theory the soul could be saved by a mere change of
beliefs and the performance of certain ceremonial observances. The
body was nothing; happiness was nothing; pleasure was often held to
be a sin; hence any amount of punishment, torture, and even death
were considered justifiable in order to produce this change and save
the soul.

On the new theory it is the body that develops, and to some extent saves, the soul. Disease, pain, and all that shortens and impoverishes life, are injurious to the soul as well as to the body. Not only is a healthy body necessary for a sound mind, but equally so for a fully-developed soul – a soul that is best fitted to commence its new era of development in the spirit world. Inasmuch as we have fully utilised and developed all our faculties – bodily, mental, and spiritual – and have done all in our power to aid others in a similar development, so have we prepared future well-being for ourselves and for them.

All this is the common knowledge and belief of Spiritualists; and I should not have thought it necessary to restate it, were it not that our creed is often misunderstood and misrepresented by outsiders, and also because it is preliminary to certain conclusions which, I think, logically follow from it, but which are not so generally accepted among us.

It seems to me that, holding these beliefs as to the future life and what is the proper and only preparation for it, we Spiritualists must feel ourselves bound to work strenuously for such improved social conditions as may render it possible for all to live a full and happy life, for *all* to develop and utilise the various faculties they possess, and thus be prepared to enter at once on the progressive higher life of the spirit-world. We *know* that a life of continuous and grinding bodily labour, in order to obtain a bare existence; a life almost necessarily devoid of beauty, of refinement, of communion with Nature; a life without adequate relaxation, and with no opportunity for the higher culture; a life full of temptation and with no cheering hope of a happy and peaceful old age, is as bad for the welfare of the soul as it is for that of the body.

If the accounts we get of the spirit world have any truth in them, the reclamation and education of the millions of undeveloped or degraded spirits which annually quit this earth, is a sore burden, a source of trouble and sorrow to those more advanced spirits who have charge of them. This burden must, for a long time to come at all events, *necessarily* be great, on account of the numbers of the less advanced races and peoples still upon the earth; but that *we*, who call ourselves *civilised*, who have learnt so much of the secret powers and mysteries of the universe, who by means of those powers could easily provide a decent and rational and happy life for our whole population – that *we* should send to the spirit world, day by day and year by year, millions of men and women, of children, and of infants, all sent there before their time through want of the necessary means of a healthy life, or by the various diseases and accidents forced upon them by the vile conditions under which alone we give them the opportunity of living at all – this is a disgrace and a crime!

I firmly believe – and the fact is supported by abundant evidence – that the very poorest class of our great cities, those that live constantly below the margin of poverty, who are without the comforts, the necessaries, and even the decencies of life, are, nevertheless, as a class, quite as good morally, and often as high intellectually, as the middle and upper classes who look down upon them as in every way their inferiors. Their condition, socially and morally, is the work of society; and in so far as they appear worse than others they are made so by society. What should we ourselves have been if we had had no education, no repose, no refined or decent homes, no means of cleanliness, which is not only next to, but is a source of, godliness; surrounded by every kind of temptation, and not unfrequently forced into crime? And a direct consequence of the millions who are compelled to lead such lives are the millions of infants who die prematurely – a slaughter a thousand times worse than that of Herod, going on year by year in our midst; surely their innocent blood cries out against our rulers, against all of *us*, who choose such rulers; and more especially against us Spiritualists, who know the higher law, if we do not work with all our strength for a radical reform.[2] [1898]

Wallace viewed the events of his own life in providential, even Panglossian, terms. What might appear to be setbacks are in fact "designed" opportunities for re-orientation.

Now, I have some reason to believe that this was the turning-point of my life, the tide that carried me on, not to fortune, but to whatever reputation I have acquired, and which has certainly been to me a never-failing source of much health of body and supreme mental enjoyment. If my brother [William] had had constant [surveying] work for me so that I never had an idle day, and if I had continued to be similarly employed after I became of age, I should most probably have become entirely absorbed in my profession, which, in its various departments, I always found extremely interesting, and should therefore not have felt the need of any other occupation or study.

I know now, though I was ignorant of it at the time, that my brother's life was a very anxious one, that the difficulty of finding remunerative work was very great, and that he was often hard pressed to earn enough to keep us both in the very humble way in which we lived. He never alluded to this that I can remember, nor did I ever hear how much our board and lodging cost him, nor ever saw him make the weekly or monthly payments. During the seven years I was with him I hardly ever had more than a few shillings for personal expenses; but every year or two, when I went home, what new clothes were absolutely necessary were provided for me, with perhaps ten shillings or a pound as pocket money till my next visit, and this, I think, was partly or wholly paid out of the small legacy left me by my grandfather. This seemed very hard at the time, but I now see clearly that even this was useful to me, and was really an important factor in moulding my character and determining my work in life. Had my father been a moderately rich man and had he supplied me with a good wardrobe and ample pocket-money; had my brother obtained a partnership in some firm in a populous town or city, or had he established himself in his profession, I might never have turned to nature as the solace and enjoyment of my solitary hours, my whole life would have been differently shaped, and though I should, no doubt, have given some attention to science, it seems very unlikely that I should have ever undertaken what at that time seemed rather a wild scheme, a journey to the almost unknown forests of the Amazon in order to observe nature and make a living by collecting.

All this may have been pure chance, as I long thought it was, but of late years I am more inclined to Hamlet's belief, when he said

> There's a divinity that shapes our ends,
> Rough-hew them how we will.[3]

Of course I do not adopt the view that each man's life, in all its details, is guided by the Deity for His special ends. That would be, indeed, to make us all conscious automata, puppets in the hands of an all-powerful destiny. But ... I have good reasons for the belief that, just as our own personal influence and expressed or unseen guidance is a factor in the life and conduct of our children, and even of some of our friends and acquaintances, so we are surrounded by a host of unseen friends and relatives who have gone before us, and who have certain limited powers of influencing, and even, in particular cases, almost of determining, the actions of living persons, and may thus in a great variety of indirect ways modify the circumstances and character of any one or more individuals in whom they are specially interested.[4] [1905] ❧

Wallace incorporated his theistic spiritualism into his science as well. Developing his interests in astronomy and astrophysics,

he derived what has come to be known as the "Anthropic Principle", which argues that the particular lucky combination of
conditions – physical, cosmological, chemical – that make life
possible on Earth is highly improbable, implying that (1) it is
inconceivable that life exists elsewhere in the universe, and (2)
some form of divine intervention occurred to ensure Earth's
suitability. Wallace even published a short book, *Is Mars Habitable?*
(1907) to rebut Percival Lowell's[5] theory that the "canali" on
Mars implied the presence of intelligent life. (His conclusion is
unequivocal: "Mars, therefore, is not only uninhabited by intelligent beings such as Mr. Lowell postulates, but is absolutely
UNINHABITABLE."). Wallace's main book on the Anthropic Principle, *Man's Place in the Universe* (1903), seems more dated than
most of his writing because it is based on an utterly flawed
cosmology.

Descending now to terrestrial physics, I have shown that, owing to
the highly complex nature of the adjustments required to render a
world habitable and to retain its habitability during the aeons of time
requisite for life-development, it is in the highest degree improbable
that the required conditions and adaptations should have occurred in
any other planets of any other suns, which might occupy an equally
favourable position as our own, and which were of the requisite size
and heat-giving power.

Lastly, I submit that the whole of the evidence I have here brought
together leads to the conclusion that our earth is almost certainly the
only inhabited planet in our solar system – and, further, that there is
no inconceivability – no improbability even – in the conception that,
in order to produce a world that should be precisely adapted in every
detail for the orderly development of organic life culminating in man,
such a vast and complex universe as that which we know exists around
us may have been absolutely required.[6]

. . .

And is it not in perfect harmony with this grandeur of design (if it
be design), this vastness of scale, this marvellous process of development through all the ages, that the material universe needed to
produce this cradle of organic life, and of a being destined to a higher
and a permanent existence, should be on a corresponding scale of
vastness, of complexity, of beauty? Even if there were no such evidence
as I have here adduced for the unique position and the exceptional
characteristics which distinguish the earth, the old idea that all the

planets were inhabited, and that all the stars existed for the sake of other planets, which planets existed to develop life, would, in the light of our present knowledge, seem utterly improbable and incredible. It would introduce monotony into a universe whose grand character and teaching is endless diversity. It would imply that to produce the living soul in the marvellous and glorious body of man – man with his faculties, his aspirations, his powers for good and evil – that this was an easy matter which could be brought about anywhere, in any world. It would imply that man is an animal and nothing more, is of no importance in the universe, needed no great preparations for his advent, only, perhaps, a second-rate demon, and a third or fourth-rate earth. Looking at the long and slow and complex growth of nature that preceded his appearance, the immensity of the stellar universe with its thousand million suns, and the vast aeons of time during which it has been developing – all these seem only the appropriate and harmonious surroundings, the necessary supply of material, the sufficiently spacious workshop for the production of that planet which was to produce, first, the organic world, and then, Man.[7] [1903] &

> We have already seen an early impact of Wallace's spiritualism on his biological thinking in his departure from Darwin on human evolution. He eventually came to embrace wholesale a theory of evolutionary teleology, to which his final major work, *The World of Life* (1910), was a paean.

& Though such a very obvious fact, it is not always kept in mind, that the entire animal world, in all its myriad manifestations, from the worm in the soil to the elephant in the forest, from the blind fishes of the ocean depths to the soaring sky-lark, depends absolutely on the equally vast and varied vegetable world for its very existence. It is also tolerably clear, though not quite so conclusively proved, that it is on the overwhelming variety of plant species, to which we have already called attention, that the corresponding variety of animal species, especially in the insect tribes, has been rendered possible.

This will perhaps be better seen by a reference to one of the best-known cases of general adaptation, which, because so common and obvious, is often overlooked or misunderstood. All lovers of a garden are apt to regard as an unmitigated evil those swarms of insects which attack their plants in spring, and in recurrent bad years become a serious nuisance and commit widespread devastation. At one time the buds or leaves of their fruit trees swarm with various kinds of caterpillars, while at others even the oak trees are so denuded of their leaves as to become an eyesore in the landscape. Many of our common vegetables, and even the grass on our lawns, are in some seasons destroyed by swarms of wire-worms which feed on their roots. Turnips, radishes, and allied plants are attacked by the turnip-fly, a small jumping beetle whose larva lives in the leaf itself, and which often swarms in millions. Then there are the aphides and froghoppers on our roses and other shrubs or flowers, and grubs which attack our apples, our carrots, and most other crops; and all these the gardener usually regards under the general term "blight," as a serious blot on the face of nature, and wonders why such harmful creatures were permitted to exist.

Most professional gardeners would be rather surprised to hear that all these insect-pests are an essential part of the world of life; that their destruction would be disastrous: and that without them some of the most beautiful and enjoyable of the living things around us would be either seriously diminished in numbers or totally destroyed. He might also be informed that he himself is a chief cause of the very evil he complains of, because, by growing in large quantities the plants the insect-pests feed upon he provides for them a superabundance of food, and enables them to increase much more rapidly than they would do under natural conditions.

Let us now consider what happens over our whole country in each recurring spring. At that delightful season our gardens and hedgerows, our orchards, woods, and copses are thronged with feathered songsters, resident and migratory, engaged every hour of the day in building their nests, hatching their eggs, or feeding and guarding their helpless offspring. A considerable proportion of these – thrushes, warblers, tits, finches, and many others – are so prolific that they have two or three, sometimes even more, families every year, so that the young birds reared annually by each pair varies from four or five up to ten or twenty, or even more.

Now, when we consider that the parents of these, to the number of perhaps fifty species or more, are all common birds, which exist in our islands in numbers amounting to many millions each, we can partially realise the enormous quantity of insect-food required to rear perhaps five or ten times that number of young birds from the egg up to full growth. Almost all of the young of the smaller birds, even when their parents are seed-eaters, absolutely require soft insect-food, such as caterpillars and grubs of various sorts, small worms, or such perfect insects as small spiders, gnats, flies, etc., which alone supply sufficient nourishment in a condensed and easily digestible form. . . .

What wonderful perfection of the senses must there be in these various parent birds; what acuteness of vision or of hearing; what rapidity of motion, and what powerful instinct of parental love, enabling them to keep up this high-pressure search for food, and of watchfulness of their nests and young, on the continuance of which, and its unfailing success, the very existence of those young and the continuance of the race depends. But all this perfect adaptation in the parent birds would be of no avail unless the insect tribes, on which alone most of them are obliged to depend, were as varied, as abundant, and as omnipresent as they actually are; and also unless vegetation were so luxuriant and abundant in its growth and so varied in its character, that it can always supply ample food for the insects without suffering any great or permanent injury to the individual plants, much less to any of the species.

By such considerations as these we learn that what we call insect-pests, when they are a little more abundant than usual in our gardens and orchards, do not exist for themselves alone as an apparently superfluous and otherwise useless part of the great world of life, but are, and must always have been throughout long past geological ages, absolutely essential for the origination and subsequent development of the most wonderful, delightful, and beautiful of all the living things around us – our garden friends and household pets, and sweet singers of the woods and fields. Without the myriad swarms of insects everywhere devouring a portion of the new and luxuriant vegetation, the nightingale and the lark, the wren, the redbreast, and the fairy-like tits and gold-crests might never have come into existence, and if the supply failed would now disappear for ever![8] [1910]   &

Wallace even found a rationale for conservation buried within his teleology.

& Already in the progress of this work I have dwelt upon the marvellous *variety* of the useful or beautiful products of the vegetable and animal kingdoms far beyond their own uses, as indicating a development for the service of man. This variety and beauty, even the strangeness, the ugliness, and the unexpectedness we find everywhere in nature, are, and therefore were intended to be, an important factor in our mental development; for they excite in us admiration, wonder, and curiosity – the three emotions which stimulate first our attention, then our determination to learn the how and the why, which are the basis of observation and experiment and therefore of all science and all philosophy. These considerations should lead us to look upon all the works of nature, animate or inanimate, as invested with a certain sanctity, to be *used* by us but not *abused*, and never to be recklessly destroyed or defaced. To pollute a spring or a river, to exterminate a bird or beast, should be treated as moral offences and as social crimes; while all who profess religion or sincerely *believe* in the Deity – the designer and maker of this world and of every living thing – should, one would have thought, have placed *this* among the first of their forbidden sins, since to deface or destroy that which has been brought into existence for the use and enjoyment, the education and elevation of the human race, is a direct denial of the wisdom and goodness of the Creator, about which they so loudly and persistently prate and preach.[9] [1910]   ঌ

He finishes the book in an appropriately grand mystical style.

ঌ   But to claim the Infinite and Eternal Being as the one and only direct agent in every detail of the universe seems, to me, absurd. If there is such an Infinite Being, and if (as our own existence should teach us) His will and purpose is the increase of conscious beings, then we can hardly be the first result of this purpose. We conclude, therefore, that there are now in the universe infinite grades of power, infinite grades of knowledge and wisdom, infinite grades of influence of higher beings upon lower. Holding this opinion, I have suggested that this vast and wonderful universe, with its almost infinite variety of forms, motions, and reactions of part upon part, from suns and systems up to plant life, animal life, and the human living soul, has ever required and still requires the continuous co-ordinated agency of myriads of such intelligences.

This speculative suggestion, I venture to hope, will appeal to some of my readers as the best approximation we are now able to formulate as to the deeper, the more fundamental causes of matter and force, of life and consciousness, and of Man himself; at his best, already "a little lower than the angels," and, like them, destined to a permanent progressive existence in a World of Spirit.[10] [1910]   ঌ

Travel

Wallace was an inveterate traveller. He spent four years (1848–52) in the Amazon, eight (1854–62) in South-east Asia,[1] and undertook a major lecture tour of North America (1886–87) as well as many journeys in and around Europe. His undiminished wanderlust contrasts strongly with Darwin's decision following his *Beagle* voyage to take root in English soil, foregoing foreign travel altogether. Wallace was once tempted to follow suit – during his traumatic journey home from the Amazon – but that temptation was short-lived: "Fifty times since I left Para [Belém] have I vowed, if I once reached England, never to trust myself more on the ocean. But good resolutions soon fade, and I am already only doubtful whether the Andes or the Philippines are to be the scene of my next wanderings."[2] As he set out on his visit to North America, Wallace, then in his mid-sixties, contemplated extending his journey "across the Pacific, lecturing in New Zealand and Australia, perhaps also in South Africa",[3] on his way home. It was only because his "voyage out was so disagreeable", making him "sick and unwell almost the whole time"[4] – he had always suffered badly from seasickness – that he decided not to undertake the round-the-world trip.

His Amazon trip famously ended in disaster when the boat carrying him and most of his collections home caught fire and sank in the middle of the Atlantic. Wallace was lucky to escape with his life, and lost much of his precious four years' accumulation of specimens and notes. Nevertheless, he still managed to produce his *Narrative of Travels on the Amazon and Rio Negro* from the "scanty materials" he had saved, supplemented with his correspondence home. He arranged with Lovell Reeve[5] for its publication "on agreement for 'half profits'".[6] But it was not a bestseller: "Only 750 copies were printed and when I returned home from the East in 1862, about 250 copies were still unsold, and there were consequently no profits to divide."[7] Darwin was unwilling to cut Wallace any slack for the loss of all his records, writing to Bates in 1861, "I was a *little* disappointed in Wallace's book on the Amazon, hardly facts enough."[8] Bates, Wallace's companion on the Amazon venture, was much more successful with his book *The Naturalist on the Amazons* (1863), and Wallace seems to have taken the fortunes of their respective books as an indictment of his writing. While working on his South-east Asia book, he wrote to Darwin:

I am at last making a beginning of a small book on my Eastern journey, which, if I can persevere, I hope to have ready by next Christmas. I am a very bad hand at writing anything like narrative. I want something to argue on, and then I find it much easier to go ahead. I rather despair, therefore, of making so good a book as Bates's, though I think my subject is better. Like every other traveller, I suppose, I feel dreadfully the want of copious notes on common everyday objects, sights and sounds and incidents, which I imagined I could never forget but which I now find it impossible to recall with any accuracy.[9] [1864]

Wallace need not have worried: Darwin, to whom, as has been noted, *The Malay Archipelago* was dedicated, [10] was enthusiastic about the work: "It seems to me excellent, and at the same time most pleasant to read. That you ever returned alive is wonderful after all your risks from illness and sea voyages, especially that most interesting one to Waigiou [Waigeo] and back. Of all the impressions which I have received from your book, the strongest is that your perseverance in the cause of science was heroic."[11] Darwin claimed he would be "astonished" if the book was not "a great success";[12] it was indeed, and Wallace was able to recall with pride that "it long continued to be the most popular of my books".[13]

Expectations

In the first letter intended for public consumption[1] that he sent from Brazil, Wallace admitted that things were not quite as he expected.

⋖ Previous to leaving England I had read many books of travels in hot countries, I had dwelt so much on the enthusiastic descriptions[2] most naturalists give of the surpassing beauty of tropical vegetation, and of the strange forms and brilliant colours of the animal world, that I had wrought myself up to a fever-heat of expectation, and it is not to be wondered at that my early impressions were those of disappointment. On my first walk into the forest I looked about, expecting to see monkeys as plentiful as at the Zoological Gardens, with humming-birds and parrots in profusion. But, for several days I did not see a single monkey, and hardly a bird of any kind, and I began to think that these and other productions of the South American forests are much scarcer than they are represented to be by travellers.[3] [1848] ⋗

However, one feature of the Brazilian environment fully lived up to Wallace's expectations and could be "fully appreciated in a single walk":[4] the virgin forest.

⋖ Here no one who has any feeling of the magnificent and the sublime can be disappointed; the sombre shade, scarce illumined by a single direct ray even of the tropical sun, the enormous size and height of the trees, most of which rise like huge columns a hundred feet or more without throwing out a single branch, the strange buttresses around the base of some, the spiny or furrowed stems of others, the curious and even extraordinary creepers and climbers which wind around them, hanging in long festoons from branch to branch, sometimes curling and twisting on the ground like great serpents, then mounting to the very tops of the trees, thence throwing down roots and fibres which hang waving in the air, or twisting round

each other form ropes and cables of every variety of size and often of
the most perfect regularity. These, and many other novel features –
the parasitic plants growing on the trunks and branches, the wonderful variety of the foliage, the strange fruits and seeds that lie rotting
on the ground – taken altogether surpass description, and produce
feelings in the beholder of admiration and awe. It is here, too, that
the rarest birds, the most lovely insects, and the most interesting
mammals and reptiles are to be found. Here lurk the jaguar and the
boa-constrictor, and here amid the densest shade the bell-bird tolls
his peal.[5] [1848]   ॐ

> Coral reefs – the marine equivalent of tropical forests – also met
> Wallace's expectations.

ॐ Passing up the harbour, in appearance like a fine river, the
clearness of the water afforded me one of the most astonishing and
beautiful sights I have ever beheld. The bottom was absolutely hidden
by a continuous series of corals, sponges, actinic [sea anemone], and
other marine productions of magnificent dimensions, varied forms,
and brilliant colours. The depth varied from about twenty to fifty feet,
and the bottom was very uneven, rocks and chasms and little hills and
valleys, offering a variety of stations for the growth of these animal
forests. In and out among them, moved numbers of blue and red and
yellow fishes, spotted and banded and striped in the most striking
manner, while great orange or rosy transparent medusa [jelly fish]
floated along near the surface. It was a sight to gaze at for hours, and
no description can do justice to its surpassing beauty and interest. For
once, the reality exceeded the most glowing accounts I had ever read
of the wonders of a coral sea. There is perhaps no spot in the world
richer in marine productions, corals, shells and fishes, than the
harbour of Amboyna [Ambon].[6] [1869]   ॐ

> Small-scale aspects of a traveller's experience can also con-
> found or live up to expectations. Wallace records that his first
> encounter with the fruit of the durian tree was inauspicious: the
> smell was so unpleasant that he deemed it inedible. But a more
> propitious re-encounter precipitated what appears to have been a
> minor love affair with the fruit. Durian continues to this day to
> provoke strong passions. To the uninitiated, virtually nothing can
> smell or taste more rank. So powerful is the smell that some
> airlines refuse to transport durian, and hotels in South-east Asia

sometimes prohibit guests from taking them into their rooms
However, to the aficionado, durian are without question the finest
food on earth. Debates rage as to what cultivar is superior and, in
season, enthusiasts will undertake what are in effect pilgrimages
through South-east Asia in pursuit of the perfect durian.

The following passage – Wallace's homage to the durian –
shows his travel writing at its best. We see him as much more
than the dispassionate scientist conscientiously recording all that
he encountered: he was also an enthusiast, and one – unusual by
Victorian standards – who occasionally allowed his excitement to
seep into his prose. Wallace's account starts in the domain of
formal botanical description but rapidly moves into a much more
subjective world.

I slept that night in the village of the Sebungow Dyaks, and the
next day reached Sarawak [Kuching], passing through a most beautiful country where limestone mountains with their fantastic forms and
white precipices shot up on every side, draped and festooned with a
luxuriant vegetation. The banks of the Sarawak River are everywhere
covered with fruit trees, which supply the Dyaks with a great deal of
their food. The Mangosteen, Lansat, Rambutan, Jack, Jambou, and
Blimbing, are all abundant; but most abundant and most esteemed is
the Durian, a fruit about which very little is known in England, but
which both by natives and Europeans in the Malay Archipelago is
reckoned superior to all others. The old traveller Linschott,[7] writing
in 1599, says: "It is of such an excellent taste that it surpasses in flavour
all the other fruits of the world, according to those who have tasted
it." And Doctor Paludanus[8] adds: "This fruit is of a hot and humid
nature. To those not used to it, it seems at first to smell like rotten
onions, but immediately when they have tasted it, they prefer it to all
other food. The natives give it honourable titles, exalt it, and make
verses on it." When brought into a house the smell is often so offensive
that some persons can never bear to taste it. This was my own case
when I first tried it in Malacca [Melaka], but in Borneo I found a ripe
fruit on the ground, and, eating it out of doors, I at once became a
confirmed Durian eater.

The Durian grows on a large and lofty forest tree, somewhat
resembling an elm in its general character, but with a more smooth
and scaly bark. The fruit is round or slightly oval, about the size of a
large cocoanut, of a green colour, and covered all over with short
stout spines the bases of which touch each other, and are consequently

somewhat hexagonal, while the points are very strong and sharp. It is
so completely armed, that if the stalk is broken off it is a difficult
matter to lift one from the ground. The outer rind is so thick and
tough, that from whatever height it may fall it is never broken. From
the base to the apex five very faint lines may be traced, over which the
spines arch a little; these are the sutures of the carpels, and show
where the fruit may be divided with a heavy knife and a strong hand.
The five cells are satiny white within, and are each filled with an oval
mass of cream-coloured pulp, imbedded in which are two or three
seeds about the size of chestnuts. This pulp is the eatable part, and its
consistency and flavour are indescribable. A rich butter-like custard
highly flavoured with almonds gives the best general idea of it, but
intermingled with it come wafts of flavour that call to mind cream-
cheese, onion-sauce, brown sherry, and other incongruities. Then
there is a rich glutinous smoothness in the pulp which nothing
else possesses, but which adds to its delicacy. It is neither acid,
nor sweet, nor juicy; yet one feels the want of none of these qualities,
for it is perfect as it is. It produces no nausea or other bad effect,
and the more you eat of it the less you feel inclined to stop. In fact to
eat Durians is a new sensation, worth a voyage to the East to
experience.

When the fruit is ripe it falls of itself, and the only way to eat
Durians in perfection is to get them as they fall; and the smell is then
less overpowering. When unripe, it makes a very good vegetable if
cooked, and it is also eaten by the Dyaks raw. In a good fruit season
large quantities are preserved salted, in jars and bamboos, and kept
the year round, when it acquires a most disgusting odour to Europeans, but the Dyaks appreciate it highly as a relish with their rice.
There are in the forest two varieties of wild Durians with much smaller
fruits, one of them orange-coloured inside; and these are probably
the origin of the large and fine Durians, which are never found wild.
It would not, perhaps, be correct to say that the Durian is the best of
all fruits, because it cannot supply the place of the subacid juicy kinds,
such as the orange, grape, mango, and mangosteen, whose refreshing
and cooling qualities are so wholesome and grateful; but as producing
a food of the most exquisite flavour, it is unsurpassed. If I had to fix
on two only, as representing the perfection of the two classes, I should
certainly choose the Durian and the Orange as the king and queen of
fruits.

The Durian is, however, sometimes dangerous. When the fruit
begins to ripen it falls daily and almost hourly, and accidents not

unfrequently happen to persons walking or working under the trees. When a Durian strikes a man in its fall, it produces a dreadful wound, the strong spines tearing open the flesh, while the blow itself is very heavy; but from this very circumstance death rarely ensues, the copious effusion of blood preventing the inflammation which might otherwise take place. A Dyak chief informed me that he had been struck down by a Durian falling on his head, which he thought would certainly have caused his death, yet he recovered in a very short time.

Poets and moralists, judging from our English trees and fruits, have thought that small fruits always grew on lofty trees, so that their fall should be harmless to man, while the large ones trailed on the ground. Two of the largest and heaviest fruits known, however, the Brazil-nut fruit (Bertholletia) and Durian, grow on lofty forest trees, from which they fall as soon as they are ripe, and often wound or kill the native inhabitants. From this we may learn two things: first, not to draw general conclusions from a very partial view of nature; and secondly, that trees and fruits, no less than the varied productions of the animal kingdom, do not appear to be organized with exclusive reference to the use and convenience of man.[9] [1869] ﻼ

City Life

Wallace was chiefly interested in the natural world, but his descriptions of city life are also evocative, benefiting no doubt from the naturalist's skills as an observer.

❧ Of all the eventful days of my life my first in Alexandria was the most striking. Imagine my feelings when, coming out of the hotel (whither I have been conveyed in an omnibus) for the purpose of taking a quiet stroll through the city, I found myself in the midst of a vast crowd of donkeys and their drivers, all thoroughly determined to appropriate my person to their own use and interest, without in the least consulting my inclinations. In vain with rapid strides and waving arms I endeavoured to clear a way and move forward; arms and legs were seized upon, and even the Christian coat-tails were not sacred from the profane Mahometans. One would hold together two donkeys by their tails while I was struggling between them, and another, forcing together their heads, would thus hope to compel me to mount upon one or both of them; and one fellow more impudent than the rest I laid flat-upon the ground, and sending the donkey staggering after him, I escaped a moment midst hideous yells and most unearthly cries. I now beckoned to a fellow more sensible-looking than the rest, and told him that I wished to walk, and would take him for a guide, and hoped now to be at rest; but vain thought! I was in the hands of the Philistines, and getting us up against a wall, they formed an impenetrable phalanx of men and brutes thoroughly determined that I should only get away from the spot on the legs of a donkey. Bethinking myself now that donkey-riding was a national institution, and seeing a fat Yankee (very like my Paris friend[1]) mounted, being like myself hopeless of any other means of escape, I seized upon a bridle in hopes that I should then be left in peace. But this was the signal for a more furious onset, for, seeing that I would at length ride, each one was determined that he alone should profit by the transaction, and a dozen animals were forced suddenly upon me and a dozen hands tried to lift me upon their respective beasts. But now my patience was exhausted, so, keeping firm hold of the bridle I had first taken with one hand, I hit right and left with the other, and calling upon my guide to do the same, we succeeded in clearing a little space around us. Now then behold your friend mounted upon a jackass in the streets of Alexandria, a boy behind holding by his tail and whipping him up, Charles[2] (who had been lost sight of in the crowd) upon another, and my guide upon a third, and off we go among a crowd of Jews and Greeks, Turks and Arabs, and veiled women and yelling donkey-boys to see the city. We saw the bazaars and the slave market, where I was again nearly pulled to pieces for "backsheesh" (money), the mosques with their elegant minarets, and then the Pasha's new palace, the interior of which is most gorgeous. . . . You may think this account is exaggerated, but it is not; the pertinacity, vigour and screams of the Alexandrian donkey-drivers no description can do justice to.[3] [1854] ੬ಽ

ੳ [Singapore.] The scene is at once so familiar and strange. The half-naked Chinese coolies, the neat shopkeepers, the clean, fat, old, long-tailed merchants, all as busy and full of business as any Londoners. Then the handsome Klings,[4] who always ask double what they take, and with whom it is most amusing to bargain. The crowd of boatmen at the ferry, a dozen begging and disputing for a farthing fare, the Americans, the Malays, and the Portuguese make up a scene doubly interesting to me now that I know something about them and can talk to them in the general language of the place. The streets of Singapore on a fine day are as crowded and busy as Tottenham Court Road, and from the variety of nations and occupations far more interesting.[5] [1856] ੬ಽ

As Wallace grew older, cities lost their appeal. They became merely emblematic of the social and economic failings of capitalism.

ੳ We passed under a fine girder bridge and the great Victoria Tubular Bridge before reaching Montreal, the appearance of which is much spoilt by factory chimneys and the usual but quite unnecessary pall of smoke. For all this unsightliness in almost every city in the world, land monopoly and competition are responsible. If each city owned its own land, it would be no one's interest to destroy its beauty and healthiness with smoke and impure water; and if every parish, district, or county owned its own land, factories would only be permitted away from centres of population, and would be so regulated as to prevent all injury or even inconvenience to those who worked in them.[6] [1905] ತಿ

Life in the Field

Wallace's priorities lay with collecting biological specimens. It was, after all, the revenue from this activity that permitted him to undertake his trips. Part of the charm of his travel writing is the way in which he weaves in information on the interplay between his scientific goals and his daily life. Here is Wallace's account of a typical entomological foray in Borneo.

To give English entomologists some idea of the collecting here, I will give a sketch of one good day's work. Till breakfast I am occupied ticketing and noting the captures of the previous day, examining boxes for ants, putting out drying-boxes and setting the insects of any caught by lamp-light. About 10 o'clock I am ready to start. My equipment is, a rug-net, large collecting-box hung by a strap over my shoulder, a pair of pliers for Hymenoptera [bees, ants, wasps], two bottles with spirits, one large and wide-mouthed for average Coleoptera [beetles], &c., the other very small for minute and active insects, which are often lost by attempting to drop them into a large mouthed bottle. These bottles are carried in pockets in my hunting-shirt, and are attached by strings round my neck; the corks are each secured to the bottle by a short string. The morning is fine, and thus equipped I first walk to some dead trees close to the house frequented by Buprestidae [jewel beetles]. As I approach I see the bright golden back of one, as he moves in sideway jerks along a prostrate trunk, – I approach with caution, but before I can reach him, whizz! – he is off, and flies humming round my head. After one or two circuits he settles again in a place rendered impassable by sticks and bushes, and when he leaves it, it is to fly off to some remote spot in the jungle. I then walk off into the swamp along the path of logs and tree-trunks, picking my way cautiously, now glancing right and left on the foliage, and then surveying carefully the surface of the smooth round log I am walking on. The first insect I catch is a pretty little long-necked Apoderus [weevil] sitting partly upon a leaf: a few paces further, I come to a place where some Curculionidae [weevils], of the genus

Mecopus, are always seated on a dry sun-shiny log. A sweep of my net captures one or two, and I go on, as I have already enough specimens of them. The beautiful Papilios [swallow-tailed butterflies], Evemon and Agamemnon, fly by me, but the footing is too uncertain to capture them, and at the same moment a small beetle flies across and settles on a leaf near me – I move cautiously but quickly on – see it is a pretty Glenea [long-horned beetle], and by a sharp stroke of the net capture it, for they are so active that the slightest hesitation is sure to lose the specimen. I now come to a bridge of logs across a little stream; this is another favourite station of the Buprestidae, particularly of the elegant Belionota sumptuosa. One of these is now on the bridge, – he rises as I approach, – flies with the rapidity of lightning around me, and settles on the handle of my net! I watch him with quiet admiration, – to attempt to catch him then is absurd; in a moment he is off again, and then settles within a yard of me; I strike with all my force, he rises at the same moment, and is now buzzing in my net, and in another instant is transferred in safety to my bottle: I wait a few minutes here in hopes that another may be heard or seen, and then go on; I pass some fallen trees, under which are always found some Curculionidae, species of Alcides and Otops, – these I sweep carefully with my net and get two or three specimens, one new to me. I now come to a large Boletus [fungus] growing on a stump, – I push my net under it, two Thyreopterae [ground beetles] run on to the top, I knock one with my hand into my net, while the other has instantly escaped into a crack in the stump and is safe for this day, but his time will come. In some distance now I walk on, looking out carefully for whatever may appear; for near half-a-mile I see not an insect worth capturing; then suddenly flies across the path a fine Longicorn [long-horned beetle], new to me, and settles on a trunk a few yards off. I survey the soft brown mud between us, look anxiously for some root to set my foot on, and then cautiously advance towards him: one more step and I have him, but alas! My foot slips off the root, down I go into the bog and the treasure escapes, perhaps a species I may never obtain again. Returning to the path, another hum salutes my ear, and the fine Cetonia [scarab beetle], Macronota Diardi, settles on a leaf near me, and is immediately secured: a little further, a yellow-powdered Buprestis is caught in the same manner.[1] [1855] ❧

Here, in a letter to his mother, he recounts his daily routine more schematically.

❧ I will tell you how my day is now occupied. Get up at half past five. Bath and coffee. Sit down to arrange and put away my insects of the day before, and set them safe out to dry. Charles mending nets, filling pincushions, and getting ready for the day. Breakfast at eight. Out to the jungle at nine. We have to walk up a steep hill to get to it, and always arrive dripping with perspiration. Then we wander about till two or three, generally returning with about 50 or 60 beetles, some very rare and beautiful. Bathe, change clothes, and sit down to kill and pin insects. Charles ditto with flies, bugs and wasps; I do not trust him yet with beetles. Dinner at four. Then to work again till six. Coffee. Read. If very numerous, work at insects till eight or nine. Then to bed.[2] [1854] ❧

A critical part of that routine appears to have been coffee.

❧ I got some of the boys to fetch me water from the river, and to bring me in a stock of fuel, and then, with coffee and cheese, roasted plantains and cassava-bread, I lived luxuriously. My coffee, however, was just finished, and in a day or two I had none. This I could hardly put up with without a struggle, so I went down to the cottage of an old Indian who could speak a little Spanish, and begged him, "por amor de Dios," to get me some coffee from a small plantation he had. There were some ripe berries on the trees, the sun was shining out, and he promised to set his little girl to work immediately. This was about ten in the morning. I went into the forest, and by four returned, and found that my coffee was ready. It had been gathered, the pulp washed off, dried in the sun (the longest part of the business), husked, roasted, and pounded in a mortar; and in half an hour more I enjoyed one of the most delicious cups of coffee I have ever tasted.[3] [1853] ❧

Like many a field biologist since, Wallace found plenty in his assistant to complain about.

❧ Charles has now been with me more than a year, and every day some such conversation as this ensues: "Charles, look at these butterflies that you set out yesterday." "Yes, sir." "Look at that one – is it set out evenly?" "No, sir." "Put it right then, and all the others that want it." In five minutes he brings me the box to look at. "Have you put them all right?" "Yes, sir." "There's one with the wings uneven, there's another with the body on one side, then another with the pin crooked. Put them all right this time." It most frequently happens that they have to go back a third time. Then all is right. If he puts up a bird, the head is on one side, there is a great lump of cotton on one side of the neck like a wen, the feet are twisted soles uppermost, or something else. In everything it is the same, what ought to be straight is always put crooked. This after twelve months' constant practice and constant teaching! And not the slightest sign of improvement. I believe he never will improve. Day after day I have to look over everything he does and tell him of the same faults. Another with a similar incapacity would drive me mad. He never, too, by any chance, puts anything away after him. When done with, everything is thrown on the floor. Every other day an hour is lost looking for knife, scissors, pliers, hammer, pins, or something he has mislaid.[4] [1855]   ઝ

And again like many a field biologist since, Wallace's interactions with local people showed him just how little he, the supposed expert, knew.

ઝ   After breakfast, we [Wallace was with Bates at this stage of his Amazon journey] loaded our old Negro [Isidora, their cook] (who had come with us to show the way) with plants that we had collected, and a basket to hold anything interesting we might meet with on the road, and set out to walk home, promising soon to make a longer visit. We reached Nazaré with boxes full of insects, and heads full of the many interesting things we had seen, among which the milk-giving tree, supplying us with a necessary of life from so new and strange a source, held a prominent place.

Wishing to obtain specimens of a tree called Caripé, the bark of which is used in the manufacture of the pottery of the country, we inquired of Isidora if he knew such a tree, and where it grew. He replied that he knew the tree very well, but that it grew in the forest a long way off. So one fine morning after breakfast we told him to shoulder his axe and come with us in search of the Caripé, – he in his usual dishabille of a pair of trousers, – shirt, hat, and shoes being altogether dispensed with in this fine climate; and we in our shirt-sleeves, and with our hunting apparatus across our shoulders. Our old conductor, though now following the domestic occupation of cook and servant of all work to two foreign gentlemen, had worked much in the forest, and was well acquainted with the various trees, could tell their names, and was learned in their uses and properties. He was of rather a taciturn disposition, except when excited by our exceeding dullness in understanding what he wanted, when he would gesticulate with a vehemence and perform dumb-show with a minuteness worthy of a more extensive audience; yet he was rather fond of displaying his knowledge on a subject of which we were in a state of the most benighted ignorance, and at the same time quite willing to learn. His method of instruction was by a series of parenthetical remarks on the trees as he passed them, appearing to speak rather to them than to us, unless we elicited by questions further information.[5] [1853]

Occasionally the behaviour of the strange foreigner was totally inexplicable.

Passing this, we got into the forest. At first the path was tolerable; soon, however, it was a mere track a few inches wide, winding among thorny creepers, and over deep beds of decaying leaves. Gigantic buttress trees, tall fluted stems, strange palms, and elegant tree-ferns were abundant on every side, and many persons may suppose that our walk must necessarily have been a delightful one; but there were many disagreeables. Hard roots rose up in ridges along our path, swamp and mud alternated with quartz pebbles and rotten leaves; and as I floundered along in the barefooted enjoyment of these, some over-hanging bough would knock the cap from my head or the gun from my hand; or the hooked spines of the climbing palms would catch in my shirt-sleeves, and oblige me either to halt and deliberately unhook myself, or leave a portion of my unlucky garment behind. The Indians were all naked, or, if they had a shirt or trousers, carried them in a bundle on their heads, and I have no doubt looked upon me as a good illustration of the uselessness and bad consequences of wearing clothes upon a forest journey.[6] [1853]

On other occasions, the presence of the strange foreigner was downright terrifying.

[During a journey into the interior of Borneo] It was evident that Europeans seldom came here, for numbers of women skeltered away as I walked through the village and one girl about ten or twelve years old, who had just brought a bamboo full of water from the river, threw it down with a cry of horror and alarm the moment she caught sight of me, turned around and jumped into the stream. She swam beautifully, and kept looking back as if expecting I would follow her, screaming violently all the time; while a number of men and boys were laughing at her ignorant terror.[7] [1869] ೞ

Or the strange foreigner was merely a curiosity.

ೞ [Later in the same journey] Early in the afternoon we reached the village of Borotói, and, though it would have been easy to reach the next one before night, I was obliged to stay, as my men wanted to return and others could not possibly go on with me without the preliminary talking. Besides, a white man was too great a rarity to be allowed to escape them, and their wives would never have forgiven them if, when they returned from the fields, they found that such a curiosity had not been kept for them to see. On entering the house to which I was invited, a crowd of sixty or seventy men, women, and children gathered around me, and I sat for half an hour like some strange animal submitted for the first time to the gaze of an inquiring public. Brass rings were here in the greatest profusion, many of the women having their arms completely covered with them, as well as their legs from the ankle to the knee. Round the waist they wear a dozen or more coils of fine rattan stained red, to which the petticoat is attached. Below this are generally a number of coils of brass wire, a girdle of small silver coins, and sometimes a broad belt of brass ring armour. On their heads they wear a conical hat without a crown, formed of variously coloured beads, kept in shape by rings of rattan, and forming a fantastic but not unpicturesque headdress.

Walking out to a small hill near the village, cultivated as a rice-field, I had a fine view of the country, which was becoming quite hilly, and towards the south, mountainous. I took bearings and sketches of all that was visible, an operation which caused much astonishment to the Dyaks who accompanied me, and produced a request to exhibit the compass when I returned. I was then surrounded by a larger crowd than before, and when I took my evening meal in the midst of a circle of about a hundred spectators anxiously observing every movement and criticising every mouthful, my thoughts involuntarily recurred to the lion at feeding time. Like those noble animals, I too was used to it, and it did not affect my appetite.[8]

. . .

Two or three [Aru Islanders] got round me and begged me for the twentieth time to tell them the name of my country. Then, as they could not pronounce it satisfactorily, they insisted that I was deceiving them, and that it was a name of my own invention. One funny old man, who bore a ludicrous resemblance to a friend of mine at home, was almost indignant. "Ung-lung!" said he, "who ever heard of such a name? – Ang-lang – Anger-lung – that can't be the name of your country; you are playing with us." Then he tried to give a convincing illustration. "My country is Wanumbai – anybody can say Wanumbai. I'm an orang-Wanumbai; but, N-glung! who ever heard of such a name? Do tell us the real name of your country, and then when you are gone we shall know how to talk about you." To this luminous argument and remonstrance I could oppose nothing but assertion, and the whole party remained firmly convinced that I was for some reason or other deceiving them.[9] [1869] ଈ

Even the worst weather could not diminish Wallace's enthusiasm for the tropics.

ଈ I am here in one of the places unknown to the Royal Geographical Society, situated in the very centre of East Sumatra, about one hundred miles from the sea in three directions. It is the height of the wet season, and the rain pours down strong and steady, generally all night and half the day. Bad times for me, but I walk out regularly three or four hours every day, picking up what I can, and generally getting some little new or rare or beautiful thing to reward me. This is the land of the two-horned rhinoceros, the elephant, the tiger, and the tapir; but they all make themselves very scarce, and beyond their tracks and their dung, and once hearing a rhinoceros *bark* not far off, I am not aware of their existence. This, too, is the very land of monkeys; they swarm about the villages and plantations, long-tailed and short-tailed, and with no tail at all, white, black, and grey; they are eternally racing about the tree-tops, and gambolling in the most amusing manner. The way they jump is amazing. They throw themselves recklessly through the air, apparently sure, with one or other of their four hands, to catch hold of something. I estimated one jump by a long-tailed white monkey, at thirty feet horizontal, and sixty feet vertical, from a high tree on to a lower one; he fell through, however, so great was his impetus, on to a lower branch, and then, without a moment's stop, scampered away from tree to tree, evidently quite pleased with his own pluck. When I startle a band, and one leader takes a leap like this, it is amusing to watch the others – some afraid and hesitating on the brink till at last they pluck up courage, take a run at it, and often roll over in the air with their desperate efforts.

Then there are the long-armed apes, who never walk or run upon the trees, but travel altogether by their long arms, swinging themselves from bough to bough in the easiest and most graceful manner possible.[10] &

"An industrious and persevering traveller"

Travelling at that time in the places Wallace travelled to was a dangerous business. Disease was the major problem, but security could be an issue as well, especially for someone like Wallace who typically travelled unaccompanied by Europeans. His own brother Herbert, who had come out with the botanist Richard Spruce and had been travelling and working on and off with Wallace, died in Brazil from yellow fever, and Wallace was seriously ill on several occasions both in the Amazon and in South-east Asia. Wallace, however, made light of the dangers.

Of real dangers I say nothing – for I do not believe that with proper precautions there is more danger in one part of the world than another. "Familiarity breeds contempt" of dangers as well as of persons – and a picture might be drawn of the terrors of London Streets – with their mad bulls and mad dogs, their garrotters, runaway cabs and falling chimneys – that would make an inhabitant of tiger infested Java or run-a-muck Macassar [Ujung Pandang] thank his stars he was not a Londoner and wonder how people could be foolhardy enough to live there.[1] [1862]

But the dangers were certainly real.

When going to bed for the night, I put out my candle, there being still a glimmering lamp burning, and, missing my handkerchief, thought I saw it on a box which formed one side of my bed, and put out my hand to take it. I quickly drew back on feeling something cool and very smooth, which moved as I touched it. "Bring the light, quick," I cried; "here's a snake." And there he was, sure enough, nicely coiled up, with his head just raised to inquire who had disturbed him.[2] [1869]

. . .

About this time we had a striking proof of the dangers of New Guinea trading. Six men arrived at the village [Goram, modern

Gorong] in a small boat almost starved, having escaped out of two praus [local South-east Asian boat, usually used for fishing], the remainder of whose crews (fourteen in number) had been murdered by the natives of New Guinea. The praus had left this village a few months before, and among the murdered men were the Rajah's son, and the relations or slaves of many of the inhabitants. The cry of lamentation that arose when the news arrived was most distressing. A score of women, who had lost husbands, brothers, sons, or more distant relatives, set up at once the most dismal shrieks and groans and wailings, which continued at intervals till late at night; and as the chief houses in the village were crowded together round that which I occupied, our situation was anything but agreeable.

It seems that the village where the attack took place (nearly opposite the small island of Lakahia) is known to be dangerous, and the vessels had only gone there a few days before to buy some tripang [sea cucumber]. The crew were living on shore, the praus being in a small river close by, and they were attacked and murdered in the day-time while bargaining with the Papuans. The six men who survived were on board the praus, and escaped by at once setting into the small boat and rowing out to sea.[3]

. . .

In the evening [while on the Aru Islands], after dark, we were suddenly alarmed by the cry of "Bajak! bajak!" (Pirates!) The men all seized their bows and spears, and rushed down to the beach; we got hold of our guns and prepared for action, but in a few minutes all came back laughing and chattering, for it had proved to be only a small boat and some of their own comrades returned from fishing. When all was quiet again, one of the men, who could speak a little Malay, came to me and begged me not to sleep too hard. "Why?" said I. "Perhaps the pirates may really come," said he very seriously, which made me laugh and assure him I should sleep as hard as I could.[4] [1869]   ❧

Wallace's major problems, however, stemmed not from hostile inhabitants but from the ever-present threat of disease in remote country. Not only was medical knowledge at the time rudimentary (for example, it was not until 1897 that Ronald Ross demon- strated that malaria is transmitted by mosquitoes), but Wallace's capacity for treating himself was additionally severely circum- scribed by what he could carry with him and by his own know- ledge of medical matters. Bouts of illness simply had to be waited

Snakes were among the many hazards Wallace faced in the course of his travels. This one, encountered in Ambon, "was about twelve feet long and very thick, capable of doing much mischief and of swallowing a dog or a child" (*The Malay Archipelago*, p. 228)

out. From the following passage, in which Wallace's habitually stiff upper lip yields the faintest of quivers, it is clear that the psychological aspect – the loneliness – of these episodes was in many ways the most difficult part to bear.

✑ I found myself quite knocked up, with headache, pains in the back and limbs. I had commenced operations that morning by taking some purgative medicine, and the next day I began taking doses of quinine, drinking plentifully cream-of-tartar water, though I was so weak and apathetic that at times I could hardly muster resolution to move myself to prepare them. . . .

I . . . made an inward vow never to travel again in such wild, unpeopled districts without some civilised companion or attendant.[5] [1853]

. . .

On leaving São Gabriel I was again attacked with fever, and on arriving at São Joaquim I was completely laid up. My Indians took the opportunity to steal a quantity of the caxaca [local alcohol] I had brought for preserving the fishes, and anything else they could lay their hands on; so I was glad, on the occasion of a slight remission of the fever, to pay their wages and send them off. After a few days, the violence of the fever abated, and I thought I was going to get over it very easily; but such was not the case, for every alternate day I experienced a great depression, with disinclination to motion: this always followed a feverish night, in which I could not sleep. The next night I invariably slept well perspiring profusely, and, the succeeding day, was able to move about, and had a little appetite. The weakness and fever, however, increased, till I was again confined to my *rédé* [hammock], – could eat nothing, and was so torpid and helpless, that Senhor L., who attended me, did not expect me to live. I could not speak intelligibly, and had not strength to write, or even to turn over in my hammock. A few days after this, I was attacked with severe ague, which recurred every two days. I took quinine for some time without any apparent effect, till, after nearly a fortnight, the fits ceased, and I only suffered from extreme emaciation and weakness. In a few days, however, the fits of ague returned, and now came every day. Their visits, thus frequent, were by no means agreeable; as, what with the succeeding fever and perspiration, which lasted from before noon till night, I had little quiet repose. In this state I remained till the beginning of February, the ague continuing, but with diminished force; and though with an increasing appetite, and eating heartily, yet gaining so little strength, that I could with difficulty stand alone, or walk across the room with the assistance of two sticks.[6] [1853] ৵

Wallace had to endure similar problems in South-east Asia. In view, however, of his belief in the power of providence – or, in his case, "the continuous co-ordinated agency of myriads of . . . intelligences"[7] – in determining one's fate, it is worth reminding ourselves that his major claim to fame, his discovery of natural selection, "flashed upon" him while he "was suffering from a sharp attack of intermittent fever [probably malaria]".[8] Perhaps, after all, disease has its advantages. But surely even Wallace could not have put a panglossian spin on the legions of insects and other pests he had to deal with.

৵ Ever since leaving Dobbo [in the Aru Islands, off the south-east coast of New Guinea] I had suffered terribly from insects, who seemed here bent upon revenging my long-continued persecution of their race. At our first stopping-place sand-flies were very abundant at night, penetrating to every part of the body, and producing a more lasting irritation than mosquitoes. My feet and ankles especially suffered, and were completely covered with little red swollen specks, which tormented me horribly. On arriving here [inland] we were delighted to find the house free from sand-flies or mosquitoes, but in the plantations where my daily walks led me, the day-biting mosquitoes swarmed, and seemed especially to delight in attacking my poor feet. After a month's incessant punishment, those useful members rebelled against such treatment and broke into open insurrection, throwing out numerous inflamed ulcers, which were very painful, and stopped me from walking. So I found myself confined to the house, and with no immediate prospect of leaving it. Wounds or sores in the feet are especially difficult to heal in hot climates, and I therefore dreaded them more than any other illness. The confinement was very annoying, as the fine hot weather was excellent for insects, of which I had every promise of obtaining a fine collection; and it is only by daily and unremitting search that the smaller kinds, and the rarer and more interesting specimens, can be obtained. When I crawled down to the river-side to bathe, I often saw the blue-winged Papilio ulysses, or some other equally rare and beautiful insect; but there was nothing for it but patience, and to return quietly to my bird-skinning, or whatever other work I had indoors. The stings and bites and ceaseless irritation caused by these pests of the tropical forests would be borne uncomplainingly; but to be kept prisoner by them in so rich and unexplored a country where rare and beautiful creatures are to be met with in every forest ramble – a country reached by such a long and tedious voyage, and which might not in the present century be again visited for the same purpose – is a punishment too severe for a naturalist to pass over in silence.[9] [1869] ತ

Insects thus represented both extremes of Wallace's experience in the field: butterflies were a constant source of delight, and biting flies a constant source, literally and figuratively, of irritation. In the Amazon, he suffered terribly from the attentions of black flies.

ತ   As it was, the torments I suffered when skinning a bird or drawing a fish, can scarcely be imagined by the unexperienced. My feet were so thickly covered with the little blood-spots produced by their bites, as to be of a dark purplish-red colour, and much swelled and inflamed. My hands suffered similarly, but in a less degree, being more constantly in motion.[10] [1853] ತ

He was forced to adopt some of the more noxious of local solutions to the insect problem.

۶ We were warned that the mosquitoes were here very annoying, and we soon found them so, for immediately after sunset they poured in upon us in swarms, so that we found them unbearable, and were obliged to rush into our sleeping rooms, which we had kept carefully closed. Here we had some respite for a time, but they soon found their way in at the cracks and key-holes, and made us very restless and uncomfortable all the rest of the night.

After a few days' residence we found them more tormenting than ever, rendering it quite impossible for us to sit down to read or write after sunset. The people here all use cow-dung burnt at their doors to keep away the "praga," or plague, as they very truly call them, it being the only thing that has any effect. Having now got an Indian to cook for us, we every afternoon sent him to gather a basket of this necessary article, and just before sunset we lighted an old earthen pan full of it at our bedroom door, in the verandah, so as to get as much smoke as possible, by means of which we could, by walking about, pass an hour pretty comfortably. In the evening every house and cottage has its pan of burning dung, which gives rather an agreeable odour; and as there are plenty of cows and cattle about, this necessary of life is always to be procured.[11] [1853] ۶

Pests represented a challenge to more than Wallace's health. One of the difficult aspects of biological collecting is ensuring the preservation of specimens. In this era of refrigeration and overnight delivery, the problems encountered by Wallace are almost unimaginable.

۶ The lean and hungry dogs [of the Aru Islands] ... were my greatest enemies, and kept me constantly on the watch. If my boys left the bird they were skinning for an instant, it was sure to be carried off. Everything eatable had to be hung up to the roof, to be out of their reach. Ali had just finished skinning a fine King Bird of Paradise one day, when he dropped the skin. Before he could stoop to pick it up, one of this famished race had seized upon it, and he only succeeded in rescuing it from its fangs after it was torn to tatters. Two skins of the large Paradisea, which were quite dry and ready to pack away, were incautiously left on my table for the night, wrapped up in paper. The next morning they were gone, and only a few scattered feathers indicated their fate. My hanging shelf was out of their reach; but having stupidly left a box which served as a step, a full-plumaged Paradise bird was next morning missing; and a dog below the house was to be seen still mumbling over the fragments, with the fine golden plumes all trampled in the mud. Every night, as soon as I was in bed, I could hear them searching about for what they could devour, under my table, and all about my boxes and baskets, keeping me in a state of suspense till morning, lest something of value might incautiously have been left within their reach. They would drink the oil of my floating lamp and eat the wick, and upset or break my crockery if my lazy boys had neglected to wash away even the smell of anything eatable. Bad, however, as they are here, they were worse in a Dyak's house in Borneo where I was once staying, for there they gnawed off the tops of my waterproof boots, ate a large piece out of an old leather game-bag, besides devouring a portion of my mosquito curtain![12] [1869]

. . .

Dorey [north coast of New Guinea] was very rich in ants. One small black kind was excessively abundant. Almost every shrub and tree was more or less infested with it, and its large papery nests were everywhere to be seen. They immediately took possession of my house, building a large nest in the roof, and forming papery tunnels down almost every post. They swarmed on my table as I was at work setting out my insects, carrying them off from under my very nose, and even tearing them from the cards on which they were gummed if I left them for an instant. They crawled continually over my hands and face, got into my hair, and roamed at will over my whole body, not producing much inconvenience till they began to bite, which they would do on meeting with any obstruction to their passage, and with a sharpness which made me jump again and rush to undress and turn out the offender. They visited my bed also, so that night brought no relief from their persecutions; and I verily believe that during my three and a half months' residence at Dorey I was never for a single hour entirely free from them. They were not nearly so voracious as many other kinds, but their numbers and ubiquity rendered it necessary to be constantly on guard against them.

The flies that troubled me most were a large kind of blue-bottle or blow-fly. These settled in swarms on my bird skins when first put out to dry, filling their plumage with masses of eggs, which, if neglected, the next day produced maggots. They would get under the wings or under the body where it rested on the drying-board, sometimes actually raising it up half an inch by the mass of eggs deposited in a few hours; and every egg was so firmly glued to the fibres of the feathers, as to make it a work of much time and patience to get them off without injuring the bird. In no other locality have I ever been troubled with such a plague as this.[13] [1853] ❧

Much of what Wallace was doing, however, was not glamorous, dangerous, or exotic, but just humdrum. He recognized that it was the quotidian hassles of life on the road that were most trying.

❧ . . . that miscellaneous lot of household furniture – bed, blankets, pots, kettles, and frying-pan, plates, dishes, and wash-basin, coffeepots and coffee, tea, sugar, and butter, salt, pickles, rice, bread and wine, pepper and curry powder, and half a hundred more odds and ends, the constant looking after which, packing and repacking, calculating and contriving, have been the standing plague of my life for the last seven years.[14] [1861] ❧

It is hardly surprising that he was occasionally homesick, as expressed in this letter from Timor.

❧ I assure you I now feel at times very great longings for the peace and quiet of home – very much weariness of this troublesome, wearisome, wandering life. I have lost some of that elasticity and freshness which made the overcoming of difficulties a pleasure, and the country and people are now too familiar to me to retain any of the charms of novelty which gild over so much that is really monotonous and disagreeable. My health, too, gives way, and I cannot now put up so well with fatigue and privations as at first. All these causes will induce me to come home as soon as possible, and I think I may promise, if no accident happens, to come back to dear and beautiful England in the summer of next year.[15] [1861] ❧

At the end of his traumatic return voyage from the Amazon, he was definitely happy to be home: "October 1. Oh, glorious day! Here we are on shore at Deal, where the ship is at anchor. Such a dinner, with our two captains[16]! Oh, beef-steaks and damson tart, a paradise for hungry sinners."[17] [1852]

Despite all the hardship, Wallace's mix of determination and scientific curiosity sustained him throughout his travels. In a letter home from the Amazon detailing his decision to venture up the largely unexplored Rio Uapés, a tributary of the Rio Negro, he supplied the perfect self-description.

My canoe is now getting ready for a further journey up to near the sources of the Rio Negro in Venezuela, where I have reason to believe I shall find insects more plentiful, and at least as many birds as here. On my return from there I shall take a voyage up the great river Uapés, and another up the Isanna, not so much for my collections, which I do not expect to be very profitable there, but because I am so much interested in the country and the people that I am determined to see and know more of it and them than any other European traveller. If I do not get profit, I hope at least to get some credit as an industrious and persevering traveller.[18] [1850]

"Tedious and unfortunate": Hazardous Voyages

Wallace endured a number of horrendous sea and river journeys with the stoicism expected of Victorians. These must have been made all the worse by his being only a weak swimmer: "This was partly due to a physical deficiency which I was unable to overcome. My legs are unusually long for my height, and the bones are unusually large. The result is that they persistently sink in the water, bringing me into a nearly vertical position, and their weight renders it almost impossible to keep my mouth above water."[1] Some journeys were merely awkward, such as this one in the Amazon.

&⅜ After four days' delay, we at length started, with a comparatively small complement of Indians, but with some extra men to assist us in passing several caxoeiras [rapids], which occur near at hand. These are the "Piréwa" (Wound), "Uacorouá" (Goat-sucker), "Maniwár" (White Ant), "Matapí" (Fishtrap), "Amána" (Rain), "Tapíracúnga" (Tapir's head), "Tapíra eura" (Tapir's mouth), and "Jacaré" (Alligator). Three of these were very bad, the canoe having to be unloaded entirely, and pulled over the dry and uneven rocks. The last was the highest; the river rushing furiously about twenty feet down a rugged slope of rock. The loading and unloading of the canoe three or four times in the course of as many hours, is a great annoyance. Baskets of farinha and salt, of mandiocca cakes and pacovas, are strewn about. Panellas [baskets] are often broken; and when there comes a shower of rain, everything has to be heaped together in a hurry, – palm-leaves cut, and the more perishable articles covered; but boxes, rédés, and numerous other articles are sure to be wetted, rendering us very uncomfortable when again hastily tumbled into the overcrowded canoe. If I had birds or insects out drying, they were sure to be overturned, or blown by the wind, or wetted by the rain, and the same fate was shared by my notebooks and papers. Articles in boxes, unless packed tight, were shaken and rumpled by not being carried evenly; so that it was an excellent lesson in patience, to bear all with philosophical serenity.[2] [1853]   &ᴗ

Other Amazon journeys were more alarming.

✍ The next day we passed Obydos, the strong current of the river, now at its height, carrying us down with great rapidity; and the succeeding night we had a tremendous storm, which blew and tossed our little vessel about in a very alarming manner. The owner of the canoe, an Indian, was much frightened; he called upon the Virgin, and promised her several pounds of candles, if she would but save the canoe; and, opening the door of the little cabin where I was sleeping, cried out in a most piteous voice, "Oh! meu amigo, estamos perdidos" (Oh! my friend, we are all lost). In vain I tried to comfort him with assurances that, as the vessel was new and strong, and not too heavily laden, there was no danger, – although the night was pitch dark, and the wind blew in the most fierce and furious gusts imaginable. We did not know whether we were in the middle of the river or near the side, and the only danger we were exposed to, was of our drifting ashore or running aground. After about an hour, however, the canoe came to a stop, without any shock whatever, and remained perfectly still, although the wind still blew. It was so dark that nothing was to be seen, and it was only by stretching his arm down over the side, that the master ascertained that we had drifted into one of the large compact beds of floating grass which, in many places, line the banks of the Amazon for hundreds of yards from the shore. Here, therefore, we were safely moored, and waited for the morning, sleeping comfortably, with the knowledge that we were out of all danger.[3] [1853] ✍

The single worst journey that Wallace experienced was his return trip from Brazil in 1852. After the *Helen* had burnt up in the middle the Atlantic, Wallace and its crew spent ten days in the leaking ship's boats before being rescued by the *Jordeson*, which in turn nearly sank. The most telling statement in Wallace's account of his nightmarish experience refers to his time adrift in an open boat: "During the night I saw several meteors, and in fact could not be in a better position for observing them, than lying on my back in a small boat in the middle of the Atlantic." Reading Wallace, it's almost possible to conclude that he views being shipwrecked as a privileged opportunity for brushing up on his astronomy!

✍ I agreed for my passage in the brig *Helen*, two hundred and thirty-five tons, Captain John Turner, whose property she was; and on the

morning of Monday, the 12th of July, we got aboard, and bade adieu
to the white houses and waving palm-trees of Pará. Our cargo consisted of about a hundred and twenty tons of india-rubber, and a
quantity of cocoa, arnotto,[4] piassaba,[5] and balsam of capivi.[6] About
two days after we left I had a slight attack of fever, and almost thought
that I was still doomed to be cut off by the dread disease which had
sent my brother and so many of my countrymen to graves upon a
foreign shore. A little calomel and opening medicines, however, soon
set me right again; but as I was very weak, and suffered much from
sea-sickness, I spent most of my time in the cabin. For three weeks we
had very light winds and fine weather, and on the 6th of August had
reached about latitude 30° 30' north, longitude 52° west.

On that morning, after breakfast, I was reading in the cabin, when
the Captain came down and said to me, "I'm afraid the ship's on fire;
come and see what you think of it," and proceeded to examine the
lazaretto, or small hole under the floor where the provisions are kept,
but no signs of fire were visible there. We then went on deck to the
forepart of the ship, where we found a dense vapoury smoke issuing
from the forecastle. The fore hatchway was immediately opened, and,
the smoke issuing there also, the men were set to work clearing out
part of the cargo. After throwing out some quantity without any
symptom of approaching the seat of the fire, we opened the after
hatchway; and here the smoke was much more dense, and in a very
short time became so suffocating, that the men could not stay in the
hold to throw out more cargo, so they were set to work pouring in
water, while others proceeded to the cabin, and now found abundance
of smoke issuing from the lazaretto, whence it entered through the
joints of the bulkhead which separated it from the hold. Attempts
were now made to break this bulkhead down; but the planks were so
thick and the smoke so unbearable that it could not be effected, as
no man could remain in the lazaretto to make more than a couple of
blows. The cabin table was therefore removed, and a hole attempted
to be cut in the cabin floor, so as to be able to pour water immediately
on the seat of the fire, which appeared to be where the balsam was
stowed. This took some time, owing to the suffocating smoke, which
also continued to pour in dense volumes out of the hatchway. Seeing
that there was now little chance of our being able to extinguish the
fire, the Captain thought it prudent to secure our own safety, and
called all hands to get out the boats, and such necessaries as we
should want, in case of being obliged to take to them. The long-boat
was stowed on deck, and of course required some time to get it afloat.

The gig was hung on davits on the quarter, and was easily let down. All now were in great activity. Many little necessaries had to be hunted up from their hiding-places. The cook was sent for corks to plug the holes in the bottoms of the boats. Now no one knew where a rudder had been put away; now the thowl-pins were missing. The oars had to be searched for, and spars to serve as masts, with proportionate sails, spare canvas, twine, cordage, tow-ropes, sail-needles, nails and tacks, carpenters' tools, etc. The Captain was looking after his chronometer, sextant, barometer, charts, compasses, and books of navigation; the seamen were getting their clothes into huge canvas bags; all were lugging about pilot-coats, blankets, south-westers, and oilskin coats and trousers; and I went down into the cabin, now suffocatingly hot and full of smoke, to see what was worth saving. I got my watch and a small tin box containing some shirts and a couple of old note-books, with some drawings of plants and animals, and scrambled up with them on deck. Many clothes and a large portfolio of drawings and sketches remained in my berth; but I did not care to venture down again, and in fact felt a kind of apathy about saving anything, that I can now hardly account for. On deck the crew were still busy at the boats; two barrels of bread were got in, a lot of raw pork, some ham and cases of preserved meats, some wine and a large cask of water. The cask had to be lowered into the boat empty, for fear of any accident, and after being securely fixed in its place, filled with buckets from those on board.

The boats, having been so long drying in a tropical sun, were very leaky, and were now half full of water, and books, coats, blankets, shoes, pork, and cheese, in a confused mass, were soaking in them. It was necessary to put two men in each, to bale; and everything necessary being now ready, the rest of the crew were called off again to pour water into the hatchways and cabin, from which rose volumes of thick yellow smoke. Now, too, we could hear in the hold the balsam bubbling, like some great boiling caldron, which told of such intense heat, that we knew the flames must soon break out. And so it was, for in less than half an hour the fire burst through the cabin-floor into the berths, and consuming rapidly the dry pine-wood, soon flamed up through the skylight. There was now a scorching heat on the quarter-deck, and we saw that all hope was over, and that we must in a few minutes be driven by the terrible element to take refuge on the scarcely less dangerous one, which heaved and swelled its mighty billows a thousand miles on every side of us. The Captain at length ordered all into the boats, and was himself the last to leave the vessel.

I had to get down over the stern by a rope into the boat, rising and falling and swaying about with the swell of the ocean; and, being rather weak, rubbed the skin considerably off my fingers, and tumbled in among the miscellaneous articles already soaking there in the greatest confusion. One sailor was baling with a bucket, and another with a mug; but the water not seeming at all to diminish, but rather the contrary, I set to work helping them, and soon found the salt-water producing a most intense smarting and burning on my scarified fingers.

We now lay astern of the ship, to which we were moored, watching the progress of the fire. The flames very soon caught the shrouds and sails, making a most magnificent conflagration up to the very peak, for the royals were set at the time. Soon after, the fore rigging and sails also burnt, and flames were seen issuing from the fore hatchway, showing how rapidly the fire was spreading through the combustible cargo. The vessel, having now no sails to steady her, rolled heavily, and the masts, no longer supported by the shrouds, bent and creaked, threatening to go overboard every minute. The main-mast went first, breaking off about twenty feet above the deck; but the foremast stood for a long time, exciting our admiration and wonder, at the time it resisted the heavy rolls and lurches of the vessel; at last, being partly burned at the bottom, it went over, more than an hour after its companion. The decks were now a mass of fire, and the bulwarks partly burnt away. Many of the parrots, monkeys, and other animals we had on board, were already burnt or suffocated; but several had retreated to the bowsprit out of reach of the flames, appearing to wonder what was going on, and quite unconscious of the fate that awaited them. We tried to get some of them into the boats, by going as near as we could venture; but they did not seem at all aware of the danger they were in, and would not make any attempt to reach us. As the flames caught the base of the bowsprit, some of them ran back and jumped into the midst of the fire. Only one parrot escaped: he was sitting on a rope hanging from the bowsprit, and this burning above him let him fall into the water, where, after floating a little way, we picked him up.

Night was now coming on. The whole deck was a mass of fire, giving out an intense heat. We determined to stay by the vessel all night, as the light would attract any ship passing within a considerable distance of us. We had eaten nothing since the morning, and had had plenty to do and to think of, to prevent our being hungry; but now, as the evening air began to get cool and pleasant, we all found we had very good appetites, and supped well on biscuits and water.

We then had to make our arrangements for the night. Our mooring ropes had been burnt, and we were thus cast adrift from the ship, and were afraid of getting out of sight of it during the night, and so missing any vessel which might chance to be attracted by its light. A portion of the masts and rigging were floating near the ship, and to this we fastened our boats; but so many half-burnt spars and planks were floating about us, as to render our situation very perilous, for there was a heavy swell, and our boats might have been in an instant stove in by coming in contact with them.

We therefore cast loose again, and kept at a distance of a quarter or half a mile from the ship by rowing when requisite. We were incessantly baling the whole night. Ourselves and everything in the boats were thoroughly drenched, so we got little repose: if for an instant we dozed off into forgetfulness, we soon woke up again to the realities of our position, and to see the red glare which our burning vessel cast over us. It was now a magnificent spectacle, for the decks had completely burnt away, and as it heaved and rolled with the swell of the sea, presented its interior towards us filled with liquid flame, a fiery furnace tossing restlessly upon the ocean.

At length morning came; the dangers of the night were past, and with hopeful hearts we set up our little masts, and rigged our sails, and, bidding adieu to the still burning wreck of our ship, went gaily bounding along before a light east wind. And then pencils and books were hunted out, and our course and distance to Bermuda calculated; and we found that this, the nearest point of land in the vast waste of waters round us, was at least seven hundred miles away. But still we went on full of hope, for the wind was fair, and we reckoned that, if it did not change, we might make a hundred miles a day, and so in seven days reach the longed-for haven.

As we had supped but scantily the night before, we had now good appetites, and got out our ham and pork, biscuit and wine and water, and made a very hearty meal, finding that even uncooked meat was not to be despised where no fire could be got to cook it with.

The day was fine and warm, and the floating seaweed, called gulf-weed, was pretty abundant. The boats still required almost incessant baling, and though we did not ship many seas, yet there was quite enough spray to keep us constantly wet. At night we got a rope fastened to the long-boat, for her to tow us, in order that we might not get separated; but as we sailed pretty equally, we kept both sails up. We passed a tolerable night under the circumstances. The next day, the 8th, was fine, gulf-weed still floated plentifully by us, and there were numerous flying-fish, some of which fell into our boats, and others flew an immense distance over the waves. I now found my hands and face very much blistered by the sun, and exceedingly sore and painful. At night two boobies, large dusky sea-birds with very long wings, flew about us. During the night I saw several meteors, and in fact could not be in a better position for observing them, than lying on my back in a small boat in the middle of the Atlantic. We also saw a flock of small birds fly by making a chirping noise; the sailors did not know what they were.

The 9th was again fine and hot, and my blistered hands were very painful. No ship appeared in sight, though we were crossing the track of the West India vessels. It was rather squally, and I passed a nervous, uncomfortable night; our boats did not, however, now leak so much, which was a great satisfaction.

The 10th was squally, and the wind veered to the southwest, so that we could not make our course for Bermuda, but were obliged to go to the north of it. The sea ran very high, and sudden gusts of wind would frequently heel us over in a manner very alarming to me. We had some heavy showers of rain, and should have liked to have caught some fresh water, but could not, as all our clothes and the sails were saturated with salt. Our position at noon was in latitude 31° 59' north, longitude 57° 22' west.

The 11th was still rough and squally. There was less gulf-weed now. The wind got still more to the westward, so that we were obliged to go nearly north. Our boats had now got swollen with the water, and leaked very little. This night I saw some more falling stars.

On the 12th the wind still kept foul, and we were getting quite out of the track of ships, and appeared to have but little chance of reaching Bermuda. The long-boat passed over some green water to-day, a sign of there being soundings, probably some rock at a moderate depth. Many dolphins swam about the boats; their colours when seen in the water are superb, the most gorgeous metallic hues of green, blue, and gold: I was never tired of admiring them.

On the 13th the wind was due west, blowing exactly from the point we wanted to go to. The day was very fine, and there were several stormy petrels, or Mother Cary's chickens, flying about us. We had now been a week in the boats, and were only halfway to the Islands, so we put all hands on short allowance of water before it was too late.

The sun was very hot and oppressive, and we suffered much from thirst.

The 14th was calm and we could not get on at all. The sun was scorching and we had no shelter, and were parched with thirst the whole day. Numerous dolphins and pilot-fish were about the boats. At night there was a very slight favourable breeze, and as we had by this time got our clothes pretty dry we slept well.

On the 15th the wind again died away, and we had another calm. The sea was full of minute *Medusae* called "blubber" by the sailors: some were mere whitish oval or spherical lumps, others were brown, and beautifully constructed like a little cap, swimming rapidly along by alternate contractions and expansions, and so expelling the water behind them. The day was very hot, and we suffered exceedingly from thirst. We were almost in despair about seeing a ship, or getting on to the Islands. At about 5pm, while taking our dinner, we saw the long-boat, which was at some distance from us, tack. "She must see a sail," said the captain, and looking round we saw a vessel coming nearly towards us, and only about five miles distant. We were saved!

The men joyfully drank the rest of their allowance of water, seized their oars, and pulled with hearty goodwill, and by seven o'clock we were alongside. The captain received us kindly on board. The men went first to the water-casks, and took long and hearty draughts, in which we joined them, and then enjoyed the almost forgotten luxury of tea. From having been so long cramped in the boats, I could hardly stand when I got on board.

That night I could not sleep. Home and all its pleasures seemed now within my grasp; and crowding thoughts, and hopes and fears, made me pass a more restless night than I should have done, had we still been in the boats, with diminished hopes of rescue. The ship was the *Jordeson*, Captain Venables, from Cuba, bound for London, with a cargo of mahogany, fustic, and other woods. We were picked up in latitude 32° 48′ north, longitude 60° 27′ west, being still about two hundred miles from Bermuda.

For several days afterwards we had fine weather and very light winds, and went creeping along about fifty miles a day. It was now, when the danger appeared past, that I began to feel fully the greatness of my loss. With what pleasure had I looked upon every rare and curious insect I had added to my collection! How many times, when almost overcome by the ague, had I crawled into the forest and been rewarded by some unknown and beautiful species! How many places, which no European foot but my own had trodden, would have been recalled to my memory by the rare birds and insects they had furnished to my collection! How many weary days and weeks had I passed, upheld only by the fond hope of bringing home many new and beautiful forms from those wild regions; every one of which would be endeared to me by the recollections they would call up, – which should prove that I had not wasted the advantages I had enjoyed, and would give me occupation and amusement for many years to come! And now everything was gone, and I had not one specimen to illustrate the unknown lands I had trod, or to call back the recollection of the wild scenes I had beheld! But such regrets I knew were vain, and I tried to think as little as possible about what might have been, and to occupy myself with the state of things which actually existed.

On the 22nd of August we saw three water-spouts, the first time I had beheld that curious phenomenon. I had much wished once to witness a storm at sea, and I was soon gratified.

Early in September we had a very heavy gale. The barometer had fallen nearly half an inch during the night; and in the morning it was blowing strong, and we had a good deal of canvas up when the captain began to shorten sail; but before it could be taken in, four or five sails were blown to pieces, and it took several hours to get the others properly stowed. By the afternoon we were driving along under double-reefed topsails. The sea was all in a foam, and dashed continually over us. By night a very heavy sea was up, and we rolled about fearfully, the water pouring completely over the bulwarks, deluging the decks, and making the old ship stagger like a drunken man. We passed an uncomfortable night, for a great sea broke into the cabin skylight and wetted us all, and the ship creaked and shook, and plunged so madly, that I feared something would give way, and we should go to the bottom after all; all night, too, the pumps were kept going, for she leaked tremendously, and it was noon the next day before she was got free of water. The wind had now abated, and we soon had fine weather again, and all hands were busy bending new sails and repairing the old ones.

We caught at different times several dolphins, which were not bad eating. I did not see so much to admire in the colours of the dying dolphin; they are not to be compared with the colours of the living fish [sic.] seen in the blue transparent water.

We were now getting rather short of provisions, owing to the increased number of mouths: our cheese and ham were finished, – then our peas gave out, and we had no more pea-soup, – next the butter came to an end, and we had to eat our biscuit dry, – our bread and pork, too, got very short, and we had to be put upon allowance. We then got some supplies from another ship; but our voyage was so much prolonged and we had adverse winds and another heavy gale, so that we were again in want, finished our last piece of meat, and had to make some scanty dinners off biscuit and water. Again we were relieved with a little supply of pork and some molasses, and so managed pretty well.

We were in the Channel on the night of the 29th of September, when a violent gale occurred, that did great damage to the shipping, and caused the destruction of many vessels much more seaworthy than our own. The next morning we had four feet of water in the hold.

On the 1st of October the pilot came on board, and Captain Turner and myself landed at Deal, after an eighty days' voyage from Pará; thankful for having escaped so many dangers, and glad to tread once more on English ground.[7] [1853] ಎ

On hearing several years later that Bates had returned safely to England from Brazil, Wallace wrote to him from Ceram (Seram), Indonesia: "Allow me to congratulate you on your safe arrival home with all your treasures; a good fortune which I trust this time is reserved for me."[8] This time Wallace's voyage home was indeed incident-free, which is more than can be said for his travel within South-east Asia. Wallace dedicates a considerable chunk of *The Malay Archipelago* to a "most tedious and unfortunate"[9] journey, which he summarizes as follows:

ಎ Looking at my whole voyage in this vessel from the time when I left Goram in May, it will appear that my experiences of travel in a native prau have not been encouraging. My first crew ran away; two men were lost for a month on a desert island; we were ten times aground on coral reefs; we lost four anchors; the sails were devoured by rats; the small boat was lost astern; we were thirty-eight days on the voyage home, which should not have taken twelve; we were many times short of food and water; we had no compass-lamp, owing to there not being a drop of oil in Waigiou [Waigeo] when we left; and to crown all, during the whole of our voyages from Goram [Gorong] by Ceram to Waigiou, and from Waigiou to Ternate, occupying in all seventy-eight days, or only twelve days short of three months (all in what was supposed to be the favourable season), we had not one single day of fair wind. We were always close braced up, always struggling against wind, tide, and leeway, and in a vessel that would scarcely sail nearer than eight points from the wind. Every seaman will admit that my first voyage in my own boat was a most unlucky one.[10] [1869] ෙ

> From this litany of disasters, the tale of the loss and eventual recovery of two of Wallace's men stands out as particularly harrowing.

ෙ   In the morning, finding it would be necessary for us to get round a rocky point [of a small uninhabited island], I wanted my men to go on shore and cut jungle-rope, by which to secure us from being again drafted away, as the wind was directly off shore. I unfortunately, however, allowed myself to be overruled by the pilot and crew, who all declared that it was the easiest thing possible, and that they would row the boat round the point in a few minutes. They accordingly got up the anchor, set the jib, and began rowing; but, just as I had feared, we drifted rapidly off shore, and had to drop anchor again in deeper water, and much farther off. The two best men, a Papuan and a Malay, now swam on shore, each carrying a hatchet, and went into the jungle to seek creepers for rope. After about an hour our anchor loosed hold, and began to drag. This alarmed me greatly, and we let go our spare anchor, and, by running out all our cable, appeared tolerably secure again. We were now most anxious for the return of the men, and were going to fire our muskets to recall them, when we observed them on the beach, some way off, and almost immediately our anchors again slipped, and we drifted slowly away into deep water. We instantly seized the oars, but found we could not counteract the wind and current, and our frantic cries to the men were not heard till we had got a long way off; as they seemed to be hunting for shell-fish on the beach. Very soon, however, they stared at us, and in a few minutes seemed to comprehend their situation; for they rushed down into the water, as if to swim off, but again returned on shore, as if afraid to make the attempt. We had drawn up our anchors at first not to check our rowing; but now, finding we could do nothing, we let them both hang down by the full length of the cables. This stopped our way very much, and we drifted from shore very slowly, and hoped the men would hastily form a raft, or cut down a soft-wood tree, and paddle out, to us, as we were still not more than a third of a mile from shore. They seemed, however, to have half lost their senses, gesticulating

Wallace's house at Bessir,
Waigeo, off New Guinea (from
The Malay Archipelago)

wildly to us, running along the beach, then going into the forest; and just when we thought they had prepared some mode of making an attempt to reach us, we saw the smoke of a fire they had made to cook their shell-fish! They had evidently given up all idea of coming after us, and we were obliged to look to our own position.[11]

. . .

Immediately on our arrival at Muka [village on Waigeo], I engaged a small boat and three natives to go in search of my lost men, and sent one of my own men with them to make sure of their going to the right island. In ten days they returned, but to my great regret and disappointment, without the men. The weather had been very bad, and though they had reached an island within sight of that in which the men were, they could get no further. They had waited there six days for better weather, and then, having no more provisions, and the man I had sent with them being very ill and not expected to live, they returned. As they now knew the island, I was determined they should make another trial, and (by a liberal payment of knives, handker-chiefs, and tobacco, with plenty of provisions) persuaded them to start back immediately, and make another attempt. They did not return again till the 29th of July, having stayed a few days at their own village of Bessir [on Waigeo] on the way; but this time they had succeeded and brought with them my two lost men, in tolerable health, though thin and weak. They had lived exactly a month on the island; had found water, and had subsisted on the roots and tender flower-stalks of a species of Bromelia,[12] on shell-fish, and on a few turtles' eggs. Having swum to the island, they had only a pair of trousers and a shirt between them, but had made a hut of palm-leaves, and had altogether got on very well. They saw that I waited for them three days at the opposite island, but had been afraid to cross, lest the current should have carried them out to sea, when they would have been inevitably lost. They had felt sure I would send for them on the first opportunity, and appeared more grateful than natives usually are for my having done so; while I felt much relieved that my voyage, though sufficiently unfortunate, had not involved loss of life.[13] [1869] ❧

"A want of harmony between man and nature": American Travels

Wallace's brief account of his visit to North America – given in his autobiography – is largely a list of people met and lectures delivered, and by no means ranks with his earlier travel writing. From the natural history standpoint, the trip disappointed him.

❧ In my journal I find this note: "During more than ten months in America, taking every opportunity of exploring woods and forests, plains and mountains, deserts and gardens, between the Atlantic and Pacific coasts, and extending over ten degrees of latitude, I never once saw either a humming-bird or a rattlesnake, or even any living snake of any kind. In many places I was told that humming-birds were usually common in their gardens, but they hadn't seen any this year! This was my luck. And as to the rattlesnakes, I was always on the look out in likely places, and there are plenty still, but they are local. I was told of a considerable tract of land not far from Niagara which is so infested with them that it is absolutely useless. The reason is that it is very rocky, with so many large masses lying about overgrown with shrubs and briars as to afford them unlimited hiding-places, and the labour of thoroughly clearing it would be more costly than the land would be worth."[1] [1905] ❧

Indeed, Wallace excoriated the landscape.

❧ ... through the greater part of North America, there results a monotonous and unnatural ruggedness, a want of harmony between man and nature, the absence of all those softening effects of human labour and human occupation carried on for generation after generation in the same simple way, and in its slow and gradual utilization of natural forces allowing the renovating agency of vegetable and animal life to conceal all harshness of colour or form, and clothe the whole landscape in a garment of perennial beauty.

Over the larger part of America everything is raw and bare and ugly, with the same kind of ugliness with which we also are defacing

Engraving from *The Geographic Distribution of Animals*, showing "the North American Prairie, with characteristic mammalia"

our land and destroying its rural beauty. The ugliness of new rows of cottages built to let to the poor, the ugliness of the mean streets of our towns, the ugliness of our "black countries" and our polluted streams. Both countries are creating ugliness, both are destroying beauty; but in America it is done on a larger scale and with a more hideous monotony.[2] [1905] ❧

Nevertheless, he was impressed by the Niagara Falls.

❧ On my way back to Washington [from Toronto] I spent four days at Niagara, living at the old hotel on the Canadian side, in a room that looked out on the great fall, and where its continuous musical roar soothed me to sleep. It was a hard frost, and the American falls had great ice-mounds below them, and ranges of gigantic icicles near the margins. At night the sound was like that of a strong, steady wind at sea, but even more like the roar of the London streets heard from the middle of Hyde Park. When in bed a constant vibration was felt. I spent my whole time wandering about the falls, above and below, on the Canadian and the American sides, roaming over Goat Island and the Three Sisters Islands far in the rapids above the Horse-shoe Fall, which are almost as impressive as the fall itself. The small Luna Island dividing the American falls was a lovely sight; the arbor-vitae trees

(*Thuya Americana*), with which it is covered, young and old, some torn and jagged, but all to the smallest twigs coated with glistening ice from the frozen spray, looked like groves of gigantic tree corals – the most magnificent and fairy-like scene I have ever beheld. All the islands are rocky and picturesque, the trees draped with wild vines and Virginia creepers, and afford a sample of the original American forest vegetation of very great interest. During these four days I was almost entirely alone, and was glad to be so. I was never tired of the ever-changing aspects of this grand illustration of natural forces engaged in modelling the earth's surface. Usually the centre of the great falls, where the depth and force of the water are greatest, is hidden by the great column of spray which rises to the height of four hundred or five hundred feet; but occasionally the wind drifts it aside, and allows the great central gulf of falling water to be seen nearly from top to bottom – a most impressive sight.[3] [1905] ❧

He found the enthusiasms and habits of the Americans curious.

❧ One morning Mr. Talbot [of the Sioux City Natural History Society] took me to see the pork-curing establishment, where, during the season, they kill a thousand hogs a day. The animals are collected in pens close to the building, with a gate opening to an inclined pathway of planks up to the top of the building. They walk up this of their own accord in a continuous procession, and at the top are caught up one after another by a chain round their hind legs, and swung on to the men who kill, scald, scrape, and cut them up; all the separate parts going through the several stages of cleaning and curing till the result is bacon, hams, barrels of pork, black puddings, sausages, and bristles, while the whole of the refuse is dried and ground up into a valuable manure. The ingenuity of the whole process is undeniable; but to go through it all, as I was obliged to do, along narrow planks and ladders slippery with blood and water, and in the warm, close, reeking atmosphere, was utterly disgusting. My friend was, however, quite amazed at my feeling anything but admiration of the whole establishment, which was considered one of the sights and glories of the city.[4] [1905]

. . .

I continued my journey to Cheyenne, across open plains of thin grass partly irrigated. Near me in the train was a lady chewing gum; I saw her at intervals for an hour, her jaws going regularly all the time, just like those of a cow when ruminating. *Not* a pleasant sight, or

conducive to beauty of expression. It must be tiring to beginners.[5]
[1905]   ɛ❧

Bizarrely this, his last major trip had, like his first, a fiery finale.
Fortunately, this time the results were not so catastrophic. He was
on the very last leg of his journey to his home at Frith Hill,
Surrey.

❧   On my way from Godalming old station to Frith Hill in a fly
[horse-drawn cab], an extraordinary event happened. Suddenly I
perceived that the driver's coat was on fire behind – actually in flames!
I called out to him. He looked round, beat it with his hands, said, "All
right, sir!" and went on. After a few minutes it began smoking again. I
called out louder, it flamed again; both overcoat, trousers, and cush-
ion were burning. Then he got down, took off his overcoat, trampled
on it, and beat out the rest. We went on. A third time it burst out in
smoke and flame. Again I shouted, and passersby called out and
stopped to look. And then at last, with their help, he finally
extinguished the conflagration. A cab-man on fire! No more curious
incident occurred during my six thousand miles of travel in America.
It originated, no doubt, from his having put a lighted pipe in his
pocket, or perhaps from a loose phosphorous match. But he did not
seem to mind it much, even when in a blaze.[6] [1905]   ɛ❧

Social Issues

"Liberty is, in my mind, a far greater and more important thing than science."

Testimony given by Wallace to the Royal Commission on Vaccination, 1890[1]

Wallace's impassioned and polemical writing on social issues ranks with his very best. Whereas many of his scientific colleagues – most notably Darwin – were happy in later life quietly to consolidate their scientific work, Wallace plunged into a second career as a socially engaged public intellectual, his scientific reputation serving as a springboard for wide-ranging forays beyond science. Always a humanitarian, he eventually became a self-declared socialist, and published on an extraordinary range of topics. Although he came increasingly to view all his work – biology, politics, spiritualism – as inter-connected parts of a whole, he was not unduly irked by those who paid homage to him as biologist but condemned his non-biological work as the output of a " 'crank' and a 'faddist' ".[2] In a letter concerning a review of his *My Life*, he wrote, "Then it points out a few things I am capable of believing, but which everybody else knows to be fallacies, and compares me to Sir I. Newton writing on the prophets! Yet of course he praises my biology up to the skies – there I am wise – everywhere else I am a kind of weak, babyish idiot! It really is most delightful!"[3]

Evolution of a Socialist

At around the age of fourteen, Wallace was living with his brother John in London.

 ... our evenings were most frequently spent at what was then termed a "Hall of Science," situated in John Street, Tottenham Court Road (now altered to Whitfield Street). It was really a kind of club or mechanics' institute for advanced thinkers among workmen, and especially for the followers of Robert Owen,[1] the founder of the socialist movement in England. Here we sometimes heard lectures on Owen's doctrines, or on the principles of secularism or agnosticism, as it is now called; at other times we read papers or books, or played draughts, dominoes, or bagatelle, and coffee was also supplied to any who wished for it. It was here that I first made acquaintance with Owen's writings, and especially with the wonderful and beneficent work he had carried on for many years at New Lanark.[2] I also received my first knowledge of the arguments of sceptics, and read among other books Paine's "Age of Reason".[3]

. . .

I have always looked upon Owen as my first teacher in the philosophy of human nature and my first guide through the labyrinth of social science. He influenced my character more than I then knew, and now that I have read his life and most of his great works, I am fully convinced that he was the greatest of social reformers and the real founder of Modern Socialism.[4] [1905]

Wallace's travels influenced more than just his biological thinking. For a start, away from the strictures of Victorian England, he had the freedom to think through his views on a range of religious, social and political issues. In a letter from Timor in 1861 to his brother-in-law, Thomas Sims, he outlined his thoughts on religious belief, and in particular on the extent to which an individual's convictions are self-determined.

❧ You allude in your last letter to a subject I never touch upon because I know we cannot agree upon it. However, I will now say a few words that you may know my opinions, and if you wish to convert me to your way of thinking, take more vigorous measures to effect it. You intimate that the happiness to be enjoyed in a future state will depend upon, and be a reward for, our belief in certain doctrines which you believe to constitute the essence of true religion. You must think, therefore, that belief is *voluntary* and also that it is *meritorious.* But I think that a little consideration will show you that belief is quite independent of our will, and our common expressions show it. We say, "I wish I could believe him innocent, but the evidence is too clear"; or, "Whatever people may say, I can never believe he can do such a mean action." Now, suppose in any similar case the evidence on both sides leads you to a certain belief or disbelief, and then a reward is offered you for changing your opinion. Can you really change your opinion and belief, for the hope of reward or the fear of punishment? Will you not say, "As the matter stands I can't change my belief. You must give me proofs that I am wrong or show that the evidence I have heard is false, and then I may change my belief"? It may be that you do get more and do change your belief. But this change is not voluntary on your part. It depends upon the force of evidence upon your individual mind, and the evidence remaining the same and your mental faculties remaining unimpaired – you cannot believe otherwise any more than you can fly.

Belief, then, is not voluntary. How, then, can it be meritorious? When a jury try a case, all hear the same evidence, but nine say "Guilty" and three "Not guilty," according to the honest belief of each. Are either of these more worthy of reward on that account than the others? Certainly you will say No! But suppose beforehand they all know or suspect that those who say "Not guilty" will be punished and the rest rewarded: what is likely to be the result? Why, perhaps six will say "Guilty" honestly believing it, and glad they can with a clear conscience escape punishment; three will say "Not guilty" boldly, and rather bear the punishment than be false or dishonest; the other three, fearful of being convinced against their will, will carefully stop their ears while the witnesses for the defence are being examined, and delude themselves with the idea they give an honest verdict because they have heard only one side of the evidence. If any out of the dozen deserve punishment, you will surely agree with me it is these. Belief or disbelief is therefore not meritorious, and when founded on an unfair balance of evidence is blameable.[5] [1861] ❧

Soon after his return from South-east Asia, Wallace extended these arguments to his assessment of the factors that affect how people act in society.[6]

❧ Mr. [J. S.] Mill truly says, that a voter is rarely influenced by "the fraction of a fraction of an interest, which he as an individual may have, in what is beneficial to the public," but that his motive, if uninfluenced by direct bribery or threats, is simply "to do right," to vote for the man whose opinions he thinks most true, and whose talents seem to him best adapted to benefit the country. The fair inference from this seems to be, that if you keep away from a man the influences of bribery and intimidation, there is no motive left but to do what he thinks will serve the public interest – in other words, "the desire to do right." Instead of drawing this inference, however, it is concluded that, as the "honest vote" is influenced by "social duty," the motive for voting honestly cannot be so strong "when done in secret, and when the voter can neither be admired for disinterested, nor blamed for selfish conduct." But Mr. Mill has not told us what motive there can possibly be to make the man, voting in secret, vote against his own conviction of what is right. Are the plaudits of a circle of admiring friends necessary to induce a man to vote for the candidate he honestly thinks the best; and is the fear of their blame the only influence that will keep him from "mean and selfish conduct," when no possible motive for such conduct exists, and when we know that, in thousands of cases, such blame does not keep him from what is much worse than "mean and selfish conduct," taking a direct bribe?

Perhaps, however, Mr. Mill means (though he nowhere says so) that "class interest" would be stronger than public interest – that the voter's share of interest in legislation that would benefit his class or profession, would overbalance his share of interest in the welfare of the whole community. But if this be so, we may assert, first, that the social influence of those around him will, in nine cases out of ten, go to increase and strengthen the ascendency of "class interests," and that it is much more likely that a man should be thus induced to vote for class interests as against public interests, than the reverse. In the second place, we maintain that any temporary influence whatever, which would induce a man to vote differently from what he would have done by his own unbiassed judgment, is bad – that a man has a perfect right to uphold the interests of his class, and that it is, on the whole, better for the community that he should do so. For, if the voter is sufficiently instructed, honest, and far-seeing, he will be convinced that nothing that is disadvantageous to the community as a whole can be really and permanently beneficial to his class or party; while, if he is less advanced in social and political knowledge, he will solve the problem the other way, and be fully satisfied that in advancing the interests of his class he is also benefiting the community at large. In neither case, is it at all likely, or indeed desirable, that the temporary and personal influence of others' opinions at the time of an election, should cause him to vote contrary to the convictions he has deliberately arrived at, under the continued action of those same influences, and which convictions are the full expression of his political knowledge and honesty at the time?

It seems to me, therefore, that if you can arrange matters so that every voter may be enabled to give his vote uninfluenced by immediate fear of injury or hope of gain (by intimidation or bribery), the only motives left to influence him are his convictions as to the effects of certain measures, or a certain policy, on himself as an individual, on his class, or on the whole community. The combined effect of these convictions on his mind will inevitably go to form his idea of "what is right" politically, that idea which, we quite agree with Mr. Mill, will in most cases influence his vote, rather than any one of the more or less remote personal interests which have been the foundation of that idea. From this point of view, I should be inclined to maintain that the right of voting is a "personal right" rather than a "public duty," and that a man is in no sense "responsible" for the proper exercise of it to the public, any more than he is responsible for the convictions that lead him to vote as he does. It seems almost absurd to say that each man is responsible to every or to any other man for the free exercise of his infinitesimal share in the government of the country, because, in that case, each man in turn would act upon others exactly as he is acted upon by them, and thus the final result must be the same as if each had voted entirely uninfluenced by others. What, therefore, is the use of such mutual influence and responsibility? You cannot by such means increase the average intelligence or morality of the country; and it must be remembered, that the character and opinions, which really determine each man's vote, have already been modified or even formed by the long-continued action of those very social influences which it is said are essential to the right performance of each separate act of voting. It appears to me that such influences, if they really produce any fresh effect, are a moral intimidation of the worst kind, and are an additional argument in favour of, rather than against, the ballot.[7] [1865] ᐧᐧ

Wallace's travels also gave him an independent yardstick with which to assess his own society. Remarkably, given the racist views of the era, Wallace found "savage" society to be in many ways superior to the "civilised" version. His conclusion in the closing pages of *The Malay Archipelago* must have made his Victorian audience squirm.

≤§ We most of us believe that we, the higher races, have progressed and are progressing. If so, there must be some state of perfection, some ultimate goal, which we may never reach, but to which all true progress must bring us nearer. What is this ideally perfect social state towards which mankind ever has been, and still is tending? Our best thinkers maintain, that it is a state of individual freedom and self-government, rendered possible by the equal development and just balance of the intellectual, moral, and physical parts of our nature, – a state in which we shall each be so perfectly fitted for a social existence, by knowing what is right, and at the same time feeling an irresistible impulse to do what we know to be right, that all laws and all punishments shall be unnecessary. In such a state every man would have a sufficiently well-balanced intellectual organization, to understand the moral law in all its details, and would require no other motive but the free impulses of his own nature to obey that law.

Now it is very remarkable, that among people in a very low stage of civilization, we find some approach to such a perfect social state. I have lived with communities of savages in South America and in the East, who have no laws or law courts but the public opinion of the village freely expressed. Each man scrupulously respects the rights of his fellow, and any infraction of those rights rarely or never takes place. In such a community, all are nearly equal. There are none of those wide distinctions, of education and ignorance, wealth and poverty, master and servant, which are the product of our civilization; there is none of that wide-spread division of labour, which, while it increases wealth, produces also conflicting interests; there is not that severe competition and struggle for existence, or for wealth, which the dense population of civilized countries inevitably creates. All incitements to great crimes are thus wanting, and petty ones are repressed, partly by the influence of public opinion, but chiefly by that natural sense of justice and of his neighbour's right, which seems to be, in some degree, inherent in every race of man.

Now, although we have progressed vastly beyond the savage state in intellectual achievements, we have not advanced equally in morals. It is true that among those classes who have no wants that cannot be easily supplied, and among whom public opinion has great influence, the rights of others are fully respected. It is true, also, that we have vastly extended the sphere of those rights, and include within them all the brotherhood of man. But it is not too much to say, that the mass of our populations have not at all advanced beyond the savage code of morals, and have in many cases sunk below it. A deficient morality is the great blot of modern civilization, and the greatest hindrance to true progress.

During the last century, and especially in the last thirty years, our intellectual and material advancement has been too quickly achieved for us to reap the full benefit of it. Our mastery over the forces of nature has led to a rapid growth of population, and a vast accumulation of wealth; but these have brought with them such an amount of poverty and crime, and have fostered the growth of so much sordid feeling and so many fierce passions, that it may well be questioned, whether the mental and moral status of our population has not on the average been lowered, and whether the evil has not overbalanced the good. Compared with our wondrous progress in physical science and its practical applications, our system of government, of administering justice, of national education, and our whole social and moral organization, remains in a state of barbarism. And if we continue to devote our chief energies to the utilizing of our knowledge the laws of nature with the view of still further extending our commerce and our wealth, the evils which necessarily accompany these when too eagerly pursued, may increase to such gigantic dimensions as to be beyond our power to alleviate.

We should now clearly recognize the fact, that the wealth and knowledge and culture of *the few* do not constitute civilization, and do not of themselves advance us towards the "perfect social state." Our vast manufacturing system, our gigantic commerce, our crowded towns and cities, support and continually renew a mass of human misery and crime *absolutely* greater than has ever existed before. They create and maintain in life-long labour an ever-increasing army, whose lot is the more hard to bear by contrast with the pleasures, the comforts, and the luxury which they see everywhere around them, but which they can never hope to enjoy; and who, in this respect, are worse off than the savage in the midst of his tribe.

This is not a result to boast of, or to be satisfied with; and, until there is a more general recognition of this failure of our civilization – resulting mainly from our neglect to train and develop more thoroughly the sympathetic feelings and moral faculties of our nature, and to allow them a larger share of influence in our legislation, our commerce, and our whole social organization – we shall never, as regards the whole community, attain to any real or important superiority over the better class of savages.[8] [1869] ❧

In particular, Wallace, who had doubtless been impressed by the communal practices he had observed in Borneo Dyak longhouses and elsewhere, addressed the issue of land in a concluding "note".

❧ We permit absolute possession of the soil of our country, with no legal rights of existence on the soil to the vast majority who do not possess it. A great landholder may legally convert his whole property into a forest or a hunting-ground, and expel every human being who has hitherto lived upon it. In a thickly-populated country like England, where every acre has its owner and its occupier, this is a power of legally destroying his fellow-creatures; and that such a power should exist, and be exercised by individuals, in however small a degree, indicates that, as regards true social science, we are still in a state of barbarism.[9] [1869] ❧

Despite consistently hewing to the political left throughout his life, it was not until 1889 that Wallace declared himself a socialist. The deciding factor was Edward Bellamy's[10] utopian novel *Looking Backward* (1888). Bellamy's tale, now largely forgotten, was in its day immensely influential: at the turn of the twentieth century it was the third most popular book in the United States (after *Uncle Tom's Cabin* and *Ben Hur*).

❧ For about ten years after I first publicly advocated land nationalization I was inclined to think that no further fundamental reforms were possible or necessary. Although I had, since my earliest youth, looked to some form of socialistic organization of society, especially in the form advocated by Robert Owen as the ideal of the future, I was yet so much influenced by the individualistic teachings of Mill and Spencer,[11] and the loudly proclaimed dogma, that without the constant spur of individual competition men would inevitably become idle and fall back into universal poverty, that I did not bestow much attention upon the subject, having, in fact as much literary work on hand as I could manage. But at length, in 1889, my views were changed once for all and I have ever since been absolutely convinced

Wallace pictured, as the frontispiece of his *Darwinism*, at about the time he became a socialist

not only that socialism is thoroughly practicable, but that it is the only form of society worthy of civilized beings, and that it alone can secure for mankind continuous mental and moral advancement, together with that true happiness which arises from the full exercise of all their faculties for the purpose of satisfying all their rational needs, desires, and aspirations.

The book that thus changed my outlook on this question was Bellamy's "Looking Backward". . . . On a first reading I was captivated by the wonderfully realistic style of the work, the extreme ingenuity of the conception, the absorbing interest of the story, and the logical power with which the possibility of such a state of society as that depicted was argued and its desirability enforced. Every sneer, every objection, every argument I had ever read against socialism was here met and shown to be absolutely trivial or altogether baseless, while the inevitable results of such a social state in giving to every human being the necessaries, the comforts, the harmless luxuries, and the highest refinements and social enjoyments of life were made equally clear. As the mere story had engrossed much of my attention, I read

the whole book through again to satisfy myself that I had not overlooked any flaw in the reasoning, and that the conclusion was as
clearly demonstrated as it at first sight appeared to be. Even as a story
I found it bore a second almost immediate perusal, a thing I never
felt inclined to give any book before (except, I think, in the case of
Herbert Spencer's "Social Statics"[12]), and during the succeeding year
I read it a third time, in order to refresh my memory on certain
suggestions which seemed to me especially admirable.

From this time I declared myself a socialist.[13] [1905]   ঌ

In 1900 Wallace published a succinct summary of his socialist
vision.

ঌ THE SOCIETY OF THE FUTURE

I am myself convinced that the society of the future will be some form
of socialism, which may be briefly defined as *the organization of labour
for the good of all.* Just as the Post Office is organized labour in one
department for the benefit of all alike; just as the railways might be
organized as a whole for the equal benefit of the whole community;
just as extensive industries over a whole country are now organized
for the exclusive benefit of combinations of capitalists; so all necessary
and useful labour might be organized for the equal benefit of all.
When a combination or trust deals with the whole of one industry
over an extensive area, there are two enormous economies; advertising, which under the system of competition among thousands of
manufacturers and dealers wastes millions annually, is all saved; and
distribution, when only the exact number of stores and assistants
needful for the work are employed, effects an almost unimaginable
saving over the scores of shops and stores in every small town,
competing with each other for a bare living. What then would be the
economy when *all* the industries of a whole country were similarly
organized for the common good; and when all absolutely useless and
unnecessary employments were abolished – such as gold and diamond
mining except to the extent needed for science and art; nine-tenths
of the lawyers, and all the financiers and stock-gamblers? It is clear
that under such an organized system three or four hours work for five
days a week by all persons between the ages of twenty and fifty would
produce abundance of necessaries and comforts, as well as all the
refinements and wholesome luxuries of life, for the whole population.

But although I feel sure that some such system as this will be adopted in the future, yet it may be only in a somewhat distant future, and the coming century may only witness a step towards it; it is important that this step should be one in the right direction. The majority of our people dislike the very idea of socialism, because they think it can only be founded by compulsion. If that were the case it would be equally repulsive to myself. I believe only in *voluntary* organization for the common good, and I think it quite possible that we require a period of true individualism – of competition under strictly equal conditions – to develop all the forces and all the best qualities of humanity, in order to prepare us for that voluntary organization which will be adopted when we are ready for it, but which cannot be profitably forced on us before we are thus prepared.

In our present society the bulk of the people have no opportunity for the full development of all their powers and capacities, while others who have the opportunity have no sufficient inducement to do so. The accumulation of wealth is now mainly effected by the misdirected energy of competing individuals; and the power that wealth so obtained gives them is often used for purposes which are hurtful to the nation. There can be no true individualism, no fair competition, without equality of opportunity for all. This alone is social justice, and by this alone can the best that is in each nation be developed and utilized for the benefit of all its citizens.[14] [1900]   ‏಄

"Robbery of the poor by the rich": The Land Problem

The publication of *The Malay Archipelago* in 1869 marks the beginning of Wallace's serious public engagement with social issues. J. S. Mill was impressed by Wallace's condemnation in the book's closing pages of the pernicious effect on society of private ownership of land, and wrote in 1870 inviting Wallace to serve on the General Committee of his Land Tenure Reform Association. The Association folded with Mill's death in 1873, but by then Wallace had learnt that his indignant comments about the failings of "civilized" with respect to "barbaric" societies could and would translate into a programme for social reform.

The need for land reform was brought into focus by the agricultural depression in Ireland through the 1870s, which culminated in the formation in 1879 of the National Land League by Michael Davitt and Charles Stewart Parnell. Dedicated to the elimination in Ireland of "landlordism" – essentially feudal exploitation of Irish farmers by absentee landlords – the League was effectively part of the Irish Home Rule movement. It promoted the passage of the Land Act in 1881 in which the "three F's" of Irish farmers were recognized – fair rent, fixity of tenure, and freedom of sale. In advocating that land be bought by the government and sold off over a 35-year period to the land's tenants, the League parted ways with an otherwise sympathetic Wallace. He objected on the grounds that such a policy would merely create a new landed class.

So soon as the new proprietors have acquired the fee simple of the land (or even before), the buying of land by the more wealthy, and the selling of it by the poorer, will, inevitably, begin again. The land would be mortgaged by the poor or improvident, and the wealthy would again accumulate large estates. Then absentee landlords and discontented tenants, rack-rents, agents, middlemen, evictions and agrarian outrages will all arise as before, till some future Government will again be asked to advance money to buy out the new landlords, and transfer the land to those who will at that time be the tenants. It is evident then that no such proposal as that of the Land League would be more than a temporary palliative applied at an enormous cost, and that we must seek in a different direction if we would effect a radical cure.[1] [1880]   ৰ

Wallace's recommendation for how the government would obtain privately owned land was ingenious.

ৰ My proposal is mainly founded upon a very simple proposition, which I think will be admitted, and which, if not capable of logical demonstration, can yet hardly be disproved. This proposition is, that whatever acts may be done by an individual without injustice or without infringing any rights which others possess or are entitled to claim in law or equity, then acts of a similar nature may be done by the State, also without injustice. In judging of the validity of this proposition, we must remember, that an individual may be actuated by purely personal motives, may be influenced by passion, by pride, or even by revenge, and yet may not go beyond what always has been admitted to be his right, while the State will, presumably, be guided in its action by a desire for the public welfare, and cannot possibly, in the particular cases here contemplated, be influenced by those lower motives which often affect the individual, and yet have never been held to impair either his legal or his moral rights.

The proposition here generally stated appears to me to be so nearly in the nature of a political axiom as to require no attempt at a formal demonstration. It will be time enough to defend it when good, or at least plausible reasons have been given why it should not be accepted. I will now proceed to its application in the present inquiry.

The right to transfer land (or other property) by will, to any successor not insane or criminal, has been allowed by most civilized nations to some extent, and by ourselves with hardly any limitations. A British landowner may leave his property to be divided among his family, or to any single member of his family. If he has no family he may leave it to any relation or to any friend; and he is not said to be unjust if he passes over some relatives and bequeaths his land either to a personal friend, or to some man of eminence, or to benefit some public institution or charity, or for any analogous purpose. Even his own immediate family – his sons and daughters, his parents, or his brothers – have no legal claim on his land, if he chooses to leave it to a more distant relation, or to a friend, or to a charity; but public opinion does, in such a case, condemn his action as more or less unjust. But whenever the choice is between remote relations and some public purpose or even personal friendship, public opinion rather applauds his freedom of choice, and it is never allowed that the more or less distant relatives who may be passed over have any right to complain of injury or robbery because the land was not left to them, even if they were the actual heirs-at-law and would have received it had the owner died intestate.

Now comes the first application of my above-stated proposition or axiom. If the personal owner of land does not rob or injure a distant relative (even if he be the heir-at-law) by making a will and otherwise disposing of his land, neither can the State be justly said to rob or injure any one if, for public purposes, it alters the law of inheritance so as to prevent the transfer of the land of intestates to any persons who are not near blood relations of the deceased. The exact degree of relationship that may be fixed upon is not of importance to the principle, expect that it must not be so narrowly limited as to interfere with what Bentham termed "just expectation." A son or a brother certainly has such just expectations, while the expectations of a third cousin or a great-grand-nephew can hardly be so termed. For the sake of illustrating the principle let us suppose that the limit of inheritance to the land of an intestate is fixed at what may be termed the second degree, that is, that it shall not pass to any more remote relative than an uncle, first-cousin, or grandchild, but when none of these exist shall devolve to the State for public purposes. No one can deny that the State could justly make such a law, when laws which disinherit acknowledged children because they are illegitimate, as well as all a man's legitimate daughters and other female relatives, have been long upheld as both just and expedient![2] [1880] &

Wallace's approach to the Irish land problem attracted a great deal of attention, and "[m]uch against [his] wishes",[3] he became in 1881 the founding president of the Land Nationalization Society (a post he held until his death in 1913). Wallace even tried to convert Darwin to the cause, recommending that he read Henry George's[4] bestselling *Progress and Poverty* (1879), in which George outlined his proposal for a "single tax" on land.[5] Darwin, in his last known letter to Wallace, was characteristically polite but non-committal: "I will certainly order 'Progress and Poverty,' for the subject is a most interesting one. But I read many years ago some books on political economy, and they produced a disastrous effect on my mind, viz. utterly to distrust my own judgement on the subject and to doubt much everyone else's judgement! So I feel pretty sure that Mr. George's book will only make my mind worse confounded than it is at present."[6]

For Wallace, the "land problem" was the bedrock upon which all of society's many ills were constructed; solve it, and those ills would be eliminated. In a number of publications, most notably *Land Nationalisation; Its Necessity and Its Aims* (1882), Wallace laid out his radical agenda. *Land Nationalisation* is dedicated to "The Working Men of England", and Wallace explains in the preface that it is not his intention to create an arcane policy document for discussion in gentleman's clubs: "To reach the landless classes – to teach them what are their rights and how to gain these rights – is the object of this work; and it was therefore necessary that it should be at once clear and forcible, moderate in bulk, and issued at a low price."[7]

The following summary of Wallace's proposals culminates in a scathing condemnation of the contemporary land tenure system that he published in a number of places.

❧ Now, I believe that the great work of this century, that which is the true preparation for the work to be done in the coming twentieth century, is not its well-meant and temporarily useful but petty and tentative social legislation, but rather that gradual reform of the political machine – to be completed, it is to be hoped, within the next six years[8] – which will enable the most thoughtful and able and honest among the manual workers to at once turn the balance of political power, and, at no distant period, to become the real and permanent rulers of the country. The very idea of such a government will excite a smile of derision or a groan of horror among the classes who have hitherto plundered and blundered at their will, and have thought they were heaven-inspired rulers. But I feel sure that the workers will do very much better; and, forming as they do the great majority of the people, it is only bare justice that, after centuries of misgovernment by the idle and wealthy, they should have their turn. The larger part of the invention that has enriched the country has come from the workers; much of scientific discovery has also come from their ranks; and it is certain that, given equality of opportunity, they would fully equal, in every high mental and moral characteristic, the bluest blood in the nation. In the organization of their trades-unions and co-operative societies, no less than in their choice of the small body of their fellow-workers who represent them in Parliament, they show that they are in no way inferior in judgment and in organizing power to the commercial, the literary, or the wealthy classes. The way in which, during the past few years, they have forced their very moderate claims upon the notice of the public, have secured advocates in the press and in Parliament, and have led both political economists and politicians to accept measures which were, not long before, scouted as utterly beyond the sphere of practical politics, shows that they have already become a power in the state. Looking forward, then, to a government by workers and largely in the interest of workers, at a not distant date, I propose to set forth a few principles and suggestions as to the course of legislation calculated to abolish pauperism, poverty, and enforced idleness, and thus lay the foundation for a true civilization which will be beneficial to all.

I. That the ownership of large estates in land by private individuals is an injustice to the workers and the source of much of their poverty and misery, is held by all the great writers I have alluded to, and has been fully demonstrated in numerous volumes. It has led directly to the depopulation of the rural districts, the abnormal growth of great cities, the diminished cultivation of the soil and reduced food-supply, and is thus at once a social evil and a national danger. Some petty attempts are now making to restore the people to the land, but in a very imperfect manner. The first and highest use of our land is to provide healthy and happy homes, where all who desire it may live in permanent security and produce a considerable portion of the food required by their families. Every other consideration must give way to this one, and all restrictions on its realization must be abolished. Hence, the first work of the people's Parliament should be, to give to the Parish and District Councils (which will by that time be in full working order) unrestricted power to take all land necessary for this purpose, so as to afford every citizen the freest possible choice of a home in which he can live absolutely secure (so long as he pays the very moderate ground rent) and reap the full reward of his labour. Every man, in his turn, should be able to choose both where he will live and how much land he desires to have, since each one is the best judge of how much he can enjoy and make profitable. Our object is that all working men should succeed in life, should be able to live well and happily, and provide for an old age of comfort and repose. Every such landholder is a gain and a safety instead of a loss and a danger to the community, and no outcry, either of existing landlords or of tenants of large farms, must be allowed to stand in their way. The well-being of the community is the highest law, and no private interests must be allowed to prevent its realization. When land can be thus obtained, co-operative communities, on the plan so clearly laid down by Mr. Herbert V. Mills in his work on "Poverty and the State" [1886], may also be established, and various forms of co-operative manufacture can be tried.

II. The next great guiding principle, and one that will enable us to carry out the resumption of the land without real injury to any individual, is, that we should recognize no rights to property in the unborn, or even in persons under legal age, except so far as to provide for their education and give them a suitable but moderate provision against want. This may be justified on two grounds. Firstly, the law allows to individuals the right to will away their property as they please, so that not even the eldest son has any vested interest, as against the power of the actual owner of the property to leave it to whom or for what purpose he likes. Now, what an individual is permitted to do for individual reasons which may be good or bad, the State may do if it considers it necessary for the good of the community. If an individual may justly disinherit other individuals who have not already a vested interest in property, however just may be their expectations of succeeding to it, *ex fortiori* the State may, partially, disinherit them for good and important reasons. In the second place, it is almost universally admitted by moralists and advanced thinkers, that to be the heir to a great estate from birth is generally injurious to the individual, and is necessarily unjust to the community. It enables the individual to live a life of idleness and pleasure, which often becomes one of luxury and vice; while the community suffers from the bad example, and by the vicious standard of happiness which is set up by the spectacle of so much idleness and luxury. The working part of the community, on the other hand, suffers directly in having to provide the whole of the wealth thus injuriously wasted. Many people think that if such a rich man *pays* for everything he purchases and wastes, the workers do not suffer because they receive an equivalent for their labour; but such persons overlook the fact that every pound spent by the idle is first provided *by* the workers. If the income thus spent is derived from land, it is *they* who really pay the rents to the landlord, inasmuch as if the landlord did not receive them they would go in reduction of taxation. If it comes from the funds or from railway shares, *they* equally provide it, in the taxes, in high railway fares, and increased price of goods due to exorbitant railway charges. Even if *all* taxes were raised by an income tax paid by rich men only, the workers would be the real payers, because there is no other possible source of annual income in the country but productive labour. If any one doubts this, let him consider what would happen were the people to resume the land as their right, and thenceforth apply the rents, locally, to establish the various factories and other machinery needful to supply all the wants of the community. Gradually all workers would be employed on the land, or in the various co-operative or municipal industries, and would themselves receive the full product of their labours. To facilitate their exchanges they might establish a token or paper currency, and they would then have little use for gold or silver. How, then, could idlers live, if these workers, in the Parliament of the country, simply declined to pay the interest on debts contracted before they were born? What good would be their much-vaunted "capital," consisting as it mostly does of mere legal power to take from the workers a portion of the product of their labours, which power would then have ceased; while their real capital – buildings, machinery, etc. – would bring them not one penny, since the workers would all possess their own, purchased by their own labour and the rents of their own land? Let but the workers resume possession of the soil, which was first obtained by private holders by force or fraud, or by the gift of successive kings who had no right to give it, and capitalists as a distinct class from workers must soon cease to exist.

III. Another principle of equal importance is to refuse to recognize the right of any bygone rulers to tax future generations. Thus all grants of land by kings or nobles, all "perpetual" pensions, and all war-debts of the past, should be declared to be legally and equitably invalid, and henceforth dealt with in such a way as to relieve the workers of the burden of their payment as speedily as is consistent with due consideration for those whose chief support is derived from such sources. Just as we are now coming to recognize that a "living wage" is due to all workers, so we should recognize a maximum income determined by the standard of comfort of the various classes of fund-holders and State or family pensioners. As a rule, these persons might be left to enjoy whatever income they now possess during their lives, and when they had relatives dependent on them the income might be continued to these, either for their lives or for a limited period according to the circumstances of each case. There would be no necessity, and I trust no inclination, to cause the slightest real privation, or even inconvenience, to those who are but the product of a vicious system; but on every principle of justice and equity it is impossible to recognize the rights of deceased kings – most of them the worst and most contemptible of men – to burthen the workers for all time in order to keep large bodies of their fellow-citizens in idleness and luxury.

By means of the principles now laid down, we may proceed to see how to deal with the present possessors of great estates, and with millionaires, whose vast wealth confers no real benefit on themselves, while it necessarily robs the workers, since, as we have seen, it has all to be provided by the workers. It will, I think, be admitted that, if a man has an income, say, of ten thousand a year, that is sufficient to supply him with every possible necessary, comfort, and rational luxury, and that the possession of one or more additional ten thousands of income would not really add to his enjoyment. But all such excessive incomes necessarily produce evil results, in the large number of idle dependents they support, and in keeping up habits of gambling and excessive luxury. Further, in the case of landed estates the management of which is necessarily left to agents and bailiffs, it leads to injurious interference with agriculture and with the political and religious freedom of tenants, to oppression of labourers, to the depopulation of villages, and other well-known evils. It will therefore be for the public benefit to fix on a maximum income to be owned by any citizen; and, thereupon, to arrange a progressive income tax, beginning with a very small tax on a minimum income from land or realized property of, say, £500, the tax progressively rising, at first slowly, afterwards more rapidly, so as to absorb all above the fixed maximum.

When a landed estate was taken over for the use of the community, the net income which had been derived from it would be paid the late holder for his life, and might be continued for the lives of such of his direct heirs as were of age at the time of passing the Act, or it might even be extended to all direct heirs living at that time. In the case of a person owning many landed estates in different counties, he might be given the option of retaining any one or more of them up to the maximum income, and that income would be secured to him (and his direct heirs as above stated) in case any of the land were taken for public use. In the case of fundholders, all above the maximum income would be extinguished, and thus reduce taxation.

The process here sketched out – by which the continuous robbery of the people through the systems of land and fundholding, may be at first greatly reduced, and in the course of one or two generations completely stopped, without, as I maintain, real injury to any living person, and for the great benefit both of existing workers and of the whole nation in the future – will, of course, be denounced as confiscation and robbery. That is the point of view of those who now benefit by the acts of former robbers and confiscators. From another, and I maintain a truer point of view, it may be described as an act of just and merciful restitution. Let us, therefore, consider the case a little more closely.

Taking the inherited estates of the great landed proprietors of England, almost all can be traced back to some act of confiscation of former owners or to gifts from kings, often as the reward for what we now consider to be disgraceful services or great crimes. The whole of the property of the abbeys and monasteries, stolen by Henry VIII., and mostly given to the worst characters among the nobles of his court, was really a robbery of the people, who obtained relief and protection from the former owners. The successive steps by which the landlords got rid of the duties attached to landholding under the feudal system, and threw the main burden of defence and of the cost of government on non-landholders, was another direct robbery of the people. Then in later times, and down to the present century, we have that barefaced robbery by form of law, the enclosure of the commons, leading, perhaps more than anything else, to the misery and destruction of the rural population. Much of this enclosure was made by means of false pretences. The general Enclosure Acts[9] declare that the purpose of enclosure is to facilitate "the productive employment of labour" in the improvement of the land. Yet hundreds of thousands of acres in all parts of the country, especially in Surrey, Hampshire, Dorsetshire, and other southern counties, were simply taken from the people and divided among the surrounding landlords, and then only used for sport, not a single pound being spent in cultivating them. Now, however, during the last twenty years, much of this land is being sold for building at high building prices, a purpose never contemplated when the Enclosure Acts were obtained. During the last two centuries more than seven millions of acres have been thus taken from the poor by men who were already rich, and the more land they already possessed the larger share of the commons was allotted to them. Even a Royal Commission, in 1869, declared that these enclosures were often made "without any compensation to the smaller commoners, deprived agricultural labourers of ancient rights over the waste, and disabled the occupants of new cottages from acquiring new rights."

Now, in this long series of acts of plunder of the people's land, we have every circumstance tending to aggravate the crime. It was robbery of the poor by the rich. It was robbery of the weak and helpless by the strong. And it had that worst feature which distinguishes robbery from mere confiscation – the plunder was divided among the individual robbers. Yet, again, it was a form of robbery specially forbidden by the religion of the robbers, a religion for which they professed the deepest reverence, and of which they considered themselves the special defenders. They read in what they call *the Word of God*, "Woe unto them that join house to house, that lay field to field, till there be no place, that they may be placed alone in the midst of the earth;" yet this is what they are constantly striving for, not by purchase only, but by robbery. Again they are told, "The land shall not be sold for ever, for the land is Mine;" and at every fiftieth year all land was to return to the family that had sold it, so that no one could keep land beyond the year of jubilee; and the reason was that no man or family should remain permanently impoverished.

Both in law and morality the receiver of stolen goods is as bad as the thief; and even if he has purchased a stolen article unknowingly, an honourable man will, when he discovers the fact, restore it to the rightful owner. Now, our great hereditary landlords know very well that they are the legal possessors of much stolen property, and, moreover, property which their religion forbids them to hold in great quantities. Yet we have never heard of a single landlord making restitution to the robbed nation of workers. On the contrary, they take every opportunity of adding to their vast possessions, not only by purchase, but by that meanest form of robbery – the enclosing of every scrap of roadside grass they can lay their hands on, so that the wayfarer or the tourist may have nothing but dust or gravel to walk upon, and the last bit of food for the cottager's donkey or goose is taken away from him.

This all-embracing system of land robbery, for which nothing is too great and nothing too little, which has absorbed meadow and forest, moor and mountain; which has secured most of our rivers and lakes, and the fish which inhabit them; which often claims the very seashore and rocky coast-line of our island home, making the peasant pay for his seaweed-manure and the fisherman for his bait of shellfish; which has desolated whole counties to replace men by sheep or cattle, and has destroyed fields and cottages to make a wilderness for deer; which has stolen the commons and filched the roadside wastes; which has driven the labouring poor into the cities, and has thus been the primary and chief cause of the lifelong misery, disease, and early death of thousands who might have lived lives of honest toil and comparative comfort had they been permitted free access to land in their native villages; – it is the advocates and beneficiaries of this inhuman system, the members of this "cruel organization," who, when a partial restitution of their unholy gains is proposed, are the loudest in their cries of "robbery!" But all the robbery, all the spoilation, all the legal and illegal filching has been on their side, and they still hold the stolen property. They made laws to justify their actions, and we propose equally to make laws which will really justify ours, because, unlike their laws which always took from the poor to give to the rich, ours will take only from the superfluity of the rich, not to give to the poor individually, but to enable the poor to live by honest work, to restore to the whole people their birthright in their native soil, and to relieve all alike from a heavy burden of unnecessary taxation. This will be the true statesmanship of the future, and will be justified alike by equity, by ethics, and by religion.[10] [1894] ॐ

In the version of the above article published in Wallace's *Studies Scientific and Social*, he inserted an additional powerful appeal.

ॐ THE INVIOLABILITY OF THE HOME

But until this great reform can be effected there is a smaller and less radical measure of relief to all tenants, which should at once be advocated and adopted by the Liberal party. It is an old boast that the Englishman's house is his castle, but never was a boast less justified by facts. In a large number of cases a working man's house might be better described as an instrument of torture, by means of which he can be forced to comply with his landlord's demands and both in religion and politics submit himself entirely to the landlord's will. So long as the agricultural labourer, the village mechanic, and the village shopkeeper are the tenants of the landowner, the parson, or the farmer religious freedom or political independence is impossible. And when those employed in factories or workshops are obliged to live, as they so often are, in houses which are the property of their employers, that employer can force his will upon them by the double threat of loss of employment and loss of a home. Under such conditions a man possesses neither freedom nor safety, nor the possibility of happiness,

except so far as his landlord and employer thinks proper. A secure
HOME is the very first essential alike of political freedom, of personal
security, and of social well-being.

Now that every worker, even to the hitherto despised and down-
trodden agricultural labourer, has been given a share of local self-
government, it is time that, so far as affects the inviolability of the
home, the landlord's power should be at once taken away from him.[11]
[1900] ઠ

At the end of a memorably titled article, "The Social Quagmire
and the Way Out of It", Wallace delivers his battle cry.

ઠ The system which permits and even encourages land monopoly
and land speculation inevitably brings about another form of slavery,
more far-reaching, more terrible in its results, than the chattel slavery
they have abolished. Let the tenement houses of New York and
Chicago, with their thousands of families in hopeless misery, their
crowds of half naked and famishing children, bear witness! These
white slaves of our modern civilization everywhere cry out against the
system of private ownership and monopoly of land, which is, from its
very nature, the robbery of the poor and landless. This system needs
no gigantic war to overthrow it; it can be destroyed without really
injuring a single human being. Only we must not waste our time and
strength in the advocacy of half-measures and petty palliatives, which
will leave the system itself to produce ever a fresh crop of evil. The
voice of the working and suffering millions must give out no uncertain
sound, but must declare unmistakably to those who claim to represent
them – Our land-system is the fundamental cause of the persistent
misery and poverty of the workers; root and branch it is wholly evil; its
fruits are deadly poison; cut it down – why cumbereth it the ground?[12]
[1893] ઠ

Wallace had a very modern-sounding vision of the benefits of
government ownership of the land. Among them he recognized
the possibility of "green belts" around towns, and the preservation
of both access to wilderness areas and historic monuments.

ઠ Other checks might be applied by local authorities, which would
tend greatly to the healthiness and enjoyability of our larger towns,
such as the interposition of belts of park and garden at certain
intervals around dense centres of population – a class of improvement

which the ruinous competition prices of land held by private owners
now renders impossible.

ENCLOSURE OF COMMONS AND MOUNTAIN WASTES AS AFFECTING THE PUBLIC

Next in importance to the power of securing pleasant and healthy
houses, the general public have most interest in the right to free
passage about the country – to roam over the commons, heaths, and
woods; to search out the grand and beautiful scenes afforded by our
rivers, moors and mountains; to have preserved for them the ruins
which are landmarks of our written history, as well as those more
ancient monuments which tell us of pre-historic ages. In each and all
of these directions they suffer injury from the powers claimed and
exercised by landlords. As we have already seen, enormous areas of
common land have been enclosed and appropriated by the surrounding owners, often without provision even of foot-paths by which the
public may enjoy any of the land they once freely roamed over. Owing
to inordinate game-preservation, the woods and copses are almost
always rigidly shut up, and thus the public are deprived of one of the
greatest enjoyments of country life – the power to wander freely under
the shade of trees, in places where the choicest wild flowers blossom,
and where the living denizens of the woods may be seen in their
native haunts. Were it not for the ancient foot-paths crossing the
country from village to village, many parts of our land would be almost
shut out from the great body of its inhabitants. Fortunately these are
tolerably numerous. But however great may be the need of fresh
centres of population, we rarely hear of new paths being formed,
while old ones are occasionally shut up or diverted, or so enclosed by
fences that all their picturesque beauty and rural enjoyability is
destroyed. . . .

THE DESTRUCTION OF ANCIENT MONUMENTS

One of the most palpable illustrations of the evil consequences of
allowing land to be the absolute property of individuals is, that it has
led to the destruction of a vast number of most interesting ancient
monuments, while the attempt of Sir John Lubbock[13] and others to
preserve those that still remain has been for some years strenuously
opposed, on the ground that it interferes with the rights of
landlords. . . .

One of the most remarkable and interesting of our very ancient monuments is Abury, or Avebury,[14] in Wiltshire, which an old antiquarian[15] declared "did as much exceed Stonehenge as a cathedral doth an ordinary parish church." The entire series of these remains presented such a colossal enigma as it would be difficult to parallel even at Karnac;[16] but this wonderful relic of the past has been for many years undergoing destruction, the great stones of which it is composed being broken up to build cottages, to make gate-posts, and even to mend the roads. "Still, even now," says Sir John Lubbock, "there is perhaps no more remarkable monument of the kind in this country, or even in Europe."[17] In the year 1875, the owner of the land on which this grand monument stands sold it unreservedly to a Building Society, by which it was lotted out in sites for cottages, and actually sold in small plots for this purpose. Fortunately, Sir John Lubbock was informed of this just in time, and succeeded in purchasing the land himself, and in persuading the villagers for a small consideration to exchange their allotments for others in an adjoining field which was just as well suited to them. Abury, the wonder of antiquarians and the enigma of the learned, was thus barely saved from complete destruction by the intervention of a private gentleman living in a remote county![18] [1882]   ❧

Grouse-shooting and other forms of "inordinate game-preservation" were not the only socially elite activities that deprived people of access to the land: Wallace echoed a modern complaint about the evils of golf's insatiable appetite for land. "Even the piece of common [in the Welsh village of Llandrindod] that was reserved for the use of the inhabitants is now used for golf-links!"[19]

Public Health

Wallace campaigned passionately for better working and living conditions. He justly prided himself on his familiarity with the working poor, both rural and urban. Recalling in his autobiography his early life in London with his brother John, a journeyman carpenter, Wallace dedicates several pages to disputing – on the grounds of personal experience – the economic facts laid out in 1884 in an influential article on "Progress of the Working Classes in the Last Century".[1] In addition, as noted earlier, Wallace was working as a surveyor in South Wales during the "Rebecca Riots" there, and thus had first-hand experience of one of the more desperate agrarian struggles in British history.

&3 As the [nineteenth] century wore on, other evils [similar to child labour in textile mills] were gradually brought to light. Children and women were found to be working underground in coal mines, under equally vile conditions as regards health and morality; and an enormous loss of life was caused by inadequate ventilation, insecure roof-propping, imperfect winding machinery, and other causes, all due to want of proper precautions by the owners of the mines. As a matter of simple justice, such owners should be held responsible to the injured person not only to the full extent of his wages and for medical attendance, but should also pay a liberal compensation for the pain suffered, and for the extra labour, expense, and anxiety to his family. But all such things are ignored in the case of poor workers, so that even the money compensation is reduced to the smallest amount possible.

It is one of the great defects of our law that deaths due to preventable causes *in any profit-making business* are not criminal offences. Till they are made so, it will be impossible to save the hundreds, or even thousands, of lives now lost owing to neglect of proper precautions in all kinds of dangerous or unhealthy trades. However costly such precautions may be, expense should not be considered when human life is risked; and the present state of the law is therefore immoral.

Notwithstanding Acts of Parliament and numerous Inspectors (whose salaries should be paid by the mine owners), explosions and other accidents underground continue to increase, the year 1910 being a record year, with its 1,775 deaths; and even the number in proportion to the workers employed is the highest for the last twenty years [i.e. since the mid-1880s].

Yet no one is punished, or even held responsible for these deaths. Surely, this shows a deplorable absence of moral feeling, both in the general public and in Parliament. The responsibility of Parliament is really criminal, since it always allows its legislation to be made ineffective by the fear of diminishing the employers' profits, thus deliberately placing money-making above human life and human well-being.

In the case of mines and quarries, Parliament is especially responsible, because the possession of the mineral wealth of our country by private individuals is itself a gross usurpation of public rights, and should have been long ago declared illegal. Whatever arguments – and they are very strong – show us that the land itself should not be private property, are ten times stronger in the case of the minerals within its bowels. The value of land increases with its proper use, but in the case of minerals, the value is absolutely destroyed. Surely, it is a crime against posterity to allow the strictly limited mineral wealth of our country to be made private property, and very largely sold to foreigners, solely to increase the wealth of individuals and to the absolute impoverishment of ourselves and our children.

I will here add one other argument which goes to the root of the matter by showing that the alleged owners of minerals have not even a legal title to them. It is, I believe, a maxim of law that public rights cannot be lost by disuse. Landed estates were, in our country, created by the Norman Conqueror to be held subject to the performance of feudal duties. Deep-seated minerals were then not known to exist, and were not (I believe) specifically included in the original grants. Except, therefore, where they have since been made private property by *Act of Parliament*, they still remain public property. I submit, therefore, that they may be both legally and equitably resumed by the Government as public property, and worked for the good of the public and of posterity. Compensation to the supposed present owners would be a matter of favour, *not of right*.

INSANITARY DWELLINGS AND LIFE-DESTROYING TRADES

The enormous difference between town and country dwellers as regards duration of life and the prevalence of zymotic [infectious] diseases has been known statistically since the era of registration, and a body of Health Officers has been set up to report upon the worst cases. The local authorities have power to compel the owners of unhealthy dwellings to put them into a sanitary condition, or even order them to be entirely rebuilt. But as many of the members of corporations and other local boards are often themselves owners of such property, or have intimate friends who are so, very little has been done to remedy the evil. Again and again, in all parts of the country, the Health Officers have duly reported, but their reports have been ignored. In some cases, where the Health Officer has been too persistent, he has been asked to resign or has been discharged. A few general facts may be here given.

By the last complete Census returns (1901), there are in England and Wales 7,036,868 tenements, and of these 3,286,526, or nearly half, have from one to four rooms only. In London, out of a total of 1,019,646 tenements, 672,030, or considerably more than half, have from one to four rooms; while there are about 150,000 tenements of only *one room*, in which are living 313,298 persons, or about two and a quarter persons in each room on the average. There are, however, about 20,000 persons living *five in a room*, and 20,000 more who have *six, seven, or eight in a room.* As most of these one-roomed tenements are either the cellars or attics of houses in the most crowded parts of large towns, where there is impure air, little light, and scanty water supply, the condition of those who dwell in them may be imagined – or rather *cannot* be imagined, except by those who have explored them.

Equally inhuman, immoral, and even criminal, is the neglect of all adequate measures to check the loss of infant life through the overwork, poverty, or starvation of the mother, together with over-crowded and insanitary dwellings. In the mad race for wealth by capitalists and employers most of our towns and cities have been allowed to develop into veritable death-traps for the poor. This has been known for the greater part of a century, yet nothing really effective has been done, notwithstanding abundant health legislation – again made useless by the dread of diminishing the excessive profits of manufacturers and slum-owners. One of the Labour newspapers calls our attention to the following facts for 1911 as to Infant mortality per 1,000 born:

	PER 1,000
Deptford, East Ward (poor)	197
Deptford, West Ward (rich)	68
Bournville Garden Village	65
St. Mary's Ward, Birmingham	331

Such facts exist all over the kingdom. They have been talked about and deplored for the last half-century at least. Who has murdered the 1,000,000 children who die annually before they are one year old? Who has robbed the millions that just survive of all that makes childhood happy – pure food, fresh air, play, rest, sleep, and proper nurture and teaching? Again we must answer, our Parliament, which occupies itself with anything rather than the immediate saving of human life and abolishing widespread human misery, the whole of which is remediable. And all for fear of offending the rich and powerful by some diminution of their ever increasing accumulations of wealth. No thinking man or woman can believe that this state of things is absolutely irremediable; and the persistent acquiescence in it while loudly boasting of our civilization, of our science, of our national prosperity, and of our Christianity, is the proof of a hypocritical lack of national morality that has never been surpassed in any former age.

A new set of evils has grown up in the various so-called "unhealthy trades" – the lead glaze in the china manufacture, the steel dust in cutlery work, and the endless variety of poisonous liquids and vapours in the numerous chemical works or processes, by which so many fortunes have been made. These, together, are the cause of a large direct loss of life, and a much larger amount of permanent injury, together with a terrible reduction in the duration of life of all the workers in such trades. Yet in one case only – that of phosphorus matches[2] – has any such injurious process of manufacture been put an end to. Wealth has been deliberately preferred to human life and happiness.[3] [1913] ❧

Wallace's most prominent public position on health matters was, at first sight, his most mistaken. He opposed vaccination against small-pox. Despite its apparently quixotic aspect, Wallace's campaign was in fact well motivated and perfectly sound: he was wrong overall, but his objections to the arguments of the medical fraternity were right. He recounts how he was gradually converted to a vaccination sceptic.

⊰ I was brought up to believe that vaccination was a scientific procedure, and that Jenner was one of the great benefactors of mankind. I was vaccinated in infancy, and before going to the Amazon I was persuaded to be vaccinated again. My children were duly vaccinated,[4] and I never had the slightest doubt of the value of the operation – taking everything on trust without any inquiry whatever – till about 1875–80, when I first heard that there were anti-vaccinators, and read some articles on the subject. These did not much impress me, as I could not believe so many eminent men could be mistaken on such an important matter. But a little later I met Mr. William Tebb,[5] and through him was introduced to some of the more important statistical facts bearing upon the subject. Some of these I was able to test by reference to the original authorities, and also to the various Reports of the Registrar-General, Dr. Farr's[6] evidence as to the diminution of small-pox *before* Jenner's time, and the extraordinary misstatements of the supporters of vaccination. Mr. Tebb supplied me with a good deal of anti-vaccination literature, especially with "Pierce's Vital Statistics," the tables in which satisfied me that the claims for vaccination were enormously exaggerated, if not altogether fallacious. I also now learnt for the first time that vaccination itself produced a disease, which was often injurious to health and sometimes fatal to life, and I also found to my astonishment that even Herbert Spencer had long ago pointed out that the first compulsory Vaccination Act had led to an increase of small-pox. I then began to study the Reports of the Registrar-General myself, and to draw out curves of small-pox mortality, and of other zymotic diseases (the only way of showing the general course of a disease as well as its annual inequalities), and then found that the course of the former disease ran so generally parallel to that of the latter as to disprove altogether any *special protective effect* of vaccination.[7] [1905] ⊱

In view of the inherent risk of vaccination – one is after all deliberately giving a healthy person a disease – Wallace demanded that the medical profession demonstrate their case in favour of mandatory vaccination. His objections were of two kinds. First he recognized that hitherto the evidence of the medical profession had been evaluated only by the medical profession; this, he recognized, was a conflict of interest. Second, and most importantly, he assessed the statistics used to support the claim that vaccination was effective, and found much of it to

TRIUMPH OF DE-JENNER-ATION.
[The Bill for the ???? agement of Small Pox was passed.]

A contemporary engraving, by E. L. Sambourne, reveals the depth of public concern about vaccination as Parliament debated public-health policy in 1898 (courtesy Wellcome Library, London)

be flawed or bogus. Wallace's statistical critique was devastatingly sound, and his analysis represents one of the first attempts to bring statistical rigour to epidemiology. His major work on the subject, a pamphlet that he printed in full in *The Wonderful Century* bears the splendid title "Vaccination a Delusion; Its Penal Enforcement a Crime: Proved by the Official Evidence in the Reports of the Royal Commission".

✐ VACCINATION AND THE MEDICAL PROFESSION

Before proceeding to adduce the conclusive evidence that now exists of the failure of vaccination, a few preliminary misconceptions must be dealt with. One of these is that, as vaccination is a surgical operation to guard against a special disease, medical men can alone judge of its value. But the fact is the very reverse, for several reasons. In the first place, they are interested parties, not merely in a pecuniary sense, but as affecting the prestige of the whole profession. In no other case should we allow interested persons to decide an important matter. Whether iron ships are safer than wooden ones is not decided by ironmasters, or by shipbuilders, but by the experience of sailors and by the statistics of loss. In the administration of medicine or any other remedy for a disease, the conditions are different. The doctor applies the remedy and watches the result, and if he has a large practice he thereby obtains knowledge and experience which no other persons possess. But in the case of vaccination, and especially in the case of public vaccinators, the doctor does not see the result except by accident. Those who get small-pox go to the hospitals, or are treated by other medical men, or may have left the district; and the relation between the vaccination and the attack of small-pox can only be discovered by the accurate registration of all the cases and deaths, with the facts as to vaccination or revaccination. When these facts are accurately registered, to determine what they teach is not the business of a doctor but of a statistician, and there is much evidence to show that doctors are bad statisticians, and have a special faculty for misstating figures.[8] [1898] ಠ

ಠ [The members of a Royal Commission on Vaccination before whom Wallace was appearing as an expert witness] were so ignorant of statistics and statistical methods that one great doctor held out a diagram, showing the same facts as one of mine, and asked me almost triumphantly how it was that mine was so different. After comparing the two diagrams for a few moments I replied that they were drawn on different scales but that with that exception I could see no substantial difference between them. The other diagram was on a greatly exaggerated vertical scale, so that the line showing each year's death-rate went up and down with tremendous peaks and chasms, while mine approximated more to a very irregular curve. But my questioner could not see this simple point; and later he recurred to it a second time, and asked me if I really meant to tell them that those two diagrams were both accurate, and when I said again that though on different scales both represented the same facts, he looked up at the ceiling with an air which plainly said, "If you will say that you will say anything!"[9] [1905] ಠ

ಠ Dr. Lettsom . . . in his evidence before the Parliamentary Committee in 1802, calculated the small-pox deaths of Great Britain and Ireland before vaccination at 36,000 annually; by taking 3000 as the annual mortality in London and multiplying by twelve, because the population was estimated to be twelve times as large. He first takes a number which is much too high, and then assumes that the mortality in the town, village, and country populations was the same as in overcrowded, filthy London![10]

. . .

Again, it is admitted by many pro-vaccinist authorities that the unvaccinated, as a rule, belong to the poorer classes, while they also include most of the criminal classes, tramps, and generally the nomad population. They also include all those children whose vaccination has been deferred on account of weakness or of their suffering from other diseases, as well as all those under vaccination age. The unvaccinated as a class are therefore especially liable to zymotic disease of any kind, small-pox included; and when, in addition to these causes of a higher death-rate from small-pox, we take account of the proved untrustworthiness of the statistics, wholly furnished by men who are prejudiced in favour of vaccination (as instanced by the declaration of Dr. Gayton, that when the eruption is so severe as on the third day to hide the vaccination marks, it affords prima facie evidence of non-vaccination . . .), we are fully justified in rejecting all arguments in favour of vaccination supported by such fallacious evidence.[11]

. . .

But even more important, as showing that vaccination has had nothing whatever to do with the decrease of small-pox, is the very close general parallelism of the line showing the other zymotic diseases, the diminution of which it is admitted has been caused by improved hygienic conditions. The decline of this group of diseases in the first quarter of this century, though somewhat less regular, is quite as well marked as in the case of small-pox, as is also its decline in the last forty years of the eighteenth century [before vaccination], strongly suggesting that both declines are due to common causes.[12]

. . .

Whether a person dies of small-pox or of some other illness is a fact that is recorded with tolerable accuracy, because the disease in fatal cases is among the most easily recognized. Statistics of "small-pox mortality" may, therefore, be accepted as reliable. But whether the patient is registered as vaccinated or not vaccinated usually depends on the visibility or non-visibility of vaccination-marks, either during the illness or after death, both of which observations are liable to error, while the latter entails a risk of infection which would justifiably lead to its omission. And the admitted practice of many doctors, to give vaccination the benefit of any doubt, entirely vitiates all such statistics, except in those special cases where large bodies of adults are systematically vaccinated or revaccinated. Hence, whenever the results of these imperfect statistics are opposed to those of the official records of small-pox mortality, the former must be rejected. It is an absolute law of evidence, of statistics, and of common sense that, when two kinds of evidence contradict each other, that which can be proved to be even partially incorrect or untrustworthy must be rejected. It will be found that all the evidence that seems to prove the value of vaccination is of this untrustworthy character. This conclusion is enforced by the fact that the more recent hospital statistics show that small-pox occurs among the vaccinated in about the same proportion as the vaccinated bear to the whole population. . . .[13] [1898] ❧

Institutional Reform

For all his radical stance on matters like land nationalization and vaccination, Wallace was no gun-toting revolutionary. His version of radicalism was genteel and somewhat bourgeois; he championed gentle reform of institutions, even the most anachronistic ones, rather than wholesale destruction. Take, for example, his recommendations for the House of Lords, which have today a familiar ring to them.

❧ A few years back, Mr. Labouchère[1] introduced a Bill into the House of Commons declaring that, after January 1, 1895, the House of Lords shall cease to exist. But it is hardly possible that such a Bill can become law, either in this Parliament or in any of its successors for the next half century, since it would require that the Peers should commit political suicide, and this they would hardly do unless an almost unanimous public opinion compelled such a course, and they considered it more dignified than submitting to actual expulsion. There is, also, as Mr. Labouchère himself acknowledges, a preliminary difficulty, in a very wide-spread impression, even among Liberals, that a second chamber is necessary, combined with an extreme diversity of opinion as to how the second chamber should be constituted. It is evident, therefore, that the abolition of the House of Lords would by no means solve the problem, but would only lead to interminable discussions on the more difficult part of the question – what kind of chamber to substitute for it. The stoppage of all useful reforms by any attempt to remodel our constitution in such a revolutionary spirit would be exceedingly unpopular; and would probably involve a longer struggle and more expenditure of parliamentary energy than the effort we recently made to give Ireland permission to manage her own affairs.[2] It may, therefore, be worth while to consider whether there is not a method by which a House of Lords may be retained in such a form as to render it a truly representative Upper Chamber, thus making it acceptable, to most Liberals, and even to many Radicals; while, by preserving its ancient name and prestige, and by giving it both greater dignity and a more important part in legislation than it now possesses, the proposed reform might be upheld as truly conservative, and receive the support of the majority of the Conservative party.

It is clear that any such fundamental reform of the British Constitution as is now advocated by advanced Liberals should proceed on the lines of evolution rather than on those of revolution. Instead of abolishing the House of Lords we must modify, reform, and elevate it; and we must do this in such a manner as, on the one hand, to bring it into general and permanent harmony with the House of Commons; while, on the other hand, it is rendered so select, so dignified, so representative of all that is best in the British Peerage, past, present, and to come, that a seat in the Upper Chamber will become a more coveted honour than the insignia of the Garter, a higher dignity than a ducal coronet. It is, I think, essential to the successful carrying out of any such great reform that it should be initiated in the House of Lords itself, and simply accepted or rejected by the House of Commons. The discussion of its principles and methods should take place in the country at large, rather than in Parliament. The peers must be well informed as to the character and amount of change that will satisfy the people and bring about that substantial harmony between the two branches of the Legislature that is essential to good government; and it is with the hope of contributing towards the peaceful settlement of this great question that I now propose to set forth what appear to me to be the main principles on which such an important reform should be founded.

The two great anomalies of the present House of Lords are first its hereditary character; and, secondly, the presence in it of the bishops of the Church of England, who thus have a voice, and often a very important influence, in making or rejecting laws which affect the whole population. Both hereditary and ecclesiastical legislators are now felt to be wholly out of place in the parliament of a people which claims to possess both political and religious freedom. They have, during the last half century, been tolerated rather from the difficulty of getting rid of them, than from any belief in the value of their services; and it has long been seen, by all but the most bigoted Conservatives, that something must soon be done to bring the Upper House into harmony with modern ideals. In these concluding years of the nineteenth century our hereditary House of Lords is an anachronism. It may be said that our hereditary Sovereign is also an anachronism; but there is this great difference – that the peers systematically use their power to prevent or delay popular legislation, which the Sovereign, at the present day, never attempts to do.

It is clear, then, that any real and effective reform of the House of Lords must, in the first place, abolish the hereditary right to legislate, and must also exclude the bishops, as such, from any share in law-making. This, of course, does not affect the hereditary succession to the peerage, which may continue at all events for the present; but it would be most advisable to discontinue the creation of new hereditary peerages. Instead of these, life-peers should be created, but always as a mode of indicating distinguished merit, whether exhibited by services to the country at large, by philanthropic labours, or exceptional achievements in the fields of science, art or literature. The object of creating these life-peers should be, to raise the character and dignity of the peerage, and thus to afford material for the selection of a new House of Lords, which should be worthy of its historic fame and be in every way fitted to take a leading part in legislating for a free and civilised people.[3] [1894] ❧

The Church of England – that "narrow religious corporation, which in no sufficient degree represents either the most culti-vated intelligence or the highest morality of our age, and which, by its dogmatic theology and resistance to progress, has become out of harmony both with the best and the least educated portion of the community"[4] – also comes in for a dose of Wallace's gentle radicalism.

❧ THE PROPOSED NATIONAL CHURCH

I will now proceed to show how it [the Church of England] can be so reformed, and how it may be made a means of national advancement more efficient than all ordinary educational machinery, because its sphere of action will be wider, and because it will carry on a higher education than that imparted by schools, not for a few years only, but throughout the entire life of all who choose to profit by it. I will first sketch out what I consider should be the status and duties of the man who will take the place of the existing clergyman as the head and representative in every parish, or district of the National Church.

First, as to his designation; he might be termed the Rector, a name to which we are already accustomed, and which does not necessarily imply a religious teacher. He should be chosen, primarily, for moral, intellectual, and social qualities of a much higher character than are now expected. Temper and disposition would be carefully considered, as his usefulness would be greatly impaired if he were not able to gain the confidence, sympathy, and friendship of his parishioners. His moral character should be unexceptionable. He should be specially trained in the laws of health and their practical application and in the principles of the most advanced political and social economy. His religion should be quite free from sectarian prejudices, and his private opinions on religious matters would be no subject for inquiry. He should, however, be of a religious frame of mind, so as to be able to work sympathetically with the clergy of the various religious bodies in his district, and excite in them neither distrust nor antagonism. He must have a fair knowledge of physiology, and of simple medicine and surgery, of the rudiments of law and legal procedure, of the principles of scientific agriculture, and of the natural history sciences, as well as of whatever is considered essential to the education of a cultivated man.

He should not be allowed to undertake the care of a parish till thirty years of age, and only after having assisted some rector in parish duties for at least five years.

The duties of the parish rector would comprise, among others, all those of the existing clergyman, *but he would never conduct religious services of any kind.* The parish church, with its appurtenances, would however be under his entire authority, in trust for the whole body of parishioners, to be used for religious services by all or any duly organized religious bodies, under such arrangements as he might find to be most convenient for all. Any religious body should be able to claim the use of the church as a right (subject to the equal rights of other such bodies), the only condition being that it should possess a permanent organization, and that its ministers should be an educated class of men, coming up to a certain standard of intellectual culture and moral character. The State might properly refuse the use of the churches to those sects whose ministers are not specially trained or well educated men, on the ground that the public teaching of religion among a civilized people is degraded by being placed in the hands of the illiterate, and that such teachers are likely to promote superstition and increase fanaticism.

The rector might himself lecture in the church on moral, social, sanitary, historical, philosophical, or any other topics which he judged most suitable to the circumstances of his parishioners. He would also allow the church to be used during the week for any purpose not inconsistent with the main objects of his position, but always having regard to religious prejudices so long as they existed, his first duty being to promote harmony and good-will, and to gain any object he might think beneficial by persuasion rather than by an abrupt exercise of authority. His knowledge of law, and his position as *ex-officio* magistrate, would enable him to settle almost all the petty disputes among his parishioners, and so greatly diminish law-suits. He would be an *ex-officio* member of the School Board, and of the governing body of any other public educational institution in his district. It would be his duty to see that new legislative enactments were brought to the notice of the persons they chiefly affected, so that no one could offend through ignorance. He might, if he pleased, visit the sick, if his services were asked for, but this would be altogether voluntary. It would be an essential part of his duty to be on good terms with the ministers of all religious sects in his district, to bring them into friendly relations with each other, and to induce them to work harmoniously together for moral and educational objects.

With a sphere of action such as is here sketched out, the rector of the parish would have far more influence than the existing clergyman can possibly have.[5] [1873] ੩

Wallace extended his secularizing campaign to what people should do on Sundays.

੩ The whole essence of the Sabbath question rests upon giving the proper meaning to the words "labour," "work," "thy work," as used in the fourth commandment. These words, as the context shows, do not refer to any particular acts, but to the work done by each one of us in the business or profession by which we live. To the summer tourist in the Alps the ascent of a mountain or the passage of a glacier is pleasure and health-giving recreation; to the guides who accompany him it is their work. A hired gardener works for his living in a garden; but though I do many of the same things as he does, to me they are not my work, but my recreation. So, a domestic servant's work is to cook or to prepare a meal, or to wait at table; but when a party go out for a picnic, light a fire, make tea, roast potatoes, arrange the meal, and help the guests, they are certainly not working but pleasuring. When a doctor attends the sick in a hospital, or the wounded on a battlefield, he is doing the work of his life; but if any one of us nurses a sick person or binds up a wound, we may be doing acts of mercy or of charity, but we are not doing "our work." Even if we take upon ourselves some of the work of others, carry a heavy load for a weary woman, or do an hour's stone-breaking to help an old rheumatic labourer, what we do ceases to be work in the true meaning of the term but is transformed into a deed of love or mercy; and such deeds are not only permissible, but even commendable, on whatever day they are done.

We have here the clue to a method by which all that needs doing for health, for enjoyment, or for charity, may be done on Sunday without any one breaking the fourth commandment. Almost all this necessary work is now done by various classes of hired servants who are employed on similar work for six days every week, and who also have not much less to do on the seventh day. To keep the Sabbath, both in the letter and the spirit, these workers must be allowed full and complete rest; they must do none of their special work on that day. All that portion of their weekly duties which is necessary for the well-being of their employers, and for the rational enjoyment of their lives, must be done by those other members of the household who have spent the week largely in idleness or in pleasure, or if in work, in work of a quite different character from that of their servants. In doing this work; in helping each other; in sharing among themselves the various household occupations which during all the week have been undertaken by others; and in doing all this in order that those others may enjoy the full and unbroken rest which their six days' continuous labour requires and deserves, each member of the family will be doing deeds of self-sacrifice and of charity (in however small a degree), and such deeds do not constitute the "work" which is so strictly forbidden on the Sabbath day.

In the ordinary middle-class household, where there are six or eight in family and two or three servants, all that is necessary may be easily done, and allow every member of the family to go to church or chapel once or oftener. In other cases there will, no doubt, be difficulties but none which may not be overcome by a little arrangement and mutual helpfulness. Where a household consists only of aged or elderly people to whom the needful operations of housework would be painful or even impossible, there are always younger relatives or friends, or even acquaintances who could, either regularly or occasionally, spend the Sunday with such old people; and there is probably not a single difficulty of this kind which could not be overcome by two or more households combining for the Sunday in such a way as to divide the work and thus render it as little irksome as possible. If it were once really felt that the thing must be done, that on no account must the commandment be broken by servants doing any of their usual work on Sunday and that the truest and most divine "service" would thus be "performed," all difficulties would vanish, and the day would become, not in name only but truly, a holy one, inasmuch as it would witness in every household deeds of true charity and mercy, because in every case they would involve some amount of personal effort and self-sacrifice.

In the larger establishments of the higher classes there would be no greater difficulty, since it would be easy to effect such a division of labour as to render the work light for each. The son or other relative who was fondest of horses and dogs would of course see after their wants on Sunday; another might undertake the fire-lighting; while the young ladies would prepare the meals and do all other really necessary domestic work. And as all visitors would be acquisitions, almost the whole of the lodging- and boarding-houses would be emptied, their occupants becoming guests at the houses of their friends and taking their share of the Sabbath day's duties. Of course the greater part of the servants thus released from their regular work would also visit their friends, and by giving some little voluntary assistance would take their part in the great altruistic movement that would characterize the day.[6] [1894]   ❧

The legal profession was another institution scrutinized by Wallace. He was unimpressed.

❧ Amid the endless discussions that have taken place as to the sphere and duties of Government, all parties are agreed that there are two great and primary functions which every efficient Government must perform if it deserve the name: it must guard the country against attack by foreign enemies; and it must make such arrangements for the administration of the laws, that every man may obtain justice – as far as possible free and speedy justice – against wilful evil-doers.

The fact that there is an absolute unanimity as to these two important functions of a good Government, while almost everything else that Governments do, or attempt to do, has been denounced by great thinkers as beyond their proper sphere of action, renders it probable that these are at all events, the primary and most important functions of the State. It may not, perhaps, be easy to determine which of these two is of the greatest importance; for even admitting that conquest by a foreign foe is an evil incalculably greater than any wrong which individuals may suffer yet the one is of so much more frequent occurrence – every member of society being daily exposed to it, while attempts at conquest occur only at distant and uncertain intervals – that repetition in the one case may make up for magnitude in the other. We are therefore pretty safe in assuming that they are of equal importance, and in confirming that it is as much the duty of Government to protect its individual subjects from wrong to person or property committed by their fellows, as to protect the entire community from foreign enemies.

But if we look around us to see how these primary duties are performed, it becomes evident, either that existing Governments do not consider these duties as equally imperative upon them (even if they are not of absolutely equal importance), or that the former duty is a very much more difficult one than the latter. In every country we find an enormous organization for the purpose of national defence, which occupies a large portion of the wealth, the skill, and the labour of the community. No cost is too great, no preparations are too tedious, in order to deter an enemy from venturing to attack us, or to secure us the victory should he be so bold as to do so. For this end we keep thousands of young and healthy men in a state of unproductive activity, or idleness; for this we pile up mountains of debt, which continue to burthen the country for successive generations. New ships, new weapons, every invention that art or science can produce, are at once taken advantage of, while the less perfect appliances of a few years ago are thrown aside with hardly a thought of the vast sums which they represent.

If we now turn to see how the other paramount duty of the State is performed, we find a very different condition of things. Here everything is antiquated, cumbrous, and inefficient. The laws are an almost unintelligible mass of patchwork which the professional study of a life is unable to master; and the mode of procedure, handed down from the dark ages, is often circuitous and ineffective, notwithstanding a number of modern improvements. It may be admitted that in criminal cases tolerably sure, if not very speedy, punishment falls on the aggressor; but the sufferer receives, in most cases, no compensation, and often incurs great expense and much trouble in the prosecution. He gets revenge, not justice. That relic of barbarism, the fixed money fine, the same for the beggar and the millionaire, though almost universally admitted to be unjust, is not yet abolished. It is, however, in cases of civil wrong that individuals find the greatest difficulty (often amounting to an absolute impossibility) of obtaining justice. This arises, not only from the enormously voluminous and intricate mass of enactments and precedents, and the tedious mode of procedure, involving grievous delay and expense to every applicant for justice, but also to the vast accumulation of cases which are allowed to come before the courts, many of which are of such a complex nature as to some extent justify the strict forms of procedure which bear so hardly on those who seek relief in much simpler cases. The result is, that it is often better for a man to put up with a palpable wrong than to endeavour to obtain redress; and the assertion that in our happy country, there is "not one law for the rich and another for the poor," though literally true, is practically the very opposite of truth, since in a large number of cases the wealthy alone can afford to pay for the means of obtaining justice.[7] [1873] &

Wallace was famously succinct on the subject of women and the vote.

& As long as I have thought or written at all on politics, I have been in favour of woman suffrage. None of the arguments for or against have any weight with me, except the broad one, which may be thus stated:– "All the human inhabitants of any one country should have equal rights and liberties before the law; women are human beings; therefore they should have votes as well as men." It matters not to me whether ten millions or only ten claim it – the right and the liberty should exist, even if they do not use it. The term "Liberal" does not apply to those who refuse this natural and indefensible right.[8] [1909] &

Public Education

Wallace's "deficient organ of language prevented [him] from ever becoming a good lecturer or having any taste for it",[1] but he nevertheless toured and lectured extensively. Perhaps because he was largely an autodidact, he was perennially concerned about public education, and the use of public funds for scientific purposes. His particular interest was museums. Not only, as a professional collector, was Wallace's career built on them, but also at that time museums were the primary means of introducing people to animals and plants from overseas – the "great group of the natural history sciences can scarcely be taught without them".[2] His first experience as a provider rather than a consumer of a museum exhibit, however, was inauspicious.

◄§ Dr. Latham[3] was at this time [1853] engaged in fitting up groups of figures to illustrate the family life and habits of the various races of mankind at the new Crystal Palace at Sydenham[4] then just completed, and he asked me to meet him there and see whether any alterations were required in a group of natives, I think, of Guiana.

I found Dr. Latham among a number of workmen in white aprons, several life-size clay models of Indians, and a number of their ornaments, weapons, and utensils. The head modellers were Italians, and Dr. Latham told me he could get no Englishmen to do the work, and that these Italians, although clever modellers of the human figure in any required attitude, had all been trained in the schools of classical sculpture, and were unable to get away from this training. The result was very curious, and often even ludicrous, a brown Indian man or girl being given the attitudes and expressions of an Apollo or a Hercules, a Venus or a Minerva. In those days there were no photographs, and the ethnologist had to trust to paintings or drawings, usually exaggerated or taken from individuals of exceptional beauty or ugliness. Under my suggestion alterations were made both in the features and pose of one or two of the figures just completed, so as to give them a little more of the Indian character, and serve as a guide in modelling others, in which the same type of physiognomy was to be preserved. I went several times during the work on the groups of South American origin, but though when completed with the real ornaments, clothing, weapons, and domestic implements the groups were fairly characteristic and life-like, yet there remained occasionally details of attitude or expression which suggested classic Greek or Italy rather than the South American savage.[5] [1905] ॐ

As Wallace himself points out, it is ironic that he was so taken with Louis Agassiz's[6] "Museum of Comparative Zoology" at Harvard University. Agassiz was an avowed anti-Darwinian, and yet Wallace much preferred his museum to London's Natural History Museum.

ॐ What ought to be exhibited to the public is a typical series of such skeletons or models [of extinct animals], so arranged as to show the progression of forms and the evolution of the more specialized types as we advance from the earlier to the later geological periods. Instead of one huge gallery, a series of moderate-sized rooms should be constructed, each to illustrate one geological epoch, with subsidiary rooms where necessary to show the successive modifications which each class or order of animals has undergone. Where only fragments of an important type have been obtained, these might be exhibited with an explanation of why they are important, and an outline drawing showing the probable form and size of the entire animal. A museum of this kind, utilizing the palaeontological treasures of the whole world, would be of surpassing interest, and would probably exceed in attractiveness and popularity all existing museums. It would offer scope for a variety of groupings of extinct and living animals, calculated, as Professor Agassiz intended his museum to do, "to illustrate the history of creation, as far as the present state of scientific knowledge reveals that history." It is surely an anomaly that the naturalist who was most opposed to the theory of evolution should be the first to arrange his museum in such a way as best to illustrate that theory, while in the land of Darwin no step has been taken to escape from the monotonous routine of one great systematic series of crowded specimens, arranged in [the] lofty halls and palatial galleries [of London's Natural History Museum], which may excite wonder, but which are calculated to teach no definite lesson.[7] [1887] ॐ

⊷ Throughout the animal kingdom, at least one or more species of every important family group should be exhibited; and in the larger and more interesting families, one or more species of each genus. The number of specimens is not, however, so important as their quality and mode of exhibiting them. A few of the more important species in each order, well illustrated by fine and characteristic specimens, would be far better than ten times the number if imperfect, badly prepared, and badly arranged. Let any one look at an artistically mounted group of fine and perfect quadruped or bird skins, which represent the living animals in perfect health and vigour, and by their characteristic attitudes and accessories tell the history of the creature's life and habits; and compare this with the immature, ragged, mangy-looking specimens one often sees in museums, stuck up in stiff and unnatural attitudes, and resembling only mummies or scarecrows. The one is both instructive and pleasing, and we return again and again to gaze upon it with delight. The other is positively repellent, and we feel that we never want to look upon it again.[8] [1869] ⊷

Wallace's vision for educating the public was often both unconventional and grand. He championed the construction of an enormous scale-model of the planet, and advocated turning parts of Epping Forest, an area of publicly owned land north-east of London, into a kind of ecological arboretum.

⊷ "How best to model the Earth," in the *Contemporary Review* (May [1896]), was a discussion of the proposal by Élisée Reclus[9] to erect an enormous model of the globe, about four hundred and twenty feet in diameter, giving a scale about one-third smaller than our ordnance maps of one inch to a mile. It was to be modelled in minute detail on the convex side, and would therefore require to be completely covered in by a building nearly six hundred feet high, and would need an elaborate system of platforms and staircases in order to see it, while only a very small portion of it could be seen at once, and accurate photographs could only be taken of very small areas. My proposal was to adopt the plan of Wyld's great globe in Leicester Square,[10] many years ago, giving all the *detailed* features on the inside surface, while the outside could be boldly modelled in some indestructible material to show all the chief physical features, which might also be coloured in fresco as naturally as possible, and would then be a grand object seen either near or at a distance, while a captive balloon would afford a splendid view of the polar regions and of all parts of the northern hemisphere. The numerous advantages of this plan are explained in some detail, and I have little doubt that it will be realized (perhaps on half the scale) some time during the present century.[11] [1905] ৯৯

৯৯ PROPOSED ILLUSTRATION OF TEMPERATE FORESTS

The plan [for Epping Forest] I have now to propose is very different from all these. It is one which would be perfectly novel, perfectly practicable, intensely interesting as a great arboricultural experiment, attractive alike to the uneducated and to the scientific, not more expensive than any other plan, and perfectly in harmony with the character of the domain as essentially "a forest." It is, briefly, to form several distinct portions of forest, each composed solely of trees and shrubs which are natives of one of the great forest regions of the temperate zone.[12] [1878] ৯৯

> Although he advocated that museums should be government-sponsored, Wallace did not favour government support for scientific research.

৯৯ The Public mind seems now to be going mad on the subject of education; the Government is obliged to give way to the clamour, and men of science seem inclined to seize the opportunity to get, if possible, some share in the public money. Art education is already to a considerable extent supplied by the State, – technical education (which I presume means education in "the arts") is vigorously pressed upon the Government, – and Science also is now urging her claims to a modicum of State patronage and support.

Now, sir, I protest most earnestly against the application of public money to any of the above specified purposes, as radically vicious in principle, and as being in the present state of society a positive wrong. In order to clear the ground let me state that, for the purpose of the present argument, I admit the right and duty of the State to educate its citizens. I uphold national education, but I object absolutely to all sectional or class education; and all the above-named schemes are simply forms of class education. The broad principle I go upon is this, – that the State has no moral right to apply funds raised by the taxation of all its members to any purpose which is not directly available for the benefit of all. As it has no right to give class preferences in legislation, so it has no right to give class preferences in the expenditure of public money. If we follow this principle, national education is not forbidden, whether given in schools supported by the State, or in museums, or galleries, or gardens, fairly distributed over the whole kingdom, and so regulated as to be equally available for instruction and amusement of all classes of the community. But here a line must be drawn. The schools, the museums, the galleries, the gardens, must all alike be *popular* (that is, adapted for and capable of being fully used and enjoyed by the people at large), and must be developed by means of public money to such an extent only as is needful for the highest attainable *popular* instruction and benefit. All beyond this should be left to private munificence, to societies, or to the classes benefited, to supply.[13] [1870]   &

In fact, he argued that government-supported scientific expeditions were inefficient, and that less costly alternatives would be much more effective.

&   I cannot avoid here referring to the enormous waste of labor and money with comparatively scanty and unimportant results to natural history of most of the great scientific voyages of the various civilized governments during the present century. All these expeditions combined have done far less than private collectors in making known the products of remote lands and islands. They have brought home fragmentary collections, made in widely scattered localities, and these have been usually described in huge folios, whose value is often in inverse proportion to their bulk and cost. The same species have been collected again and again, often described several times over under new names, and not unfrequently stated to be from places they never inhabited. The result of this wretched system is that the productions of some of the most frequently visited and most interesting islands on the globe are still very imperfectly known, while their native plants and animals are being yearly exterminated; and this is the case even with countries under the rule or protection of European governments. Such are the Sandwich Islands [Hawaii], Tahiti, the Marquesas, the Philippine Islands, and a host of smaller ones; while Bourbon [Réunion] and Mauritius, St. Helena, and several others have only been adequately explored after an important portion of their productions has been destroyed by cultivation or the reckless introduction of

goats and pigs. The employment in each of our possessions, and those
of other European powers, of a resident naturalist at a very small
annual expense, would have done more for the advancement of
knowledge in this direction than all the expensive expeditions that
have again and again circumnavigated the globe.[14] [1880]   ✎

Capitalism and Empire

Economics

Having grown up poor (". . . my father passed the last years of his life in comparative freedom from worry about money matters, because these had reached such a pitch that nothing worse was to be expected"[1]), having spent twelve years among "savages" overseas, and having developed an instinctive sympathy for the underdog, it was perhaps inevitable that Wallace would become a critic of Victorian imperial social order.

A theme to which Wallace returns again and again is the discrepancy between the potential social benefits of technological progress and the reality of an expanding urban underclass. Like the scientists involved in the manufacture of the atomic bomb a few decades later, Wallace had realized that scientific advances and technological progress do not necessarily result in the improvement of the lot of the average person. Indeed, he was sure that innovation had, overall, had a detrimental effect by facilitating capitalism and increasing the polarization of wealth.

During the whole of the nineteenth century there was a continuous advance in the application of scientific discovery to the arts, and especially in the invention and application of labour-saving machinery; and our wealth has increased to an equally marvellous extent. Various estimates which have been made of the increase in our wealth-producing power show that, roughly speaking, the use of mechanical power has increased more than a hundredfold during the century; yet the result has been to create a limited upper class, living in unexampled luxury, while about one-fourth of our whole population exists in a state of fluctuating penury, often sinking below what has been termed "the margin of poverty." Of these, many thousands are annually drawn into the gulf of absolute destitution, dying either from direct starvation, or from diseases produced by their employment, and

BAD TIMES:

AN ESSAY ON THE PRESENT DEPRESSION OF TRADE,
TRACING IT TO ITS SOURCES IN ENORMOUS FOREIGN
LOANS, EXCESSIVE WAR EXPENDITURE, THE INCREASE
OF SPECULATION AND OF MILLIONAIRES, AND THE
DEPOPULATION OF THE RURAL DISTRICTS;
WITH SUGGESTED REMEDIES.

BY

ALFRED RUSSEL WALLACE, LL.D.

"In regretting the depopulation of the country I inveigh against the
increase of our luxuries; and here too I expect the shout of
modern politicians against me."
GOLDSMITH, *Dedication to Deserted Village.*

London:

MACMILLAN AND CO.

1885

The title page of Wallace's short 1885 book on socioeconomic matters. Wallace entered the essay "in competition for the Pears prize of one hundred guineas for the best essay on the present depression of trade. It did not obtain the prize, and it is therefore submitted to the judgement of the public – and more especially the working classes" (from the Preface to *Bad Times*)

rendered fatal by want of the necessaries and comforts of a healthy existence.

But during this long period, while wealth and want were alike increasing side by side, public opinion was not sufficiently educated to permit of any effectual remedy being applied for the extirpation of this terrible social disease. The workers themselves had not visualised its fundamental causes – land monopoly and the competitive system of industry, giving rise to an ever-increasing private capitalism which, to a very large extent, controlled the legislature. This rapid growth of wealth through the increase of the various kinds of manufacturing industry led to a still greater increase of middlemen engaged in the distribution of its products, from wealthy merchants, through various grades of tradesmen and petty shopkeepers who supplied the daily wants of the whole community. To these must be added the innumerable parasites of the ever-increasing wealthy classes; the builders of their mansions and their factories; the makers of their furniture and clothing, of their costly ornaments and their children's toys; the vast body of their immediate dependents, from their managers, their agents, commercial travellers and clerks, through various grades of domestic servants, grooms and game-keepers, butlers and housekeepers, down to stable-boys and kitchen-maids, all deriving their means of existence from the wealth daily produced in mines, factories and workshops. This was apparently due primarily, if not exclusively, to the capitalists themselves as the employers of labour, without whose agency and supervision it was believed that all productive labour would cease, bringing ruin and starvation to the whole population. Thus, a vast mass of public opinion was created, all in favour of the capitalists as the employers of labour and the true source of the creation of wealth.

To those who lived in the midst of this vast industrial system, or were a part of it, it seemed natural and inevitable that there should be rich and poor; and this belief was enforced on the one hand by the clergy, and on the other by the political economists, so that religion and science agreed in upholding the competitive and capitalistic system of society as being the only rational and possible one. Hence, till quite recently, it was believed that the abolition of poverty was entirely outside the true sphere of governmental action. It was, in fact, openly declared and believed that poverty was due to economic causes over which governments had no power; that wages were kept down by the "iron law" of supply and demand; and that any attempts to find a remedy by Acts of Parliament only aggravated the disease. This was the doctrine held, even by such great men as W. E. Gladstone and Sir William Harcourt,[2] together with the dogma that it was a government's duty to buy in the cheapest market, in order to protect the taxpayer. It was the doctrine also which converted the misnamed "guardians" of the poor into guardians of the ratepayers' interests, and led to that rigid and unsympathetic treatment of the very poor which made the workhouse more dreaded than the jail, and which to this very day leads many of the most destitute to die of lingering starvation, or to commit suicide, rather than apply for relief or enter the gloomy portals of the workhouse.[3] [1913] &

In a letter written from South-east Asia in response to urgings that he return, Wallace revealed his own aspirations. Wealth was not one of them.

& Your ingenious arguments to persuade me to come home are quite unconvincing. I have much to do yet before I can return with satisfaction of mind; were I to leave now I should be ever regretful and unhappy. That alone is an all-sufficient reason. I feel that my work is *here* as well as my pleasure; and why should I not follow out my vocation? As to materials for work at home, you are in error. I have, indeed, materials for a life's study of entomology, as far as the forms and structure and affinities of insects are concerned; but I am engaged in a wider and more general study – that of the relations of animals to space and time, or, in other words, their geographical and geological distribution and its causes. I have set myself to work out this problem in the Indo-Australian Archipelago, and I must visit and explore the largest number of islands possible, and collect materials from the greatest number of localities, in order to arrive at any definite results. As to health and life, what are they compared with peace and happiness? and happiness is admirably defined in the *Family Herald* as to be best obtained by "work with a purpose, and the nobler the purpose the greater the happiness." But besides these weighty reasons there are others quite as powerful – pecuniary ones. I have not yet made enough to live upon, and I am likely to make it quicker here than I could in England. In England there is only one way in which I could live, by returning to my old profession of land-surveying. Now, though I always liked surveying, I like collecting better, and I could never now give my whole mind to any work apart from the study to which I have devoted my life. So far from being angry at being called an enthusiast (as you seem to suppose), it is my pride and glory to be worthy to be so called. Who ever did anything good or great who was not an enthusiast? The majority of mankind are enthusiasts only in one thing – in money-getting; and these call others enthusiasts as a term of reproach because they think there is something in the world better than money-getting. It strikes me that the power or capability of a man in getting rich is in an inverse proportion to his reflective powers and in direct proportion to his impudence. It is perhaps good to be rich, but not to get rich, or to be always trying to get rich, and few men are less fitted to get rich, if they did try, than myself.[4] [1859]

He rationalized his disdain for inherited wealth.

How often we hear the remark upon such cases, "He is nobody's enemy but his own." But this is totally untrue, and every such spendthrift is really a worse enemy of society than the professional burglar, because he lives in the midst of an ever-widening circle of parasites and dependents, whose idleness, vice, and profligacy are the direct creation of his misspent wealth. He is not only vicious himself, but he is a cause of vice in others. Perhaps worse even than the vice is the fact that among his host of dependents are many quite honest people, who live by the salaries they receive from him or the dealings they have with him, and the self-interest of these leads them to look leniently upon the whole system which gives them a livelihood. Innumerable vested interests thus grow up around all such great estates, and the more wastefully the owner spends his income the better it seems to be for all the tradesmen and mechanics in the district. But the fundamental evil is the kind of sanctity we attach to property, however accumulated and however spent. Hence no real reform is ever suggested; and those who go to the root of the matter and see that the evil is in the very fact of inheritance itself, are scouted as socialists or something worse. The inability of ordinary political and social writers to follow out a principle is well shown in this matter. It is only a few years since Mr. Benjamin Kidd[5] attracted much attention to the principle of "equality of opportunity" as the true basis of social reform, and many of the more advanced political writers at once accepted it as a sound principle and one that should be a guide for our future progress. Herbert Spencer, too, in his volume on "Justice,"[6] lays down the same principle, stating, as "the law of social justice" that "each individual ought to receive the benefits and evils of his own nature and consequent conduct; neither being prevented from having whatever good his actions normally bring him, nor allowed to shoulder off on to other persons whatever ill is brought to him by his actions." This, too, has, so far as I am aware never been criticized or objected to as unsound, and, in fact, the arguments by which it is supported are unanswerable. Yet no one among our politicians or ethical writers has openly adopted these principles as a guide for conduct in legislation, or has even seen to what they inevitably lead. Stranger still, neither Mr. Kidd nor Herbert Spencer followed out their own principle to its logical conclusion, which is, the absolute condemnation of unequal inheritance. Herbert Spencer even declares himself in favour of inheritance as a necessary corollary of the right of property rightfully acquired; and he devotes a chapter to "The Rights of Gift and Bequest." But he apparently did not see, and did not discuss the effect of this in neutralizing his "law of social justice," which it does absolutely. . . .

It is in consequence of *not* going to the root of the matter, and *not* following an admitted principle to its logical conclusion, that the idea prevails that it is only the *misuse* of wealth that produces evil results. But a little consideration will show us that it is the inheritance of wealth that is wrong in itself, and that it necessarily produces evil. For if it is right, it implies that *inequality* of opportunity is right, and that "the law of social justice" as laid down by Herbert Spencer is *not* a just law. It implies that it is *right* for one set of individuals, thousands or millions in number, to be able to pass their whole lives without contributing anything to the well-being of the community of which they form a part, but on the contrary keeping hundreds, or perhaps thousands, of their fellow men and women wholly engaged in ministering to *their* wants, *their* luxuries, and *their* amusements. Taken as a whole, the people who thus live are no better in their nature – physical, moral, or intellectual – than other thousands who, having received no such inheritance of accumulated wealth, spend their whole lives in labour, often under exhausting, unhealthful, and life-shortening conditions, to produce the luxuries and enjoyments of others, but of which they themselves rarely or more often *never* partake. Even leaving out of consideration the absolute vices due to wealth on the one hand and to poverty on the other, and supposing both classes to pass fairly moral lives, who can doubt that *both* are injured morally, and that *both* are actually, though often unconsciously, the causes of ever-widening spheres of demoralization around them? If there is one set of people who are tempted by their necessities to prey upon the rich, there is a perhaps more extensive class who are in the same way driven to prey upon the poor. And it is the very *system* that produces and encourages these terrible inequalities that has also led to the almost incredible result, that the ever-increasing power of man over the forces of nature, especially during the last hundred years, while rendering easily possible the production of all the necessaries, comforts, enjoyments, and wholesome luxuries of life for every individual, have yet, as John Stuart Mill declared, "not diminished the toil of any worker," but even, as there is ample evidence to prove, has greatly increased the total mass of human misery and want in every civilized country in the world.

And yet our rulers and our teachers – the legislature, the press, and the pulpit alike – shut their eyes to all this terrible demoralization in our midst, while devoting all their energies to increasing our already superfluous and injurious wealth-accumulations, and in turn compelling other peoples, against their will, to submit to our ignorant and often disastrous rule.[7] [1905] ᴥ

> Millionaires symbolized for Wallace everything wrong with society. He must have enjoyed (mis-)quoting Adam Smith to prove his point.

ᴥ   Between 1862 and 1873 1 find that 162 persons died with fortunes of over a quarter of a million. In the next ten years they had increased to 208 persons who had died with fortunes of over a quarter of a million. This is an increase of over 29 per cent. The detailed figures show still more remarkable results, because they show that the increase was still more rapid in very great fortunes over a million. In addition to that a very considerable number of great landowners have died who paid no probate duty, but whose capitalised fortunes have been from one to five millions sterling each. We have not the exact figures, but still we know that their fortunes have been of late increasing, owing to the increase of our large towns and the enormous increase of ground rents which have arisen in them. The main result is, that a few, that is comparatively few, have become much richer than they ever were before; and it appears to me that it is a demonstrable fact that, when those who are very rich suddenly become more numerous and still richer, without any increased power of wealth-creation independent of labour, then, as a necessary result, those who are poor become poorer.

This principle was laid down very clearly by Adam Smith, strange to say, in the very first sentence of his *Wealth of Nations*, but I do not know that much attention has been paid to it. The sentence is this. He says:

> "The actual labour of every nation is the fund which originally supplies it with all the necessaries and conveniences of life which it actually consumes, and which consists always either in the immedi-

ate produce of that labour or in what is purchased with that produce from other nations."[8]

This lays down a proposition perfectly clear, that there is no other source whatever of wealth in the country than the produce of the labour of its people. Hence it follows absolutely and indisputably, that if a larger proportion of that wealth goes to the few, a smaller proportion must remain with the many. As some people may not clearly see the bearing of this statement of Adam Smith, let me just illustrate it by a few particular cases. It is quite evident that all the wealth of the country is produced by labour, or by the use of labour and capital combined, and everybody who gets wealth must get a portion of this total amount. There is no other source from which he can get it. Whether he obtains it in the form of rent or from the taxes it comes exactly to the same thing, it can only come out of the produce of labour. In the same manner, whether he gets it in payment of wages or remuneration for professional services, those who pay it can only have got it, directly or indirectly, by labour. Consequently the fact is indisputable, that the produce of our labour measures the whole available wealth produced by us in the country, and that wealth has to be distributed by various methods among the whole community. Consequently, if it is clearly proved, as I think it is – to prove it in detail would require a much more complete examination of the statistics of the country, but I am sure it can be proved – that the large body of the very rich have been steadily growing richer, then it follows as a logical result that the remaining body, or at least a portion of the remaining body, must have been growing poorer in proportion.[9] [1886]   ह⇒

Speculation on the stock market disgusted Wallace.

⇐§ Our Stock Exchanges ... are used largely for pure gambling, which, owing to its vast extent and being carried out under business forms, is perhaps more ruinous than any other. But this form of gambling goes on unchecked, and is generally accepted as quite honest business. Yet ordinary betting on races and other forms of direct gambling are hypocritically condemned as immoral and criminal.[10] [1913]   ह⇒

He dismissed "fictitious" (paper) wealth in the same way as he did millionaires and what he termed "hereditary plutocrats"[11] – those who exist on hereditary wealth.

✑ REAL AND FICTITIOUS WEALTH

Nothing is more certain than that wealth – real wealth – is continually used up and destroyed, and as continually reproduced by fresh labour. All articles of food, of clothing, of furniture, and even tools, machines, dwellings, books, works of art, and ornaments, are either wholly or partly used up day by day or year by year, and as continually reproduced; and it is those which are most continually consumed and reproduced which are, pre-eminently, beneficial wealth. If, then, a man acquires a large surplus of this real wealth beyond what he can consume himself, it must be either profitably consumed by others or be wasted by natural decay. In either case it soon ceases to exist. But our fiscal and legislative arrangements enable a man to change this *perishable* wealth into securities which bring him a *permanent* income – an income supposed to be perpetual, but at all events lasting long after the wealth of which it is the symbol, and which is supposed to produce it, has totally disappeared. Incomes thus derived constitute a tax or tribute on the community, for which it receives nothing in return, and may truly be termed fictitious wealth.

To show that this is so, and at the same time to exhibit clearly not only the evil but the inherent absurdity of these arrangements, let us consider a few indisputable facts. Every year much surplus wealth is accumulated by individuals and is invested as reproductive capital. This invested capital goes on producing an income which, it is supposed, is and ought to be permanent; and as more and more surplus wealth is continually produced and invested, it is evident that the permanent incomes thus derived will continually increase, and thus in each successive generation a larger and larger number of persons will be enabled to live in idleness on incomes derived from this invested capital. But this is a state of things that evidently carries with it its own destruction, since a time must come when the number of idle persons living on "independent incomes," and the aggregate of those incomes, will be so great that the working portion of the population will be ground down to penury in the attempt to pay them, and this will more quickly come about from the tendency of a large idle and luxurious class to withdraw labour from beneficial work to the production of useless luxuries. The end will inevitably be the worst kind of revolution, brought about by the determination of the labouring poor no longer to support the burthen of an ever-increasing class of idle rich.[12] [1900]   ✑

As to ways in which to bring about change, Wallace, despite being in many ways a doctrinaire socialist, had his own ideas. For example, he felt that strikes were ineffective.

❧ Has not the time come when the workers should cease to employ so rude, inefficient, and wasteful a method of improving their condition as by means of STRIKES? In most cases a strike effects little or nothing of a permanent nature, nothing but what may be lost within a year or two, nothing that tends to raise the whole body of the workers in any country. The strike may have been an essential weapon in the past – perhaps the only weapon the worker possessed. Now, however, all the higher grades of workers are better educated, better organised, and have higher ideals. They have learnt the benefits of co-operation and of union; they have accumulated funds which may be reckoned by millions; and to waste those funds in keeping thousands and tens of thousands of men idle during a strike is one of those economic and social blunders which, in their effects, are often worse than crimes. Instead of keeping men idle for months, in order to obtain a small and perhaps temporary advance in wages or reduction of working hours, would it not be wiser to adopt a totally different method, one which would be much more dreaded by the employers, because it would tend to produce a permanent, instead of a temporary, rise of wages. That method is, *competition with the employers instead of strikes against them*; and it is to be effected by saving and accumulating all the money now spent in keeping men *idle*, and, as occasion arises, using it for the purpose of acquiring shops and tools by which the unemployed in each trade may be gradually absorbed and kept at *work*. Then, step by step, wage-earners would be withdrawn from employers' shops or factories to work in those of their union. Even if, at first, some of these shops were not able to pay full rates of wages, still the men would earn *something* instead of *nothing*, and they could hardly earn less than the usual pay during a strike.[13] [1899] ❧

Despite having strong views on how best to effect change, Wallace was happy to endorse any approach – even strikes – that he saw to be advancing the cause of the working man. In an interview given in his ninetieth year, he was in fact positively effusive on the subject of strikes:

❧ "And now about this strike," says Dr. Wallace, his eyes smiling, his voice chuckling with an almost schoolboy enthusiasm. "It's a grand thing. Oh, a really grand thing! First the Railway Strike. And now the Coal Strike.[14] Splendid – the whole thing – splendid!" He laughs quietly, but blithely, for a moment. Then, still smiling, but with greater earnestness and a deep gratitude, he adds: "I'm glad I've lived to see it. I'm glad of that!

". . . These big strikes, really big strikes, bring home to people the fact of their dependence on the working man – the poor working man! They are apt to criticise him; to say he is one thing and another; to find fault with him; to be angry with him; in their hearts to hate and despise him. Well, a strike like this says to the self-satisfied and criticising public: The working man may be everything you say he is, your criticisms may be entirely just, you alone may be virtuous, thrifty, industrious, and logical; nevertheless, you depend upon the man you criticise, you absolutely depend upon him for your very life!

"And that is a good thing for them to know. Because it must lead the more intelligent of them to perceive the importance, the enormous importance, of so ordering the national life that those upon whom they depend, absolutely and entirely, for the necessities of existence are properly provided for, are living in conditions which leave no reasonable room for complaint. It makes us all think about the working man, the poor working man!

"What do these strikes mean? They mean that the working classes are living their lives conscious of injustice, burdened by a sense of grievance. Every strike reveals some definite wrong. Now, is that not a perilous state of things? How can a community be safe, how can any national life be wholesome and strong where the vast majority of the population is convinced that it is being cheated and wronged by the minority? Little strikes only serve to make us uneasy. The big strikes bring home to us the essential truth – that we depend upon the working classes, that our whole national position is in their hands, and that these essential people in the State are conscious of injustice, conscious of wrong. Now, is that a rational state of things?"[15] [1912] ❧

Wallace's view on nationalization was also somewhat quirky. Here, in a letter written in 1908 to the editor of *The New Age*, he lays out how his ideas would apply to the railways.

❧ As it seems to be a very general idea, even among Socialists, that the two operations – nationalisation and purchase – must go together, will you allow me space to point out that, while nationalisation is in the highest degree advisable, and may be effected at once by a very simple enactment, purchase is equally unadvisable and unnecessary, and had far better be left till a much later period, when in all probability some general method of dealing with similar claims to other forms of nationally produced wealth may be found practicable.

I may take it for granted that every reader of *The New Age* recognises the advantages to the public, as individuals, in the whole of the railways being worked with the sole view of the maximum of use and enjoyment of the people, so far as is consistent with the safety and well-being of the great army of employees, which will itself tend to secure the safety of the public; while to the nation, this complete unity of organisation and management will be of incalculable advantage as a safeguard against foreign invasion.

But these, and many other collateral advantages will accrue, just as certainly, and even more rapidly, by the State taking over the fixed and rolling stock of the whole of the railways, to be managed and worked in the public interest, while continuing to pay to the present owners of the railways – the shareholders and possessors of every kind of railway stock – that proportion of the net profits to which they are now equitably entitled.

It is, I believe, generally estimated that the economies which would be effected by the co-ordination of the whole system would amount to many millions annually, and this great saving would all be expended in reduced fares, better services, higher wages, and shorter hours of work, by which all shareholders and employees, as well as the whole of the public, would greatly benefit.

But the increased facilities to all who use the railways, and the abolition of the needless and often irritating restrictions of most of the existing managements, would certainly lead to a large increase of traffic, and thus render any considerable discharge of existing railway employees unnecessary, while the position of all would be much improved.[16] [1908]   ॐ

Globalization

Wallace's objections to globalization remind us just how topical much of his work remains.

ॐ Admitting that free trade will necessarily benefit a country materially, it does not follow that it will be best for that country to adopt it. Man has an intellectual, a moral, and an æsthetic nature; and the

exercise and gratification of these various faculties is thought by some
people to be of as much importance as cheap cotton, cheap silk, or
cheap claret. We will suppose a small country to be but moderately
fertile, yet very beautiful, with abundance of green fields, pleasant
woodlands, picturesque hills, and sparkling streams. The inhabitants
live by agriculture and by a few small manufactures, and obtain some
foreign necessaries and luxuries by means of their surplus products.
They have also abundance of coal and of every kind of metallic ore,
which pervade their whole country but which they have hitherto
worked only on a small scale for the supply of their own wants. They
are a happy and a healthy people; their towns and cities are comparatively small; their whole population enjoy pure air and beautiful
scenery, and a large proportion of them are engaged in healthy
outdoor occupations. But now the doctrines of free trade are spread
among them. They are told that they are wasting their opportunities:
that other nations can supply them with various articles of food and
clothing far cheaper than they can supply themselves; while they, on
the other hand, can supply half the world with coal and iron, lead
and copper, if they will but do their duty as members of the great
comity of nations, and develop those resources which nature has so
bountifully given them. Visions of wealth and power float before
them; they listen to the voice of the charmer; they devote themselves
to the development of their natural resources; their hills and valleys
become full of furnaces and steam-engines; their green meadows are
buried beneath heaps of mine-refuse or destroyed by the fumes from
copper-works; their waving woods are cut down for timber to supply
their mines and collieries; their towns and cities increase in size, in
dirt, and in gloom; the fish are killed in their rivers by mineral
solutions, and entire hill-sides are devastated by noxious vapours; their
population is increased from ten millions to twenty millions, but most
of them live in "black countries" or in huge smoky towns and, in
default of more innocent pleasures, take to drink: the country as a
whole is more wealthy, but, owing to the large proportion of the
population depending upon the fluctuating demands of foreign trade,
there are periodically recurring epochs of distress far beyond what
was ever known in their former condition.

With this example of the natural effects of carrying out the essential
principles of free trade, another people in almost exactly similar
circumstances determine that they prefer less wealth and less population, rather than destroy the natural beauty of their country and give
up the simple, healthful, and natural pleasures they now enjoy. They

accordingly, by the free choice of the people in Parliament assembled, forbid by high duties the exportation of any minerals, and even regulate the number of mines that shall be worked, in order that their country shall not be changed into a huge congeries of manufactories. A balance is thus kept up between different industries, all of which are allowed absolutely free development so long as they do not interfere with the public enjoyment, or cause any permanent deterioration to the water, the soil, or the vegetation of the country. They are in fact protectionists, for the purpose of preserving the beauty and enjoyability of their native land for themselves and for their posterity. Free trade would destroy these, and give them instead cheaper wine and silk, stale eggs instead of fresh, and butter ingeniously manufactured from various refuse fats. They prefer nature to luxury. They prefer intellectual and æsthetic pleasures, with fresh air and pure water, to an endless variety of cheap manufactures. Are they morally or intellectually wrong in doing so?[17] [1879]   &

War and Imperialism

As a pacifist and a socialist, Wallace recognized that the elimination of war would require a popular movement. It is perhaps fortunate that he – utopian, optimist, and supreme believer in the essential goodness of his fellow human – died when he did. The advent of World War I, a little more than a year after his death, would have been too horribly disillusioning.

& It is, I think, clear that no hope of a complete solution – hardly even of amelioration – is to be expected from the ruling classes, urged on as they are on the one hand by those who are ever seeking for place and power, or for official appointments in newly-acquired territories, and on the other hand by the military class, who ever seek to justify their existence and the enormous burden they are to the nation by obtaining for it extensions of territory or military glory, and with either of these an extension of their own influence. It is, therefore, the *people*, and the people alone, that must be relied upon to banish militarism and war, and for this end every possible effort must be made to educate and enlighten them, not only as to the horrors and iniquity of war, but as to the utter inadequacy and worthlessness of almost all the causes for which wars are waged. They must be shown that all modern wars are dynastic; that they are caused by the ambition, the interests, the jealousies, and the insatiable greed of power of their rulers, or of the great mercantile and financial classes which have power and influence over their rulers; and that the results of war are *never* good for the people, who yet bear all its burthens.[18] [1899] ও৶

The evils of warfare are self-evident. Wallace's views on colonialism, however, were not so straightforward: he was only semi-enlightened by today's standards, but was extraordinarily progressive for his day. As we have seen, he was not a racist and held high opinions of non-Europeans, but he nevertheless persisted in using terms like "lower races" and in believing that the technological prowess of Western Europeans guaranteed their expansionist agendas at the expense of indigenous peoples – "that the higher – the more intellectual and moral – must displace the lower and more degraded races".[19] He propounded a brand of pacifist international politics whose hallmark was a well-meaning paternalism.

৶ Robert Blatchford[20] [comes to] ... the one conclusion: We must increase Army, Navy, and Home Defences, and be prepared to fight all the world. Not one word about there being any alternative to all this blood-and-iron bluster and defiance; not one syllable to show that the writer is a great Socialist teacher, a believer in the goodness of human nature and the brotherhood of man. "But," he replies by his heading, "this is very good in theory, and very true, but it is not Practical Politics. The danger is urgent. Tell us, ye Labour leaders, what you propose to do *now*."

I am not a Labour leader, but I hope I am a true friend of Labour and a true Socialist; and I will now state the case as it appears to me, and suggest what, in my opinion, is the only course of action worthy of Socialism or politic for Labour, and, besides, the only course which has the slightest chance of succeeding *in the long run*: in one word, the only RIGHT course.

It is a notorious and undeniable fact that we – that is, our Governments – are, with a few exceptions, hated and feared by almost all other Governments, especially those of the Great Powers. Is there no cause for this? Surely we know there is ample cause. We have either annexed or conquered a larger portion of the world than any other Power. We have long claimed the sovereignty of the sea. We hold islands and forts and small territories offensively near the territories of other Powers. We still continue grabbing all we can. In disputes with the powerful we often give way; with the weak and helpless, or those we think so, we are – allowing for advance in civilisation – bloody, bold, and ruthless as any conqueror of the Middle Ages. And with it all we are sanctimonious. We profess religion. We claim to be more moral than other nations, and to conquer and govern and tax and plunder weaker peoples for *their* good! While robbing them we actually claim to be benefactors! And then we wonder, or profess to wonder, why other Governments hate us! Are they not fully justified in hating us? Is it surprising that they seek every means to annoy us, that they struggle to get navies to compete with us, and look forward to a time when some two or three of them may combine together and thoroughly humble and cripple us? And who can deny that any just Being, looking at all the nations of the earth with impartiality and thorough knowledge, would decide that we deserve to be humbled, and that it might do us good?

Now the course I recommend as the only true one is, openly and honestly, without compulsion and without vain-glory, to do away with many of the offences to other peoples, and to treat all subject peoples and all foreign Powers on exactly the same principles of equity, of morality, and of sympathy, as we treat our friends, acquaintances, and neighbours with whom we wish to live on friendly terms.

And, to begin with, and to show that our intentions are genuine, I would propose to evacuate Gibraltar, dismantle the fortress, and give it over to Spain; Crete and Cyprus should be free to join Greece; Malta, in like manner, would be given the choice of absolute self-government under the protection of Britain, or union with Italy. But the effect of these would be as nothing compared with our giving absolute internal self-government to Ireland, with protection from attack by any foreign Power; and the same to the Transvaal and Orange Free State;[21] and this last we should do "in sackcloth and ashes," with full acknowledgment of our heinous offences against liberty and our plighted word.

Now we come to India, which our friend Blatchford seems to consider the test case. And so it is; for if ever there was an example of a just punishment for evil deeds, it is in the fact that, after a century of absolute power, we are still no nearer peace and plenty and rational self-government in India than we were half-a-century ago, when we took over the government from the "[East India] Company" with the promise to introduce home-rule as soon as possible. And now we have a country in which plague and famine are chronic – a country which we rule and plunder for the benefit of our aristocracy and wealthy classes, and which we are, therefore, in continual dread of losing to Russia.

If we had honestly kept our word, if we had ruled India with the one purpose of benefiting its people, had introduced home-rule throughout its numerous provinces, states, and nations, settling disputes between them, and guarding them from all foreign attack, we should by now have won the hearts of its teeming populations, and no foreign Power would have ventured to invade a group of nations so united and so protected. Such a position as we might have now held in India – that of the adviser, the reconciler, and the powerful protector of a federation of self-governing Native States – would be a position of dignity and true glory very far above anything we can claim to-day.

But, it will be replied, all this is foolish talk; it will be a century before the British people will be persuaded to give up its possessions and its power; and, in the meantime, if we do not defend ourselves we shall not have the opportunity of being so generous, hardly shall we keep our own liberties. I have not so low an opinion of my countrymen as to believe that they really *wish* to keep other peoples subject to them against their will: that they are really *determined* to go on denying that freedom to others which is so dear a possession to themselves. And if there is not now a majority who would agree to act at once as I suggest, I am pretty confident that there is, even now, a majority who would acknowledge that such action is theoretically just, and that they would be willing to do it by degrees, and as soon as it is safe, etc. To look forward to it, in fact, as an ideal to be realised at some future time.

Now, what I wish to urge is, that it is of the most vital importance to us, now, that all who agree with me that there can be no national honour or glory apart from justice and mercy, and that to take away people's liberty and force our rule upon them against their will is the greatest of all national crimes, should take every opportunity of making their voices heard. If, for instance, every Socialist in our land, and I hope a very large proportion of workers and advanced thinkers who may not be Socialists, would agree to maintain this as one of their fundamental principles, to be continually brought before the people through the Press and on the platform, to be urged on the Government at every opportunity, and to be made a condition of our support of every advanced Parliamentary candidate, we should create a body of ethical opinion and feeling that would not only be of the highest educational value at home, but which would influence the whole world in their estimate of us. It would show them that though our Government is bad – as all Governments are – yet the people at heart are honest and true, and that it will not be very long before the people will force their Governments to be honest also.

This, I submit, *would* be really "practical politics." At the present day we *have* got so far as this – that none of the Great Powers wages a war of aggression and conquest against another Power without some quarrel or some colourable pretence of injury. But surely the fact of there being such a party as I have outlined, and especially if it would (as I think it certainly could) compel the next Government to make some of the smaller concessions here indicated and adopt the general principle of respecting the liberties of even the smallest nationalities, would so reduce the amount of envy and hatred with which we are now regarded as to considerably reduce the danger of combined aggression upon us.[22] [1904]   ❧

American imperialism was as roundly condemned as the British variety.

❧ In conclusion, I again emphasise the fact, that we, as a nation, have no right whatever to claim any superiority as regards our treatment of those less civilised people with whom we come in contact. Our conduct towards the Boers and Zulus in South Africa, the Burmese, and many of the hill tribes on our Indian frontier, and the Chinese in our wars growing out of the opium trade, has been certainly not better than what the Americans have done or are likely to do in Cuba and the Philippines. But many of us have always protested against our own unfair dealings with those inferior races, and have denounced the conduct of our Governments as unworthy of a civilised and professedly Christian people. And if we now venture to express our disappointment that our American kinsfolk are apparently following our bad example, it is because, in the matter of the rights of every people to govern themselves, we had looked up to them as being about to show us the better way, by respecting the aspirations towards freedom, even of less advanced races, and by acting in accordance with their own noble traditions and Republican principles.[23] [1899]   ❧

Wallace's paternalism, however, caused him to adopt some paradoxical positions. When in Borneo (1854–56), he spent a great deal of time as the guest of Sir James Brooke, the "White

Rajah" of the northern province of Sarawak. Brooke, an "adventurer" in the true Victorian sense, had essentially annexed Sarawak as his own personal fiefdom. The future president of the Land Nationalization Society, however, was undisturbed, referring subsequently to Brooke's regime as "a splendid and almost unrivalled success in the art of government".[24] In fact, Wallace had been impressed by the wealth of colonial opportunities available in the Tropics during his first major trip overseas, to the Amazon.

When I consider the excessively small amount of labour required in this country, to convert the virgin forest into green meadows and fertile plantations, I almost long to come over with half-a-dozen friends, disposed to work, and enjoy the country; and show the inhabitants how soon an earthly paradise might be created, which they had never even conceived capable of existing.

It is a vulgar error, copied and repeated from one book to another, that in the tropics the luxuriance of the vegetation overpowers the efforts of man. Just the reverse is the case: nature and the climate are nowhere so favourable to the labourer, and I fearlessly assert, that here, the "Primeval" forest can be converted into rich pasture and meadow land, into cultivated fields, gardens, and orchards, containing every variety of produce, with half the labour, and, what is of more importance, in less than half the time than would be required at home, even though there we had clear, instead of forest, ground to commence upon.[25] [1853]

Wallace even wrote an entire article – with the non-ironic title "White Men in the Tropics" – to demonstrate that Caucasians were not constitutionally incapable of living in a hot climate.

The fact is that white men *can* live and work anywhere in the tropics, *if they are obliged*, and unless they are obliged they will not, as a rule, work even in the most temperate regions. Hence, wherever there are inferior races, the white men get these to work for them, and the kinds of work performed by these inferiors become *infra dig.* for the white man. This is the real reason why the myth, as to white men not being able to work in the tropics, has been spread abroad. It applies in most cases to agricultural work only, because natives can usually be got to do this kind of work, while that of the skilled mechanics has usually to be done by white men. And another reason is that it is only by getting cheap labour in quantity that fortunes can be made in most tropical countries. But when people come to recognize that the fortune-makers, whether by gold mining, speculating or any of the various forms of thinly-veiled slavery, are not by any means the happiest, the healthiest or the wisest men, whereas those who really *work*, under the best conditions, so as to receive the whole produce of their labour, may be both healthy and happy, and usually live longer and enjoy life more, and by working in association may obtain all the necessaries and comforts of existence – then the enormous advantage of living in the best parts of the tropics will become evident. For not only is nature so much more productive that equal amounts of produce may be obtained with half or perhaps a quarter of the labour required in northern lands, but the essentials of a happy and an easy life are so much fewer in number. Houses may be slighter and far less costly: clothing may be reduced to less than half what is required here; fuel is only wanted for cooking; while the enjoyability of the early morning hours is so great that everybody rises before the sun, and thus comparatively little artificial light is required. When all this is fully realized we may hope to see co-operative colonies established in many tropical lands, where families of the same grade of education and refinement may so live as really to enjoy the best that life can give them. Thus only, in my opinion, can the best use be made of the tropics.[26] [1899] ॐ

> Wallace struggled to find a middle road between the imperial values of the Victorian society he belonged to and the respect he had gained for non-European peoples in the course of his travels. The product was a paradoxical mix of paternalism towards, and esteem for, indigenous peoples. However, it was Wallace's insistence that "the coloured races are men of fundamentally the same nature as ourselves" that set him apart from his era, even from the progressive liberals of the day. For example, his friend and colleague T. H. Huxley, as good a liberal as any and active in the anti-slavery movement, declared that the "higher places in the hierarchy of civilisation will assuredly not be within the reach of our dusky cousins".[27] The following extract illustrates how Wallace was at once both of his time and ahead of it.

ॐ So long as we possess colonies in which a considerable native population still exists we should, I think, always retain our guardian-ship of those natives in order to protect them from the oppression and cruelty which always occurs when a young, and mainly wealth-seeking community has absolute power over them. Where these natives are numerous and energetic, and are rapidly acquiring our education, our religion, and the outward form at all events of our civilisation, things cannot remain as they are. What the ultimate condition of such mixed communities may be it is difficult to say, but, whatever the future may have in store for us, it is certain that a method which recognises that the coloured races are men of fundamentally the same nature as ourselves, and which aims at developing the best that is in them, by granting them some at least of the elementary rights of men and citizens, is more likely to bring about a satisfactory solution of this difficult problem, than that system of contemptuous superiority and denial of all political and social claims that has hitherto so largely prevailed.[28] [1906] ❧

Coda: Wallace and Darwin

By the end of his life Wallace was famous, widely recognized as "one of the supreme figures in modern science."[1] Today he scarcely merits more than a footnote in biology textbooks: he is remembered merely as the obscure naturalist whose feverish insight put an end to Darwin's lifelong habit of procrastination, and as the discoverer of the eponymous line that divides the Asian and Australasian faunal regions. Why, in less than a century, has Wallace's reputation dwindled almost to nothing?

Paradoxically, what was in many ways Wallace's finest hour – the discovery of natural selection – may have contributed to his neglect by posterity. It has caused him to be forever bracketed with Darwin, but not as an equal; he has been condemned always to play Watson to Darwin's Holmes. Even Wallace's biographers have been sucked into the Darwin vortex: a 1966 biography, *Darwin's Moon*,[2] gives Darwin star billing as the Sun.

As we have seen, there is a school of thought[3] that holds that evolutionary biology's laurels belong rightfully to Wallace, not Darwin.[4] Most commentators, however, are happy to consider Darwin the senior partner in the pairing on the grounds of his long engagement with the problem, but, even if this were allowed, a junior partner surely merits, say, 30% of the kudos, not, as in this case, 1%. Today Darwin is lionized, and Wallace all but forgotten. Almost every word Darwin wrote is in print; almost every word Wallace wrote is out of print.

Despite the obvious overlaps between their careers – formative years spent immersed in the natural history of distant regions, and later years in the upper echelons of Victorian science – Darwin and Wallace were very different. How did those differences translate into the profound contrast in the ways in which they are remembered? The two men pursued very different career strategies. Wallace, as we have seen, was socially engaged. He viewed the prominence won him by his biological work as an opportunity to diversify and address what turned out to be virtually every issue under the sun. Darwin, on the other hand, knuckled down. In the twenty-three years between the publication of the *Origin* and his death, he published ten books, each one building in some way on the platform furnished by that seminal work. His final book, published in 1881, *The Formation of Vegetable Mould, through the Action of Worms, with Observations on Their Habits*, is the ultimate illustration of the Darwin strategy. His theory of evolution is based on extrapolation: he borrowed the uniformitarianism of the geologists to argue that processes that have minor effects on a day-to-day basis can have major consequences over long periods. Thus the subtle action of natural selection may be barely discernible from one generation to the next, but over a few thousand generations significant changes will occur. So too it is with the impact of earthworms on landscapes: only over long periods will their soil-churning activities be noticeable. Darwin stuck to his theme to the very end; he consolidated. Wallace, in contrast, went off at tangents.

Moreover, Wallace wrote no *Origin of Species*. The announcement of the Darwin–Wallace theory at the Linnaean Society in July 1858 actually occasioned very little comment. The then president of the Society, Thomas Bell, summarized the year's events, "The year has not, indeed, been marked by any of those striking discoveries which at once revolutionize, so to speak, the department of science on which they bear."[5] It was the *Origin* that created the controversy, that put natural selection on the map. Wallace's major work on evolution, *Darwinism* (1889), came thirty years later, and is anyway more of a textbook than – to use Darwin's own characterization of the *Origin* – "one long argument".[6]

Not only did Wallace depart from the biological straight and narrow, he was also wrong – wildly so – on a number of issues, and in public. His unfortunate endorsement of phrenology[7] is the most startling example, but – in most people's eyes – his embrace of spiritualism and outspoken support for the "Anti-Vaccination League" furnish others. Darwin made mistakes too, but they were generally only manifest to relatively small audiences of specialists. For example, after publishing the *Origin*, he became increasingly wed to the theory of inheritance of acquired characters, and even introduced his own version of it, Pangenesis. Wallace too made biological errors. Most biologists today would agree with Darwin, and not Wallace, that female choice is an important factor in sexual selection. It was not, however, these in-house technical miscalls that cost Wallace; it was the high-profile public claims. In essence, Wallace paid a price for taking intellectual risks. Darwin – as his procrastination *vis-à-vis* publishing on evolution amply illustrates – was risk-averse.

The lesson history is apparently teaching us here – that "knuckling down" is rewarded by the makers of reputations, while a willingness to take on diverse issues is not – is unfortunate because it suggests that scientists should remain narrowly focused on their particular corners of science, and not venture out into the public domain. For Wallace, however, considerations of reputation and posterity were irrelevant. His priorities – social justice and the search for truth – made him the prototypical socially engaged scientist. Thus Wallace's review of the achievements of the nineteenth century, *The Wonderful Century*, went beyond merely lauding the scientific and technological developments of the era to assessing the impacts that those developments had had on ordinary people. In the penultimate chapter, "The Demon of Greed", he pointed out that many aspects of the industrial revolution had made people's lives worse, not better. Today, as technology comes to play an ever greater role in our lives, there is a pressing need for latter-day Wallaces – scientists willing to forsake their ivory towers (or their biotech companies) and become publicly engaged with social issues. Neglected though he may be, Wallace's example is nevertheless inspirational.

Part of the standard Wallace mythology is the suggestion that he was subtly discriminated against by the Victorian scientific elite for reasons of class, but this was surely only a very minor factor. T. H. Huxley, the most voluble and dynamic member of that elite, was himself born above a village butcher's shop. More significant, I suspect, was Wallace's want of social skills. At a gawky six foot one, he was shy, awkward, and diffident.

❧ Up to middle age, and especially during the first decade after my return from the East, I was so much disinclined to the society of uncongenial and commonplace people that my natural reserve and coldness of manner often amounted, I am afraid, to rudeness. I found it impossible, as I have done all my life, to make conversation with such people, or even to reply politely to their trivial remarks. I therefore often appeared gloomy when I was merely bored. I found it impossible, as some one had said, to tolerate fools gladly; while, owing to my deficient language-faculty, talking without having anything to say, and merely for politeness or to pass the time, was most difficult and disagreeable. Hence I was thought to be proud, conceited, or stuck-up.[8] [1905] ð

That Wallace was strongly opinionated and also inclined to disregard social niceties translated, as he grew older, into a simple lack of tact. He brought his characteristic bluntness to even the most exalted of social situations: on visiting Alfred Tennyson in 1884, he lambasted the House of Lords and hereditary peerage in general despite Tennyson's having earlier that year been made a peer.[9]

Wallace's chronic inability to land a job was probably a reflection in part of his social awkwardness. But it went beyond that. Having made his mind up on what was right (natural selection, spiritualism, socialism), he would doggedly – and brilliantly – defend that position, regardless of the data to hand. Thus his critical facility was fully engaged when he went overboard in savaging Samuel Haughton for criticizing Darwin,[10] and fully suspended when he testified in court in support of a fraudulent spiritualist medium, Henry Slade.[11] I think this trait – twisting the data to fit his worldview – can even be seen in some of his biological arguments. Take, for example, his response to Lord Kelvin's (erroneous) claim[12] that the planet was too young for evolution to have occurred: he contrived an ingenious and plausible scenario involving accelerated rates of evolutionary change in the early phases of earth history. Wallace's view is wrong, and, it turns out, unnecessary, and we cannot help being a little suspicious of a scientist who has such an uncanny ability to paper over every crack – whether real or not – in his pet theory. Wallace's behaviour occasionally hinted at zealotry.

That Wallace was so self-effacing no doubt also contributed to the rapid dimming of his star.

ð I cannot understand why you or anyone should care about my being an F.R.S. [Fellow of the Royal Society, Britain's highest scientific accolade], because I have really done so little of what is usually considered scientific work to deserve it. I have for many years felt almost ashamed at the amount of reputation and honour that has been awarded me.[13] [1893] ð

Even Wallace's autobiography lacks the self-congratulation characteristic of the genre – it includes, for instance, an account of "certain marked deficiencies in my mental equipment".[14] Wallace's gentlemanly deference to Darwin is famous:[15] not only did he not complain about Darwin's arrangements for joint publication on natural selection, which were made without consulting Wallace, but he entitled his major later work on their joint discovery *Darwinism*.

Despite his best efforts, however, Wallace is today undergoing a minor renaissance. His neglected grave in Broadstone, Dorset, has been renovated, and the Linnean Society – at which the Darwin–Wallace joint paper was presented in 1858 – has finally hung a Wallace portrait beside Darwin's. A number of new biographies and critical works have, or are about to, come out. If, as he believed he would be, Wallace is still around and in a position to view these goings-on, I suspect he is not very interested. Humanitarian that he was, he would have valued more the simple testimony of the people who lived and worked with him during his travels: Old Jeronymo, a "poor wanderer" in the Amazon, who wanted Wallace to know that he was grateful for, and "still has the shirt that Senhor Alfredo gave him";[16] the Aru Islanders who implored Wallace to "stay here a year or two";[17] and Ali, Wallace's faithful local assistant during his years in Southeast Asia, who in 1907, 45 years after he had last seen Wallace, still proudly introduced himself to an American biologist visiting the Moluccan Islands as "Ali Wallace".[18]

Notes

Editor's Note

1. Charles H. Smith, ed., *Alfred Russel Wallace: An Anthology of His Shorter Writings*, Oxford University Press, Oxford, 1991. Updated at: http://www.wku.edu/~smithch/bibintro.htm.
2. Ibid.

A Biographical Sketch

1. Aged twenty-one, Wallace's "very imperfect school training, added to my shyness and want of confidence, must have caused me to appear a very dull, ignorant, and uneducated person to numbers of chance acquaintances." ML I 224.
2. Ibid. 7.
3. Ibid. 10.
4. Ibid. 12.
5. Ibid.
6. Ibid. 13.
7. Ibid. 29.
8. Ibid. 46–7.
9. Ibid. 70.
10. Ibid. 44.
11. Ibid. 65.
12. Ibid. 58.
13. Ibid. 51.
14. Mungo Park (1771–1806), explorer who drowned in the Niger after being attacked.
15. Dixon Denham (1786–1828) and Hugh Clapperton (1788–1827) reached Lake Chad and explored the Sudan.
16. Produced 1830–39 by Thomas Hood (1799–1845), poet and humorist.
17. ML I 74.
18. Ibid. 108.
19. Ibid. 137.
20. ML I 141–2. Undated letter, probably 1839/40.
21. Ibid. 109. Founded in 1826 by Henry Brougham, an influential advocate of formal education for the poor and of mass adult education, the Society for the Diffusion of Useful Knowledge aimed at

"imparting useful information to all classes of the community, particularly to such as are unable to avail themselves of experienced teachers, or may prefer learning by themselves".

22. Ibid. 133.

23. Ibid. 123–4.

24. Ibid. 109–10.

25. Ibid. 194–5.

26. Ibid. 223.

27. Ibid. 229–30.

28. Ibid. 239.

29. Henry Walter Bates (1825–92), Wallace's travelling companion in the Amazon, made important contributions to evolutionary biology, including his theory of mimicry.

30. ML I 236–7.

31. Ibid. 239.

32. Ibid. 242.

33. Ibid. 241.

34. The building currently houses the offices of the Neath Museum (until recently it housed the museum itself). A plaque on the outside states that Wallace lived in Neath 1841–48.

35. ML I 247.

36. Initial disturbances took place in 1839, and escalated between 1842 and 1844. The rioters took as their motto the text of Genesis 24:52: "And they blessed Rebecca, and said to her, '. . . may your descendants possess the [toll]gate of those who hate them!'" James Moore ('Wallace's Malthusian Moment: The Common Context Revisited', in Bernard Lightman, ed., *Victorian Science in Context*, University of Chicago Press, Chicago, 1997, pp. 300–3) has suggested that it was this early exposure to Malthusian rigours of rural existence that was the key both to Wallace's insight that yielded natural selection and to his politics.

37. ML I 253.

38. Charles Lyell (1797–1875) was arguably the most influential scientist of his generation. His *The Principles of Geology* (1830–33) established and popularized the geological doctrine of uniformitarianism, whereby the grand geological changes of the past are accounted for by extrapolating over time from current-day geological processes. Both Darwin and Wallace recognized the impact of Lyell's thinking on their ideas.

39. Darwin's *Journal of Researches into the geology and natural history of the various countries visited by H.M.S. Beagle* (1839) is more commonly referred to as *The Voyage of the Beagle*.

40. Alexander von Humboldt (1769–1859), German naturalist-explorer, travelled extensively in South and Central America between 1799 and 1804. His *Personal Narrative of a Journey to the Equinoctial Regions of the*

New Continent comprised the final three volumes (and most readable part) of his encyclopaedic account of the region.

41. ML I 255–6.
42. William Edwards (1822–1909), American lepidopterist, author of *The Butterflies of North America* (1868).
43. Edward Doubleday (1811–49), lepidopterist, author of *The Genera of Diurnal Lepidoptera* (1846–50).
44. ML I 264.
45. Ibid. 266.
46. Ibid.
47. Peter Raby, *Alfred Russel Wallace*, Chatto and Windus, London, 2001, pp. 44–5.
48. ML I 275.
49. Ibid. 281.
50. Ibid. 282.
51. Ibid. 283–4.
52. Captain Turner skippered the *Helen*.
53. ML I 310–11. Wallace's full account of the *Helen* incident is given on pp. 291–9.
54. Ibid. 306.
55. Ibid. 321.
56. MA xii.
57. Ibid.
58. Letter to H. W. Bates, Singapore, April 30 1856. ML I 351.
59. Now called, courtesy taxonomic revision, *Trogonoptera brookiana*.
60. ML I 357.
61. See pp. 50–62.
62. ML I 363–4.
63. MA 280.
64. Parting words to Wallace from "'Sunshine,' an Indian girl" encountered by Wallace at a seance in April 1896. Wallace records that "These three words were spoken very impressively, and I wrote them at once in a small note-book with capital letters." ML II 390. Because this anthology presents most of Wallace's major contributions in science and elsewhere, this section is merely a chronological outline of his work. Detailed information is to be found in the sections that follow.
65. ML II 1. For health reasons, Wallace was unable to take up Darwin's initial invitation. Instead he visited Down House a few months later in the summer of 1862. See Raby, *Alfred Russel Wallace*, pp. 164 and 168.
66. Sir Joseph Hooker (1817–1911), who succeeded his father as director of Kew Gardens in 1865, was the leading British botanist of the nineteenth century, and longtime member of Darwin's inner circle.
67. Herbert Spencer (1820–1903), philosopher and polymath. Spencer coined the phrase "survival of the fittest" and obsessively applied

evolutionary ideas to every aspect of life. (Wallace commented [ML II 33] on "the often unexpected way in which he would apply the principles of evolution to the commonest topics of conversation".) Today Spencer is chiefly remembered as the father of social Darwinism.

68. See pp. 176–190.
69. See pp. 89–102.
70. See pp. 319ff.
71. See pp. 309ff.
72. ML II 197.
73. See pp. 117ff.
74. ML II 129.
75. Ibid. 134.
76. Given for distinguished service in the armed forces or in science, art, literature, or the promotion of culture. The order is limited to twenty-four members.
77. Letter to Mrs Fisher (née Buckley), Charles Lyell's one-time secretary, 1908. M 447.
78. In fact, Wallace was only invited to participate as a pall-bearer at the last minute, and very much as an afterthought. See Adrian Desmond and James Moore, *Darwin*, Michael Joseph, London, 1991, p. 669.
79. In 1871 Wallace wanted to move out of London: "I had a great longing for life in the country where I could devote much of my time to gardening and rural walks." ML II 90.
80. Mr Ridsdale, a fellow spiritualist (Raby, *Alfred Russel Wallace*, p. 193).
81. ML II 360–1.
82. Ibid. 362.
83. Ibid. 363.
84. Ibid. 363–4.
85. Ibid. 364.
86. Ibid. 365.
87. Ibid. 374.
88. Ibid. 376.
89. The *Oxford English Dictionary* gives: "A frame or sledge on which a traitor was drawn through the streets to execution."
90. ML II 371.
91. ML I 415.
92. See pp. 354.
93. ML I 416.
94 Letter to Wallace, January 5 1880. M 249.
95. ML II 378.
96. One-time secretary of Sir Charles Lyell.
97. ML II 378.
98. Letter to Wallace, January 7 1881. M 257.
99. Letter to Darwin, January 8 1881. Ibid. 258.
100. Letter to George Silk, Singapore, January 20 1862. Ibid. 70–1.

101. Lewis Leslie, an auctioneer. Wallace's fiancée was Marion. (Raby, *Alfred Russel Wallace*, p. 172.)
102. ML I 409–11.
103. Ibid. 411–12. In fact, Annie Mitten was twenty, not eighteen, when she married Wallace (Raby, *Alfred Russel Wallace*, p. 187).
104. Letter to Darwin, July 9 1881. M 260.
105. ML II 191.
106. Ibid. 228.
107. Ibid. 229.
108. Letter to Alfred Russell, May 11 1900. M 395.
109. Letter to Mr E. Smedley, August 26 1913. Ibid. 347.
110. Letter to J. W. Marshall, March 6 1894. Ibid. 436.

Science

Evolution

1. ML I 257.
2. Anonymously published by Robert Chambers (1802–71) in 1844, *The Vestiges* was a bestselling outline of evolution. Scientifically sloppy and short on mechanistic detail, it was nevertheless important in introducing a broad Victorian audience to evolutionary ideas.
3. ML I 254.
4. Ibid. 257.
5. Thomas Malthus (1766–1834) first published his *Essay on the Principle of Population* anonymously in 1798. He pointed out that natural populations tend to increase faster than their resources. Both Darwin and Wallace acknowledged the importance of Malthus in their independent formulations of natural selection.
6. ML I 232.
7. Darwin was one of Wallace's customers, buying a pair of Balinese birds (James Moore, 'Wallace's Malthusian Moment: The Common Context Revisited', in Bernard Lightman, ed., *Victorian Science in Context*, University of Chicago Press, Chicago, 1997, p. 293).
8. Spencer St John, *Life of Sir James Brooke*, Smith, Elder, London, 1890.
9. William Swainson (1789–1855), English naturalist and bird illustrator.
10. Charles-Lucien Bonaparte (1803–57), a nephew of Napoleon, was one of the nineteenth century's leading ornithologists.
11. ML I 354.
12. Ibid. 355.
13. Letter from "Tunantins, Upper Amazon," November 19 1856. M 52.
14. Letter, May 1 1857. Ibid. 107.
15. Edward Blyth (1810–73), naturalist who, from 1841 to 1862, was curator of the museum of the Asiatic Society in Calcutta.
16. Letter, December 22 1857. M 109.

17. Letter to Bates, Amboyna, January 4 1858. Ibid. 54.

18. Carolus Linnaeus (1707–78), Swedish naturalist and founder of modern biological classification.

19. Edward Forbes (1815–54), naturalist and geologist. Wallace wrote to Bates from Amboyna (Ambon) on January 4 1858 that "[i]t was the promulgation of Forbes's theory which led me to write and publish ['The Sarawak Law'], for I was annoyed to see such an ideal absurdity put forth when such a simple hypothesis will explain *all the facts*" (M 54). Forbes's "absurdity" was his "Polarity Theory", which held that biodiversity was divinely ordained to be high early and late in earth's history.

20. Wallace is using "antitype" as a euphemism for "ancestor".

21. Hugh Strickland (1811–53), naturalist and geologist, killed by a train while studying geological strata in a railway cutting. In the 1841 page that Wallace refers to, Strickland discredited the then popular quinarian school of classification which posited that organisms were related to each other via clusters of circles. Strickland, H. E., "On the true method of discovering the natural system in zoology and botany", *Annals and Magazine of Natural History*, vol. 6, pp. 184–94, 1841.

22. Lovell Reeve (1814–65), malacologist.

23. *The Principles of Geology* (1830–33).

24. *Annals and Magazine of Natural History*, vol. 16 (2nd Series), pp. 184–96, 1855.

25. "On the Habits of the Orang-Utan of Borneo", *Annals and Magazine of Natural History*, vol. 18 (2nd Series), pp. 26–32, 1856. Reprinted in Charles H. Smith, ed., *Alfred Russel Wallace: An Anthology of His Shorter Writings*, Oxford University Press, Oxford, 1991 (excerpt pp. 13–14).

26. See Lewis McKinney, *Wallace and Natural Selection*, Yale University Press, New Haven and London, 1972.

27. Jean Baptiste, Chevalier de Lamarck (1744–1829), French naturalist and author of the most scientifically respectable pre-Darwinian theory of evolution. Lamarck's theory depended on a bogus theory of genetics: the inheritance of characters acquired in the course of an individual's lifetime, such that a blacksmith's child should inherit the highly developed upper arms of its father.

28. ML I 360–3.

29. Now extinct. The last one died in the Cincinnati Zoological Garden in 1914.

30. Sir Richard Owen (1804–92), influential Victorian zoologist and anatomist who headed the Natural History Museum. He was a prominent opponent of Darwin.

31. *Journal of the Proceedings of the Linnean Society: Zoology*, vol. 3, pp. 53–62, 1858.

32. Letter from Darwin to Charles Lyell, June 18 (or May 18?) 1858.

Quoted in John L. Brooks, *Just Before the Origin: Alfred Russel Wallace's Theory of Evolution*, Columbia University Press, New York, 1984, p. 262.

33. Ibid.

34. Arnold C. Brackman, *A Delicate Arrangement: The Strange Case of Charles Darwin and Alfred Russel Wallace*, Times Books, New York, 1980.

35. As Wallace himself recognized, writing to Darwin (May 29 1864), "All the merit I claim is the having been the means of inducing *you* to write and publish at once." M 131.

36. ML I 363.

37. Letter to his mother, October 6 1858. M 57.

38. Letter to George Silk, November 1858. ML I 365.

39. Letter, also to George Silk, September 1 1860. Ibid. 372.

40. Letter to Bates, Ternate, December 24 1860. Ibid. 374.

41. George-Louis Leclerc, Comte de Buffon (1707–88), French naturalist and encyclopaedist who advanced evolutionary ideas and argued that the earth was older than suggested by scripture.

42. Erasmus Darwin (1731–1802), Charles's grandfather, who, in a series of rather whimsical epic poems, advanced evolutionary ideas akin to Larmarck's.

43. Robert Chambers (1802–71), anonymous author of *The Vestiges of Creation* (1844), a work that, despite its scientific weaknesses, did much to popularize evolutionary ideas among the Victorians.

44. M 91–5.

45. WL 109–10.

46. Phidias (fifth century BC), Greek sculptor.

47. Michael Faraday (1791–1867), chemist and physicist.

48. WL 113–15.

49. Anton Kerner von Marilaun (1831–98), Austrian botanist and author of *The Natural History of Plants* (1886–91).

50. Richard Kearton (1862–1928), naturalist.

51. WL 121–3.

52. Ibid. 112–13.

53. Ibid. 133.

54. Letter to Wallace, July 1867. M 154.

55. Rev. Samuel Haughton (1821–97), Irish physicist, geologist, and biologist.

56. Quoted in Wilma George, *Biologist Philosopher: A Study of the Life and Writings of Alfred Russel Wallace*, Abelard-Schuman, London, 1964, p. 62.

57. ML II 87.

58. Sir John Lubbock (1834–1913), politician and author of many books on natural history and on human prehistory.

59. Asa Gray (1810–88), American botanist, professor of natural history at Harvard 1842–73, and one of Darwin's strongest supporters. Wallace visited him in the course of his North American trip in 1886.

60. "Remarks on the Rev. S. Haughton's Paper on the Bee's Cell, And on

the Origin of Species", *Annals and Magazine of Natural History*, vol. 12 (3rd Series), pp. 303–9, 1863. Reprinted in Smith, ed., *Alfred Russel Wallace* (excerpt pp. 333–5).

61. ML II 88.

62. Ibid. 89.

63. Letter to Wallace, January 2 1881. M 256.

64. ML II 22.

65. Ibid. 16.

66. Review of *The Descent of Man and Selection in Relation to Sex* by Charles Darwin, *Academy*, vol. 2, pp. 177–83, 1871.

67. ML II 17–19.

68. "A Theory of Birds' Nests: Showing the Relation of Certain Sexual Differences of Colour in Birds to Their Mode of Nidification", *Journal of Travel and Natural History*, vol. 1, pp. 73–89, 1868. Reprinted in NSTN (excerpt pp. 122–4).

69. Letter to E. B. Poulton, November 28 1889. M 302.

70. Letter to C. Lyell, February 20 1868. ML I 422.

71. "Darwin and Darwinism [review of *Charles Darwin and the Theory of Natural Selection* by Edward B. Poulton, 1986]", *Nature*, vol. 55, pp. 289–90, 1897. My attention was first drawn to this excerpt by Charles Smith's website, www.wku.edu/~smithch/quotes.htm.

72. Sir Francis Galton (1822–1911), a cousin of Charles Darwin who founded the field of (and coined the term) eugenics.

73. August Weissman (1834–1914), German biologist responsible for recognizing the fundamental separation of the germ line (reproductive cells) and soma (the cells that make up the body of an organism).

74. ML II 21–2.

75. "Are Individually Acquired Characters Inherited? II", *Fortnightly Review*, vol. 53 (n.s.), pp. 655–8, 1893. Reprinted in SSS I (excerpt pp. 329–32).

76. Sir Edward Poulton (1856–1943), Oxford entomologist.

77. Letter to E. B. Poulton, July 27 1907. M 333.

78. Alfred Bennett (1833–1902), botanist.

79. "Natural Selection – Mr. Wallace's Reply to Mr. Bennett", *Nature*, vol. 3, pp. 49–50, 1870 (p. 50).

80. "The Origin of Species and Genera", *Nineteenth Century*, vol. 7, pp. 93–106, 1880. Reprinted in SSS I (excerpt pp. 302–3).

81. Paul Janet (1823–99), French philosopher.

82. Letter to Darwin, July 2 1866. M 140–1.

83. Letter, July 5 1866. Ibid. 144.

84. ML I 401.

85. Pieter Cramer (1721–76), a Dutch merchant with an interest in butterflies.

86. William Hewitson (1806–78), entomologist and illustrator.

87. 'The Malayan Papilionidae or Swallow-tailed Butterflies, As Illustrative of the Theory of Natural Selection", in CTNS, pp. 130–3, 1870.

88. James Cowles Prichard (1786–1848), physician and ethnologist, author of *Researches into the Physical History of Man* (1813), which argued for a single human species. Wallace misspells his name throughout his work.
89. Birdwing butterflies, belonging to the family Papilionidae.
90. "Malayan Papilionidae", pp. 141–7.
91. John Gray (1800–75), naturalist and keeper of zoology at the British Museum (1840–74).
92. "Malayan Papilionidae", pp. 157–62.
93. Ibid., 165–6.
94. A sub-family of butterflies, including, for example, the Monarch Butterfly, within the family Nymphalidae.
95. Thomas Horsfield (1773–1859), American naturalist, author of *Zoological Researches in Java, and the Neighbouring Islands* (1824).
96. Willem de Haan (1801–55), Dutch zoologist.
97. "Malayan Papilionidae", pp. 179–85.
98. Letter to Wallace, February 23 1867. M 147.
99. John Jenner Weir (1822–94), civil servant (ultimately Accountant and Controller-General of the Customs Service) and naturalist.
100. Arthur Butler (1844–1925).
101. ML II 3–6.
102. Ibid. 201.
103. Rev. Alfred Eaton (1845–1929).
104. D 105–6.

Biogeography

1. A 326.
2. ML I 256.
3. A 327.
4. Johann Baptist von Spix (1781–1826) was the principal zoologist on the Bavarian Expedition to Brazil, 1817–20.
5. *Proceedings of the Zoological Society of London*, vol. 20, pp. 107–10, 1852 (excerpt pp. 109–10).
6. John Gray (1800–75), naturalist and keeper of zoology at the British Museum (1840–74).
7. ML I 377.
8. Ibid. 336.
9. Ibid. 327.
10. MA 13.
11. Letter to Bates, Amboyna (Ambon), January 4 1858. M 55.
12. ML I 356.
13. P. L. Sclater, "On the General Distribution of the Members of the Class Aves", *Journal of the Proceedings of the Linnaean Society*, vol. 2, pp. 130–45, 1858. Philip Sclater (1829–1913), ornithologist and origi-

nator of the "global biogeographic zones" theory popularized (and modified) by Wallace.

14. *Journal of the Proceedings of the Linnaean Society: Zoology*, vol. 4, pp. 172–84, 1860. Reprinted in Charles H. Smith, ed., *Alfred Russel Wallace: An Anthology of His Shorter Writings*, Oxford University Press, Oxford, 1991 (excerpt pp. 236–7).
15. GD I 389.
16. Letter to Darwin, January 9 1880. M 250.
17. Letter to Wallace, November 3 1880. Ibid. 252.
18. Ibid.
19. IL 3–6.
20. Sclater's scheme, proposed in 1858, included Palearctic (N. Eurasia), Aeteopica (Africa), Indica (Oriental), Neotropica (S. America), Nearctica (N. America), and Australiana (Australia, New Guinea, New Zealand).
21. ML II 94.
22. "Animals and their Native Countries", *Nineteenth Century*, vol. 5, pp. 247–59, 1879. Reprinted in SSS I (excerpt pp. 282–4).
23. WL 14.
24. Ibid. 103.

Natural History and Conservation

1. ML I 194.
2. NSTN 240.
3. Letter to J. W. Marshall, September 23 1892. M 307.
4. A 52.
5. Ibid. 115–16.
6. Ibid. 118–19.
7. Ibid. 34.
8. Ibid. 166.
9. Ibid. 221–2.
10. Wallace preferred to use the local name for the orang.
11. Indigenous group in Borneo.
12. On Wallace's love for the durian, which is an acquired taste, see pp. 266–9.
13. MA 44–6.
14. "On the Habits of the Orang-Utan of Borneo", *Annals & Magazine of Natural History*, vol. 18 (2nd Series), pp. 26–32, 1856 (excerpt p. 32).
15. Letter home, June 1855. ML I 343–5.
16. MA 35.
17. ML I 394.
18. "Narrative of Search After Birds of Paradise", *Proceedings of the Zoological Society of London*, vol. 1862, pp. 153–61, 1862. Reprinted ML I (excerpt p. 394).

19. George R. Gray (1808–72), entomologist and ornithologist at the British Museum; author of *Genera of Birds* (1844–9).
20. MA 252–3.
21. Ibid. 339–40.
22. Ibid. 354.
23. Ibid. 408.
24. Ibid. 423–4.
25. Ibid. 257–8.
26. Ibid. 328.
27. "The Disguises of Insects", *Hardwicke's Science Gossip*, vol. 3, pp. 193–8, 1867. Reprinted in SSS I (excerpt pp. 191–3).
28. "On the Physical Geography of the Malay Archipelago", *Journal of the Royal Geographical Society*, vol. 33, pp. 217–34, 1863. Reprinted in Charles H. Smith, ed., *Alfred Russel Wallace: An Anthology of His Shorter Writings*, Oxford University Press, Oxford, 1991 (excerpt p. 377).
29. WL 83–6.
30. "Free-Trade Principles and the Coal Question", letter to the editor, *Daily News*, September 16 1873. Reprinted in SSS II (excerpt pp. 142–3).
31. WC 83–4.
32. "Topics of the Time: Landscape-Gardeners Needed for America", *The Century*, vol. 34, no. 2, June 1887, p. 313.
33. "English and American Flowers. I", *Fortnightly Review*, vol. 50 (n.s.), pp. 525–34, 1891. Reprinted in SSS I (excerpt pp. 209–10).
34. "English and American Flowers. II. Flowers and Forests of the Far West", *Fortnightly Review*, vol. 50 (n.s.), pp. 796–810, 1891. Reprinted in ibid. (excerpt p. 234).
35. Letter to Lord Avebury, June 23 1908. M 398.
36. See Alfred Crosby, *Ecological Imperialism: The Biological Expansion of Europe, 900–1900*, Cambridge University Press, Cambridge, 1986.
37. GD I 44.

Geography, Geology, and Glaciology

1. IL 502.
2. ML I 54–5.
3. A 248–9.
4. "Inaccessible Valleys: A Study in Physical Geography", *Nineteenth Century*, vol. 33, pp. 391–404, 1893. Reprinted in SSS I (excerpt p. 27).
5. Alfred Wegener (1880–1930), German meteorologist, advanced his theory of continental drift in 1915; it was translated, as *Origin of Continents and Oceans*, in 1924.
6. IL 84.
7. Jacques Boucher (de Crèvecoeur) de Perthes (1788–1868), French archaeologist.

8. Hugh Falconer (1808–65), botanist and palaeontologist.

9. Sir Joseph Prestwich (1812–96), geologist. A wine merchant until he was sixty, he became Oxford Professor of Geology in 1874.

10. Sir John Evans (1823–1908), industrialist and archaeologist.

11. A limestone cave near Torquay, Devon. Originally investigated by the Rev. J. McEnery 1825–29, but his findings were disregarded until subsequent research by William Pengelly in 1865–80.

12. WC 129–32. Wallace's speculations about humans being derived early in the Great Ape lineage are wrong; humans and chimpanzees shared a common ancestor about five millions years ago.

13. William Thomson, Lord Kelvin (1824–1907), physicist and inventor.

14. Letter to Wallace, July 12 1871. M 220.

15. James Croll (1821–90), self-trained physicist whose astronomical theory of Ice Age causation was embraced by Wallace.

16. IL 202–3.

17. William J. Sollas (1849–1936), Professor of Geology at the University of Oxford.

18. IL 212–14.

19. Ibid. 215–18.

20. Louis Agassiz (1807–73). He founded the Harvard University Museum of Comparative Zoology, which so impressed Wallace, but remained a vocal anti-evolutionist.

21. William Buckland (1784–1856). Professor of Geology at Oxford University; later, Dean of Westminster.

22. Croll's theory was mathematically formalized and rather taken over by the Serbian mathematician Milutin Milankovitch (1879–1958).

23. WC 127–9.

Humans

"Uncivilised people"

1. Letter May 1855. M 45.

2. ML I 288–9.

3. "The Native Problem in South Africa and Elsewhere", *Independent Review*, vol. 11, pp. 174–82, 1906. Reprinted in Charles H. Smith (ed.), *Alfred Russel Wallace: An Anthology of His Shorter Writings*, Oxford University Press, Oxford, 1991 (excerpt p. 168).

4. A 83–4.

5. Ibid. 193–4.

6. Karl Wilhelm von Humboldt (1767–1835), German statesman and philologist, elder brother of Alexander von Humboldt of South American explorations fame.

7. James Cowles Prichard (1786–1848), physician and ethnologist, author of *Researches into the Physical History of Man* (1813), which

argued for a single human species, Wallace misspells his name throughout his work.

8. MA 15.

"A being apart": Human Evolution

1. ML II 17.
2. Letter to Wallace, May 28 1864. M 127.
3. It is also his most popular. By using available citation databases, Charles Smith has compiled a list of Wallace's most cited publications (http://www.wku.edu/~smithch/cited.htm). This 1864 paper is Wallace's greatest hit.
4. Eocene: 57–35 million years ago; Miocene: 23–5 million years ago.
5. Denise: Middle Pleistocene skeletal remains from La Denise, France. Engis: a partial skull from the Engis caves in the Meuse valley, Belgium, discovered and described (1833) by Philippe-Charles Schmerling (1791–1836).
6. From "On the Characters, Principles of Division and Primary Groups of the Class Mammalia" (1858). Owen deleted this passage from later publications of his essay (see http://aleph0.clarku.edu/huxley/guide7.html).
7. *Journal of the Anthropological Society of London*, vol. 2, pp. clviii–clxx, 1864.
8. Charles Smith, *Alfred Russel Wallace: An Anthology of His Shorter Writings*, Oxford University Press, Oxford, 1991, p. 31.
9. "Sir Charles Lyell on Geological Climates and the Origin of Species", 1867–68, and *Elements of Geology*, 1865, both by Sir Charles Lyell, *Quarterly Review*, vol. 126, pp. 359–94, 1869. Partially reprinted in Smith, ed., *Alfred Russel Wallace* (excerpt p. 33).
10. Letter to Wallace, March 27 1869. M 197.
11. Samuel Morton (1799–1851), American anatomist whose erroneous measurements of the skull capacity buttressed nineteenth-century racism. Morton's results purporting to show racial differences in brain size are specious because they fail to take into account differences in body size among groups.
12. A Bronze Age settlement beside Lake Zurich, investigated by Swiss naturalist Ferdinand Keller, who published his report in 1854.
13. From "On Some Fossil Remains of Man" (1863).
14. Cave site in the Dordogne valley of France, where in 1868 Cro Magnon man was first described by Louis Lartet.
15. Paul Broca (1824–80), French surgeon and anthropologist.
16. Georges Cuvier (1769–1832), French zoologist and anatomist, and champion of the geological doctrine of "Catastrophism" whereby the planet undergoes periodic major upheavals.
17. Daniel O'Connell (1775–1847), Irish politician.

18. A member of an indigenous group of Chile/Argentina.
19. A Polynesian from the Marquesas, a Pacific archipelago.
20. A Cambridge University student excelling in maths.
21. A member of the Munda people of North-east India.
22. A group of several tribes that subsisted mainly on roots.
23. Albert S. Bickmore (1839–1914) founded the American Museum of Natural History in 1869.
24. A Burmese family with congenital hypertrichosis lanuginosa, a hereditary condition of excessive hairiness. In Burma they were revered; later in America they were exhibited by P. T. Barnum.
25. A member of the Nama people, now mainly in Namibia, but previously in the Cape of Good Hope region.
26. Thomas Laycock (1812–76), author of *Mind and Brain* (1860).
27. CTNS 332–71 (excerpt pp. 332–60).
28. ML I 224.
29. "The Expressiveness of Speech, or Mouth-gesture as a Factor in the Origin of Language", *Fortnightly Review*, vol. 58 (n.s.), pp. 528–43, 1895. Reprinted in SSS II (excerpt pp. 135–7).

Human Improvement

1. See pp. 315.
2. Herbert Spencer, "Progress: Its Law and Cause", *Westminster Review*, vol. 67, pp. 244–67, 1857 (excerpt p. 244).
3. Letter to R. Meldola, June 1893. M 308.
4. ML II 33.
5. According to a letter to Darwin (October 1 1867; M 155), the name "Herbert" was also a reference to Wallace's younger brother, who died of yellow fever on the Amazon.
6. "Human Selection", *Fortnightly Review*, vol. 48 (n.s.), pp. 325–37, 1890. Reprinted in SSS I (excerpt p. 509).
7. Diane Paul, *Controlling Human Heredity: 1865 to the Present*, Humanities Press, Atlantic Highlands, NJ, 1998, p. 75.
8. Interview, *The Millgate Monthly*, August 1912. Reprinted in Charles H. Smith, ed., *Alfred Russel Wallace: An Anthology of His Shorter Writings*, Oxford University Press, Oxford, 1991 (excerpt p. 177).
9. See pp. 77ff.
10. The Eugenics Education Society, founded in 1907, renamed the Eugenics Society in 1926, and, since 1989, the Galton Institute.
11. Grant Allen (1848–99), Canadian novelist and literary polymath.
12. Emanuel Swedenborg (1688–1772), Swedish mystic, theologian, and scientist.
13. SEMP 141–53.

Spiritualism and Metaphysics

1. Martin Fichman, "Science in Theistic Contexts: A Case Study of Alfred Russel Wallace on Human Evolution", *Osiris*, vol. 16 (2nd Series), 2001, pp. 227–50.
2. G. K. Chesterton, "Alfred Russel Wallace", *English Illustrated Magazine*, vol. 30, pp. 420–2, 1904 (p. 420).

"Strange doings": Conversion

1. Named after Austrian physician Franz Anton Mesmer (1734–1815), the first investigator of what he called "animal magnetism".
2. The analysis of mind and character by the study of the external shape and contours of the skull, founded by Franz Joseph Gall (1758–1828). Huxley, "when I [Wallace] once asked him *why* he did not accept phrenology as a science, replied at once, 'Because, owing to the varying thickness of the skull, the form of the outside does not correspond to that of the brain itself, and therefore the comparative development of different parts of the brain cannot be determined by the form of the skull'" (WC 182).
3. Recounted in the Book of Esther. Esther, wife of the Persian king Xerxes I, conspires with an exiled Jew, Mordecai, to prevent the extermination of the Jews.
4. The fourteenth chapter of the Book of Daniel (from the Apocrypha/ Septuagint). The first half recounts the story of the Babylonian idol Bel, ministered to by priests who secretly consume food left for it, thus deceiving the king and the people. Daniel reveals the fraud, and priests and idol are destroyed by the king. The second half of the passage tells of a dragon – a reptile – worshipped as a god; Daniel kills it and is thrown to the lions. The prophet Habakkuk is brought miraculously to the den by an angel to feed him.
5. ML I 226–7.
6. Robert Dale Owen (1801–77), son of Robert Owen, the social reformer who influenced Wallace. Robert Dale Owen moved to the US in 1825, and was a journalist, member of the US Congress, US Ambassador to India, and a spiritualist.
7. ML I 88.
8. In Leicester, working as a teacher, 1843–45, Wallace boarded with the school's headmaster, Rev. Abraham Hill.
9. Letter to his brother-in-law, Thomas Sims, Delli (Dili), Timor, March 15 1861. M 65–7.
10. Frederick Engels (*Dialectics of Nature*, International Publishers, New York, 1940 [1883]) also encountered Mr Hall:

    Now it happens that I also saw this Mr. Spencer Hall in the winter of 1843–44 in Manchester. He was a very mediocre charlatan. . . .

While we with our frivolous scepticism thus found that the basis of magnetico-phrenological charlatanry lay in a series of phenomena which for the most part differ only in degree from those of the waking state and require no mystical interpretation, Mr. Wallace's "ardour" led him into a series of self-deceptions, in virtue of which he confirmed Gall's [phrenological] map of the skull in all its details and noted a mysterious relation between operator and patient. Everywhere in Mr. Wallace's account, the sincerity of which reaches the degree of naïveté, it becomes apparent that he was much less concerned in investigating the factual background of charlatanry than in reproducing all the phenomena at all costs. Only this frame of mind is needed for the man who was originally a scientist to be quickly converted into an "adept" by means of simple and facile self-deception. (pp. 298, 300–1)

11. George Combe (1788–1858), phrenologist, published *The Constitution of Man* in 1828.
12. Rev. Abraham Hill, headmaster of the Leicester Collegiate School (and Wallace's employer).
13. ML I 232–6.
14. ML II 275.
15. Ibid. 81–2. On dentistry, Wallace concludes, "My experience of modern dentists is that they all want to improve upon nature, and care nothing for the comfort of those who are to use the teeth" (ibid. 83).
16. ML I 257–60.
17. WC 193.
18. A 120.
19. Ibid. 84–5.
20. Modern spiritualism can be dated from March 1848 when a nine-year-old, Kate Fox, living in upstate New York, had "intelligent communications ... with the unknown cause of the mysterious knockings" (MMS 152–3). Thirty years afterwards, she confessed that she had produced the sounds by snapping her big toe inside her shoe.
21. MMS 131–5.

"To excite to inquiry": Spiritualism and Science

1. Sir William Crookes (1832–1919), author of *Select Methods of Chemical Analysis* (1871).
2. Sir Oliver Lodge (1851–1940), a pioneer in wireless telegraphy.
3. Gullible though Wallace may in retrospect appear, it is worth noting that he did not adopt "alternative" worldviews indiscriminately. In a letter to a Mr. E. Smedley dated October 2 1911 (M 441) he wrote:

   I am quite astonished at your wasting your money on an advertising astrologer. In the horoscope sent you there is not a single definitive fact that would apply to you any more than to thousands of other men.

4. Letter to Wallace, November 1866. M 418.

5. William Carpenter (1813–85), English biologist known for his *Principles of General and Comparative Physiology* (1839). He and Wallace were "near neighbours" when Wallace was living in London, but they fell out over "a rather acute controversy upon mesmerism and clairvoyance" (ML II 43).

6. Sir David Brewster (1781–1868), Scottish physicist.

7. Daniel Home (1833–86), Scots-American spiritualist medium. He was the inspiration for Robert Browning's sceptical poem "Mr Sludge, the Medium" (1864).

8. John Tyndall (1820–93), Irish physicist and alpinist. He became Royal Institution Professor in 1854.

9. G. H. Lewes (1817–78), writer and intellectual. Unhappily married, he started a lifelong affair with George Eliot in 1854.

10. ML II 349–50.

11. Sir Edwin Ray Lankester (1847–1929), zoologist and editor of the monumental *Treatise on Zoology* (1900–9).

12. Henry Slade (1836–1905).

13. Richard Milner, "Charles Darwin and Associates, Ghostbusters", *Scientific American*, vol. 275(4), 1996, pp. 96–101.

14. Felix Anton Dohrn (1840–1909), German biologist with expertise on the evolution of the arthropods.

15. David Friedrich Strauss (1808–74), German theologian whose *Leben Jesu* (1835; translated by George Eliot in 1846) sought to demonstrate that the Gospels are a collection of myths.

16. Carl Vogt (1817–95), German materialist philosopher and zoologist.

17. MMS vi–viii.

18. Letter to John Tyndall, May 8 1868. ML II 292–3.

19. Wallace seems to have overlooked the reversal of the "r" and the "n" in the backwards version of "Henry".

20. Robert Chambers (1802–71), anonymous author of *The Vestiges of Creation* (1844), a work that, despite its scientific weaknesses, did much to popularize evolutionary ideas among the Victorians.

21. John Elliotson (1791–1868), English physician and founder of the Phrenological Society.

22. William Gregory (1803–58), Professor of Chemistry at Edinburgh University.

23. Robert Hare (1781–1858), American chemist and author of *Spiritualism Demonstrated* (1855).

24. James Gully (1808–83), English physician who administered Darwin's "water cure" at his Malvern spa.

25. John Worth Edmonds (1816–74), American spiritualist and writer.

26. Letter to the editor, *The Times* (London), January 4 1873.

27. William Harvey (1578–1657), discoverer of the circulation of the blood.

28. Edward Jenner (1749–1823), discoverer of small-pox vaccination.

29. Thomas Young (1773–1829), English physicist, physician and Egyptologist. Discoverer of the undulatory nature of light.

30. When repeating the same argument elsewhere (MMS 18–19), Wallace wisely places less emphasis on mesmerism as a previously reviled but now generally accepted scientific theory.

31. Sir Humphry Davy (1778–1829), chemist, the leading British scientist of his day.

32. Sir John Herschel (1792–1871), astronomer.

33. "On the Attitude of Men of Science Towards the Investigators of Spiritualism", in Hudson Tuttle and J. M. Peebles, eds, *The Year-Book of Spiritualism for 1871*, William White & Co., Boston 1871, pp. 28–31 (excerpt pp. 30–1).

34. MMS 43.

35. "On the Attitude" (excerpt p. 29).

36. David Hume, *Enquiries Concerning Human Understanding*, L. A. Selby-Bigge, ed., 3rd edn, Oxford: Oxford University Press, 1975 [1748], pp. 115 and 114, respectively.

37. Ibid. p. 114. Emphasis added by Wallace.

38. Roger Boyle, First Earl of Orrery (1621–79), Irish soldier and statesman.

39. Valentine Greatorex (1629–83), Irish faith healer.

40. Henry More (1614–87), English philosopher and theologian interested in demonstrating the compatibility between reason and faith.

41. Joseph Glanvill (1636–80), English philosopher and clergyman, author of *Sadducismus Triumphatus* (edited and published posthumously in 1681 by Henry More) in which he attacked the rationalizing scepticism of those who denied the existence of ghosts and other apparitions. Wallace misspells his name here.

42. Girolamo Savonarola (1452–98), Italian political and religious reformer.

43. Alban Butler (1710–73), English hagiographer whose *The Lives of Saints* (1756–59) makes no distinction between fact and fiction. This did not apparently worry Wallace.

44. "An Answer to the Arguments of Hume, Lecky, and others, against Miracles", *The Spiritualist*, vol. 1, pp. 113c–16b, 1870. Reprinted with minor revisions in MMS (excerpt pp.1–8).

A World Viewed through the Lens of Spiritualism

1. "Why Live a Moral Life? The Answer of Rationalism", in Charles A. Watts, ed., *The Agnostic Annual 1895*, W. Stewart & Co., London 1895, pp. 6–12. Reprinted in SSS II (excerpt pp. 382–3).

2. "Spiritualism and Social Duty", *Light*, vol. 18, pp. 334–6, 1898. Reprinted in SSS II (excerpt pp. 521–4).

3. Act V, Scene 2.

4. ML I 195–6.
5. Percival Lowell (1855–1916), American astronomer who postulated the existence of Pluto, and, less successfully, the presence of intelligent beings on Mars.
6. MPU 306.
7. Ibid. 317–18.
8. WL 131–5.
9. Ibid. 300–1.
10. Ibid. 431.

Travel

1. Some of Wallace's descriptive writing on natural history can be found in the "Natural History and Conservation" section of this anthology.
2. Letter to Richard Spruce, this part dated October 5 1852. ML I 310.
3. ML II 107.
4. Ibid.
5. Lovell Reeve (1814–65) was both naturalist and publisher.
6. ML I 322.
7. Ibid.
8. Cited in Wilma George, *Biologist Philosopher: A Study of the Life and Writings of Alfred Russel Wallace*, Abelard-Schuman, London, 1964, p. 14.
9. Letter to Darwin, January 2 1864. M 124.
10. The dedication was to "Charles Darwin ... not only as a token of personal esteem and friendship but also to express my deep admiration for his genius and his works". Darwin deemed this "a *great* honour, and this is nothing more than the truth". (Letter to Wallace, January 22 1869. Ibid. 191.)
11. Letter to Wallace, March 22 1869. Ibid. 194.
12. Ibid.
13. ML I 386. *The Malay Archipelago* was particularly popular with Joseph Conrad: "Wallace's *Malay Archipelago*, apparently one of [Conrad's] favourite books, supplied him with a number of details for *Lord Jim*, *The Rescue* and other novels; and Stein's appearance, his apprenticeship as watchmaker and his butterfly collecting are based on Wallace himself" (Jocelyn Baines, *Joseph Conrad: A Critical Biography*, Weidenfeld and Nicolson, London, 1969, p. 254).

Expectations

1. Written in "fulfilment of a promise I made before I left Neath" (ML I 269) to the members of the Mechanics' Institute there in early 1849.
2. W. H. Edwards, author of *A Voyage Up the River Amazon* (John Murray,

London, 1847), which strongly influenced Wallace and Bates, was inclined to whimsical descriptions: ". . . vast numbers of trees add their tribute of beauty, and the flower-domed forest from its many coloured altars ever sends heavenward worshipful incense. Nor is this wild luxuriance unseen or unenlivened. Monkeys are frolicking through festooned bowers, or chasing in revelry over the wood arches. Squirrels scamper in ecstasy from limb to limb, unable to contain themselves for joyousness . . ." (p. 29; quoted in Sandra Knapp, *Footsteps in the Forest: Alfred Russel Wallace in the Amazon*, Natural History Museum, London, 1999, p. 8).

3. ML I 269–70.
4. Ibid. 270.
5. Ibid. 270–1.
6. MA 226.
7. Jan Huygen van Linschoten (1563–1611), Dutch traveller, who, after working as assistant to the Archbishop of Goa (1583–89), published extensively on Asia.
8. Dr. Bernard Paludanus (1550–1633), traveller.
9. MA 56–8.

City Life

1. A "stout, good-humoured American" (ML I 325) who happened to be in the railway compartment with Wallace and his friend George Silk on a trip to the continent in 1853.
2. Charles Allen, Wallace's field assistant, was sixteen at the time of this Alexandrian adventure.
3. Letter to George Silk, steamship *Bengal*, Red Sea, March 26 1854. M 37–8.
4. Indian settlers in Malaysia.
5. Letter to brother-in-law Thomas Sims, March 1856. M 50.
6. ML II 189.

Life in the Field

1. Letter from Si Munjon Coal Works, Sarawak, April 8 1855, *Zoologist*, vol. 13, pp. 4803–7 (excerpt pp. 4805–7). My attention was drawn to this passage by Charles Smith's Wallace website, www.wkv.edu/~smithch/home.htm.
2. May 28 1854, Singapore. M 40.
3. A 181.
4. Letter to his sister, Mrs Sims, Sadong River, Borneo, June 25 1855. M 47.
5. A 22.
6. Ibid. 149.

7. MA 49.
8. Ibid. 52.
9. Ibid. 351.
10. Letter to George Silk, Lobo Roman, Sumatra, December 22 1861. ML I 380–1.

"An industrious and persevering traveller

1. Notes for an address to the Linnaean Society quoted by Wilma George, *Biologist Philosopher: A Study of the Life and Writings of Alfred Russel Wallace*, Abelard-Schuman, London, 1964, p. 48.
2. MA 267.
3. Ibid. 286.
4. Ibid. 345.
5. A 225–6.
6. Ibid. 234.
7. WL 431.
8. ML I 361–2.
9. MA 353.
10. A 213.
11. Ibid. 99–100.
12. MA 357.
13. Ibid. 390–1.
14. Letter to his mother, Sourabaya (Surabaya), Java, July 20 1861. M 67–8.
15. Letter to brother-in-law Thomas Sims, Delli (Dilli), Timor, March 15 1861. Ibid. 64.
16. Capt. Turner of the shipwrecked *Helen* and Capt. Venables of the *Jordeson*.
17. Letter to botanist Richard Spruce commenced on board the *Jordeson*, September 19 1852. ML I 309.
18. Letter written to undisclosed correspondent, Guia on the Rio Negro, late 1850. ML I 285.

"Tedious and unfortunate": Hazardous Voyages

1. ML I 179.
2. A 242–3.
3. Ibid. 266.
4. [Annatto.] Orange–red dye obtained from the seed coat of a tropical tree.
5. Fibre derived from leaf stalks of palm trees.
6. Resinous substance used for medicinal purposes.
7. A 271–9.
8. November 25 1859. M 58.

9. ML I 370.
10. MA 418.
11. Ibid. 394–5.
12. Plant family to which the pineapple belongs.
13. MA 400.

"A want of harmony between man and nature": American Travels

1. ML II 191.
2. Ibid. 193–4.
3. Ibid. 127–8.
4. Ibid. 149–50.
5. Ibid. 156.
6. Ibid. 200–1.

Social Issues

1. *The Vaccination Inquirer and Health Review*, vol. 12, pp. 164–8, 1891 (p. 168).
2. Letter to Sir William Barrett, physicist and spiritualist, February 17 1901. M 436.
3. Letter to Mrs Fisher (née Buckley), Charles Lyell's one-time secretary, November 7 1905. Ibid. 450.

Evolution of a Socialist

1. Robert Owen (1771–1858), social and educational reformer, author of *A New View of Society* (1813).
2. A model community, with improved housing and working conditions, associated with a Scottish cotton mill.
3. Thomas Paine's (1737–1809) *The Age of Reason* (1794–96) was a powerful attack on conventional religion. Excerpt from ML I 87.
4. Ibid. 104.
5. Letter to Thomas Sims, Delli (Dili), Timor, March 15 1861. M 65–6.
6. I thank Charles Smith for pointing out to me the relationship between these two arguments.
7. "Public Responsibility and the Ballot", *Reader*, vol. 5, p. 517, 1865.
8. MA 455–7.
9. Ibid. 458.
10. Edward Bellamy (1850–98), American novelist and social reformer.
11. Though decrying the uneven distribution of resources in society, Mill and Spencer nevertheless objected to increases in state intervention on the grounds that such action compromised the rights of the individual.
12. Published in 1851. Wallace was strongly influenced by this work.

All my readers know the name of our great philosophic thinker and writer, Herbert Spencer, but they are perhaps not aware that to him is primarily due the formation of the Land Nationalization Society. In 1853, soon after I returned from my travels in the Amazon Valley, I read his book on "Social Statics", and from it first derived the conception of the radical injustice of private property in land. His irresistible logic convinced me once for all, and I have never since had the slightest doubt upon the subject. ("Herbert Spencer and the Land Question", SSS II 333).

13. ML II 266–7.
14. "True Individualism: The Essential Preliminary of a Real Social Advance", SSS II 510–20 (excerpt pp. 512–13).

"Robbery of the poor by the rich": The Land Problem

1. "How to Nationalize the Land: A Radical Solution to the Irish Land Problem", *Contemporary Review*, vol. 38, pp. 716–36, 1880. Reprinted in SS II (excerpt p. 267).
2. Ibid. (excerpt pp. 268–70).
3. ML II 240.
4. Henry George (1839–97), American journalist and reformer.
5. Letter to Darwin, July 9 1881. M 260.
6. Darwin to Wallace, July 12 1881. Ibid. 261.
7. LN viii.
8. Wallace is writing in 1894 looking forward to the end of the century.
9. A series of Acts of Parliament (largely late eighteenth/early nineteenth century) promoting the partitioning and appropriation by individuals of common land.
10. "Economic and Social Justice", in Andrew Reid, ed., *Vox Clamantium: The Gospel of the People*, A. D. Innes & Co., London, pp. 166–97, 1894. Reprinted in SSS II (excerpt pp. 440–52).
11. SSS II 443.
12. *Arena*, vol. 7, pp. 395–410 and 525–42. Reprinted in SSS II (excerpt pp. 430–1). The biblical quotation is from Luke, 13:7, from the parable of the barren fig tree.
13. Sir John Lubbock (1834–1913), naturalist and politician, who in 1882 sponsored the "Ancient Monuments Act". He was made "Baron Avebury".
14. Avebury is the largest stone circle in the world: it is 427 metres in diameter and covers an area of some 28 acres (11.5 hectares).
15. John Aubrey (1626–97), antiquarian and folklorist.
16. Karnak, ancient Egyptian temple complex near Luxor.
17. *Nineteenth Century*, March 1877.
18. LN 126–30.
19. ML I 151.

Public Health

1. By Sir Robert Giffen, published as a pamphlet by G. Bell and Sons, London. Gladstone apparently lauded it as a "masterly paper". ML I 81.
2. The "match girls' strike" led by Annie Besant at Bryant & May's factory in the East End of London in 1888, protesting poor pay and dangerous conditions, was a major early success for the trade union movement.
3. SEMP 52–60
4. In later life, Wallace advised his son William to use a home remedy when a small-pox epidemic threatened, suggesting (in a letter, November 1 1903) that he "take a hot bath for about 20 minutes, then drink a half pint of tolerably strong salt and water" (cited in Peter Raby, *Alfred Russel Wallace*, Chatto and Windus, London, 2001, p. 267).
5. William Tebb (1839–1914), leading member of anti-vaccination lobby.
6. William Farr (1807–83), superintendent of the statistical department of the registrar-general.
7. ML II 351–2.
8. WC 213–323 (pp. 222–223).
9. ML II 352–3.
10. WC 225.
11. Ibid. 241.
12. Ibid. 247.
13. Ibid. 303–4.

Institutional Reform

1. Henry du Pré Labouchère (1831–1912), radical journalist and politician.
2. Gladstone failed to get a Home Rule Bill through the House of Commons in 1886. A second Gladstone-sponsored Home Rule Bill was passed by the Commons in 1893, but scuppered by the House of Lords.
3. "How to Preserve the House of Lords", *Contemporary Review*, vol. 65, pp. 114–22, 1894. Reprinted in SSS II (excerpt pp. 223–5).
4. "Disestablishment and Disendowment: With a Proposal for a Really National Church of England", *Macmillan's Magazine*, vol. 27, pp. 498–507, 1873. Reprinted in SSS II (excerpt p. 252).
5. Ibid. (excerpt pp. 240–2).
6. "A Suggestion for Sabbath-keepers", *Nineteenth Century*, vol. 36, pp. 604–11, 1894. Reprinted in SSS II (excerpt pp. 367–9).
7. "Limitation of State Functions in the Administration of Justice", *Contemporary Review*, vol. 23, pp. 43–52, 1894. Reprinted and revised in SSS II (excerpt pp. 150–2).

8. Letter from Wallace read at a meeting in support of female suffrage quoted in *The Times*, February 11 1909.

Public Education

1. ML I 415.
2. "Museums for the People", *Macmillan's Magazine*, vol. 19, pp. 244–50, 1869. Reprinted in SSS II (excerpt p. 2).
3 Robert Latham (1812–88), ethnographer and linguist.
4 The huge glasshouse structure constructed by Joseph Paxton for the Great Exhibition of 1851 in Hyde Park was moved after the exhibition to Sydenham in south London.
5. ML I 322–3.
6. Louis Agassiz (1807–73), Swiss naturalist credited with discovering the Ice Ages.
7. "American Museums: The Museum of Comparative Zoologys, Harvard University", *Fortnightly Review*, vol. 42 (n.s.), pp. 347–59, 1887. Reprinted in SSS II (excerpt pp. 36–7).
8. "Museums for the People", *Macmillan's Magazine*, vol. 19, pp. 244–50, 1869. Reprinted in SSS II (excerpt p. 8).
9. Jean Jacques Élisée Reclus (1830–1905), French geographer and radical.
10. James Wyld's Great Globe (1851) had a 10,000-square foot representation of the world painted inside. Scaffolding allowed visitors to climb up and view any part of the world they wished.
11. ML II 214–15.
12. "Epping Forest", *Fortnightly Review*, vol. 24 (n.s.), pp. 628–45, 1878. Reprinted in SSS II (excerpt p. 82).
13. Letter to the Editor on "Government Aid to Science", *Nature*, vol. 1, pp. 288–9, 1870. Reprinted ML II 55–9.
14. Footnote, IL 7.

Capitalism and Empire

1. ML I 13.
2. Sir William Harcourt (1827–1904), succeeded Gladstone as leader of the Liberal party in 1893.
3. RD 2–6.
4. Letter to Thomas Sims, brother-in-law, April 1859. ML I 367–8.
5. Benjamin Kidd (1858–1916), social theorist.
6. Herbert Spencer, *Principles of Ethics, Part IV*, Appleton and Co., New York, 1891.
7. ML I 173–6.
8. Wallace misquotes Smith, his main departure being the substitution (twice) of "actual" for "annual". The correct quotation is: "The annual

labour of every nation is the fund which orignally supplies it with all the necessaries and conveniences of life which it annually consumes, and which consist always either in the immediate produce of that labour, or in what is purchased with that produce from other nations."

9. "The Depression of Trade, its Causes and its Remedies", in *The Claims of Labour*, Co-operative Printing Co., Edinburgh, pp. 112–54, 1886. Reprinted in SSS II (excerpt pp. 202–3).

10. SEMP 67.

11. "Interest-bearing Funds Injurious and Unjust", SSS II (excerpt p. 264).

12. Ibid. 256–7.

13. "The Inefficiency of Strikes: Is There Not a Better Way?" in Joseph Edwards, ed., *The Labour Annual: 1899*, Joseph Edwards, Wallasey, Cheshire, 1899, p. 105. Reprinted in Charles H. Smith, ed, *Alfred Russel Wallace: An Anthology of His Shorter Writings*, Oxford University Press, Oxford (excerpt pp. 169–70).

14. The national railway (August 1911) and national coal (March 1912) strikes took place respectively over the issues of union recognition and a minimum wage.

15. "The Great Strike – And After", interview by Harold Begbie, *The Daily Chronicle*, March 13 1912. Interview reproduced in full at: www.wku.edu/~smithch/5749.htm.

16. "Nationalisation, Not Purchase, of Railways", letter in *The New Age*, vol. 3, pp. 417–18, 1908. Reprinted in Smith, ed., *Alfred Russel Wallace* (excerpt pp. 175–6).

17. "Reciprocity the True Free Trade", *Nineteenth Century*, vol. 5, pp. 638–49 , 1879. Reprinted in SSS II (excerpt pp. 171–2).

18. "The Causes of War, and the Remedies", *The Clarion*, July 8 1899. Reprinted in SSS II (excerpt p. 390).

19. "The Origin of Human Races and the Antiquity of Man Deduced From the Theory of 'Natural Selection'", *Journal of the Anthropological Society of London*, vol. 2, pp. clviii–clxx, 1864 (p. clxix).

20. Robert Blatchford (1851–1943), prominent Fabian socialist, and founding editor of the socialist newspaper, *The Clarion*. He later disavowed socialism.

21. The second Boer War had ended in 1902.

22. "Practical Politics", a letter to *The Clarion*, September 30 1904. Reprinted in ML II (excerpt pp. 224–6).

23. "America, Cuba, and the Philippines", letter to *The Daily Chronicle*, January 19 1899.

24. *Australasia*, 3rd edn, Stanford's Companion of Geography and Travel: Edward Stanford, London, 1883, p. 375.

25. A 230–1.

26. "White Men in the Tropics", *The Independent*, vol. 51, pp. 667–70, 1899. Reprinted in SSS II (excerpt pp. 105–6).

27. "Emancipation – Black and White", in E. Rhys, ed., *Lectures and Lay*

Sermons [1871], Everyman's Library, J. M. Dent & Co., London, 1926, reprint, p. 115.

28. "The Native Problem in South Africa and Elsewhere", *Independent Review*, vol. 11, pp. 174–82, 1906. Reprinted in Smith, ed., *Alfred Russel Wallace* (excerpt p. 188).

Coda

1. Obituary of Wallace (anon.), *Current Opinion* (New York), vol. 56, pp. 32–3, 1914 (p. 32).
2. Amabel Williams-Ellis, *Darwin's Moon: A Biography of Alfred Russel Wallace*, Blackie, London, 1966.
3. Arnold C. Brackman, *A Delicate Arrangement: The Strange Case of Charles Darwin and Alfred Russel Wallace*, Times Books, New York, 1980. John L. Brooks, *Just Before the Origin: Alfred Russel Wallace's Theory of Evolution*, Columbia University Press, New York, 1984.
4. See pp. 62–3.
5. "Presidential Address", *Proceedings of the Linnaean Society*, vol. 1858–59, p. viii, 1859.
6. "Recapitulation and Conclusion", final chapter of *On the Origin of Species*, John Murray, London, 1859.
7. See p. 230.
8. ML II 382.
9. Peter Raby, *Alfred Russel Wallace*, Chatto and Windus, London, 2001, p. 233.
10. See pp. 74–6.
11. See pp. 236–7.
12. See p. 159ff.
13. Letter to Sir William Thiselton-Dyer (who succeeded Joseph Hooker as director of Kew Gardens in 1885), January 17 1893. M 445–6.
14. ML I 224.
15. See p. 63ff.
16. Letter to Richard Spruce from Joaõ [*sic*] Antonio de Lima, Portuguese trader, São Joaquim, Rio Negro, June 7 1853. ML I 313.
17. MA 352.
18. Jane R. Camerini, "Wallace in the Field", *Osiris*, vol. 11 (2nd Series), 1996, p. 54.

Bibliography

Wallace's Major Publications

Listed below are all of Wallace's books and a selection of his more important articles. For a full Wallace bibliography, see Charles H. Smith, ed., *Alfred Russel Wallace: An Anthology of His Shorter Writings*, Oxford University Press, Oxford, 1991, or Charles Smith's Wallace website, http://www.wku.edu/~smithch/bibintro.htm. When quotations used in the anthology have come from an edition other than the first, the relevant edition is given as well as the first, along with the abbreviation used in the text. In general, I have tried to use the most widely available edition of any given work. Thus, for example, I have avoided using the original 1853 edition of *Travels on the Amazon* – only 750 copies were printed and it is accordingly difficult to find – and used instead a more easily found second edition published in 1889. Page numbers used in the text and notes therefore refer to that edition. Note that Wallace often changed the titles of articles when reprinting them.

Abbreviations

A	*A Narrative of Travels on the Amazon and Rio Negro*, 1853
CTNS	*Contributions to the Theory of Natural Selection*, 1870
D	*Darwinism*, 1889
GD I & II	*The Geographical Distribution of Animals*, 1876
IL	*Island Life*, 1880
LN	*Land Nationalisation*, 1882
M	J. Marchant, ed., *Alfred Russel Wallace: Letters and Reminiscences*, 1916
MA	*The Malay Archipelago*, 1869
ML I & II	*My Life*, 1905
MMS	*On Miracles and Modern Spiritualism*, 1875/96
MPU	*Man's Place in the Universe*, 1903
NSTN	*Natural Selection and Tropical Nature*, 1891

RD   *The Revolt of Democracy*, 1913
SEMP   *Social Environment and Moral Progress*, 1913
SSS I & II *Studies Scientific and Social*, 1900
TNOE   *Tropical Nature, and Other Essays*, 1878
WC   *The Wonderful Century*, 1898
WL   *The World of Life*, 1910

Bibliography

"On the Monkeys of the Amazon", *Proceedings of the Zoological Society of London*, vol. 20, pp. 107–10, 1852.

A Narrative of Travels on the Amazon and Rio Negro, With an Account of the Native Tribes, and Observations on the Climate, Geology, and Natural History of the Amazon Valley, Reeve & Co., London, 1853.
  [A. Second Edition. Minerva Library of Famous Books, Ward, Lock & Co., London, New York, and Melbourne, 1889]

Palm Trees of the Amazon and Their Uses, John Van Voorst, London, 1853.

"On the Law Which Has Regulated the Introduction of New Species ['Sarawak Law']", *Annals and Magazine of Natural History*, vol. 16 (2nd Series), pp. 184–96, 1855.

"On the Habits of the Orang-Utan of Borneo", *Annals and Magazine of Natural History*, vol. 18 (2nd Series.), pp. 26–32, 1856.

"On the Tendency of Varieties to Depart Indefinitely from the Original Type ['Ternate Paper']", *Journal of the Proceedings of the Linnaean Society: Zoology*, vol. 3, pp. 53–62, 1858.

"On the Zoological Geography of the Malay Archipelago", *Journal of the Proceedings of the Linnaean Society: Zoology*, vol. 4, pp. 172–84, 1860.

"On the Physical Geography of the Malay Archipelago", *Journal of the Royal Geographical Society*, vol. 33, pp. 217–34, 1863.

"Remarks on the Rev. S. Haughton's Paper on the Bee's Cell, And on the Origin of Species", *Annals and Magazine of Natural History*, vol. 12 (3rd Series), pp. 303–9, 1863.

"The Origin of Human Races and the Antiquity of Man Deduced from the Theory of 'Natural Selection'", *Journal of the Anthropological Society of London*, vol. 2, pp. clviii–clxx, 1864.

"On the Phenomena of Variation and Geographical Distribution as Illustrated by the Papilionidae of the Malayan Region", *Transactions of the Linnaean Society of London*, vol. 25, part I, pp. 1–71, 1865.

"Public Responsibility and the Ballot", *Reader*, vol. 5, p. 517, 1865.

"Mimicry, and Other Protective Resemblances Among Animals", *Westminster Review*, vol. 32 (n.s.), pp. 1–43, 1867.

"A Theory of Birds' Nests: Showing the Relation of Certain Sexual Differences of Colour in Birds to Their Mode of Nidification", *Journal of Travel and Natural History*, vol. 1, pp. 73–89, 1868.

The Malay Archipelago; The Land of the Orang-Utan and the Bird of Paradise; A Narrative of Travel With Studies of Man and Nature, 2 volumes, Macmillan & Co., London, 1869. [MA. Tenth Edition, reprinted, Periplus Editions, Singapore, 2000]

"Museums for the People", *Macmillan's Magazine*, vol. 19, pp. 244–50, 1869.

"Sir Charles Lyell on Geological Climates and the Origin of Species", *Quarterly Review*, vol. 126, pp. 359–94, 1869.

Contributions to the Theory of Natural Selection. A Series of Essays, Macmillan & Co., London & New York, 1870. [CTNS. Second Edition, Macmillan & Co., London and New York, 1871]

"The Limits of Natural Selection as Applied to Man", in CTNS, pp. 332–71, 1870.

"On the Attitude of Men of Science Towards the Investigators of Spiritualism", in Hudson Tuttle and J. M. Peebles, eds, *The Year-book of Spiritualism for 1871*, William White & Co., Boston, pp. 28–31, 1871.

"Disestablishment and Disendowment: With a Proposal for a Really National Church of England", *Macmillan's Magazine*, vol. 27, pp. 498–507, 1873.

"Free-trade Principles and the Coal Question", *The Daily News*, September 16 1873.

"Limitation of State Functions in the Administration of Justice", *Contemporary Review*, vol. 23, pp. 43–52, 1873.

"Spiritualism and Science [letter to the Editor]", *The Times*, January 4 1873.

On Miracles and Modern Spiritualism. Three Essays, James Burns, London, 1875. [MMS Third Edition, George Redway, London, 1896]

The Geographical Distribution of Animals; With A Study of the Relations of Living and Extinct Faunas as Elucidating the Past Changes of the Earth's Surface, 2 volumes, Macmillan & Co., London, 1876. [GD I & II]

"The Colours of Animals and Plants. I. – The Colours of Animals", *Macmillan's Magazine*, vol. 36, pp. 384–408, 1877; "The Colours of Animals and Plants. II. – The Colours of Plants", *Macmillan's Magazine*, vol. 36, pp. 464–71, 1877.

"Epping Forest", *Fortnightly Review*, vol. 24 (n.s.), pp. 628–45, 1878.

Tropical Nature, and Other Essays, Macmillan & Co., London and New York, 1878. [TNOE]

"Animals and Their Native Countries", *Nineteenth Century*, vol. 5, pp. 247–59, 1879.

Australasia [i.e. modern Australasia and South-east Asia], Stanford's Compendium of Geography and Travel: Edward Stanford, London, 1879.

"Reciprocity the True Free Trade", *Nineteenth Century*, vol. 5, pp. 638–49, 1879.

"How to Nationalize the Land: A Radical Solution of the Irish Land Problem", *Contemporary Review*, vol. 38, pp. 716–36, 1880.

Island Life: Or, The Phenomena and Causes of Insular Faunas and Floras, Including a Revision and Attempted Solution of the Problem of Geological Climates, Macmillan & Co., London, 1880. [IL. Reprinted Prometheus Books, Amherst, NY, 1997]

"The Origin of Species and Genera", *Nineteenth Century*, vol. 7, pp. 93–106, 1880.

Land Nationalisation; Its Necessity and Its Aims; Being a Comparison of the System of Landlord and Tenant With That of Occupying Ownership in Their Influence on the Well-being of the People, Trübner & Co., London, 1882. [LN. Third Edition, W. Reeves, London, 1883]

"The Debt of Science to Darwin", *Century Magazine*, vol. 25, pp. 420–32, 1883.

"The 'Why' and the 'How' of Land Nationalisation. I.", *Macmillan's Magazine*, vol. 48, pp. 357–68, 1883; II., *Macmillan's Magazine*, vol. 48, pp. 485–93, 1883.

Bad Times: An Essay on the Present Depression of Trade, Tracing It to Its Sources in Enormous Foreign Loans, Excessive War Expenditure, the Increase of Speculation and of Millionaires, and the Depopulation of the Rural Districts; With Suggested Remedies, Macmillan & Co., London & New York, 1885.

"American Museums. The Museum of Comparative Zoology, Harvard University", *Fortnightly Review*, vol. 42 (n.s.), pp. 347–59, 1887.

Darwinism: An Exposition of the Theory of Natural Selection With Some of Its Applications, Macmillan & Co., London & New York, 1889. [D. Second Edition, Macmillan & Co., London and New York, 1889]

"Human Selection", *Fortnightly Review*, vol. 48 (n.s.), pp. 325–37, 1890.

Natural Selection and Tropical Nature: Essays on Descriptive and Theoretical Biology, Macmillan & Co., London and New York, 1891. [NSTN]

"English and American Flowers. I.", *Fortnightly Review*, vol. 50 (n.s.), pp. 525–34, 1891; "English and American Flowers. II. Flowers and Forests of the Far West", *Fortnightly Review*, vol. 50 (n.s.), pp. 796–810, 1891.

"Human Progress: Past and Future", *Arena*, vol. 5, pp. 145–59, 1892.

"H. W. Bates, the Naturalist of the Amazons [obituary]", *Nature*, vol. 45, pp. 398–9, 1892.

"The Permanence of the Great Oceanic Basins", *Natural Science*, vol. 1, pp. 418–26, 1892.

"Are Individually Acquired Characters Inherited? I.", *Fortnightly Review*, vol. 53 (n.s.), pp. 490–8, 1893; II., *Fortnightly Review*, vol. 53 (n.s.), pp. 655–8, 1893.

"Inaccessible Valleys: A Study in Physical Geography", *Nineteenth Century*, vol. 33, pp. 391–404, 1893.

"The Social Quagmire and the Way Out of It. I. The Farmers", *Arena*, vol. 7, pp. 395–410, 1893; "The Social Quagmire and the Way Out of It. II. Wage-workers", *Arena*, vol. 7, pp. 525–42, 1893.

"Economic and Social Justice", in Andrew Reid, ed., *Vox Clamantium: The Gospel of the People*, A. D. Innes & Co., London, pp. 166–97, 1894.

"How to Preserve the House of Lords", *Contemporary Review*, vol. 65, pp. 114–22, 1894.

"A Limitation of State Functions in the Administration of Justice", *Contemporary Review*, vol. 23, pp. 43–52, 1894.

"A Suggestion to Sabbath-keepers", *Nineteenth Century*, vol. 36, pp. 604–11, 1894.

"What Are Zoological Regions?", *Nature*, vol. 49, pp. 610–13, 1894.

"The Expressiveness of Speech, or Mouth-gesture as a Factor in the Origin of Language", *Fortnightly Review*, vol. 58, n.s., pp. 528–43, 1895.

"The Problem of Utility: Are Specific Characters Always or Generally Useful?", *Journal of the Linnaean Society: Zoology*, 25, pp. 481–96, 1896.

"The Proposed Gigantic Model of the Earth", *Contemporary Review*, vol. 69, pp. 730–40, 1896.

"Darwin and Darwinism [review of *Charles Darwin and the Theory of Natural Selection* by Edward B. Poulton, 1896]", *Nature*, vol. 55, pp. 289–90, 1897.

"Spiritualism and Social Duty", *Light*, vol. 18, pp. 334–6, 1898.

The Wonderful Century: Its Successes and Its Failures, Swan Sonnenschein & Co., London, 1898. [WC]

"America, Cuba, and the Philippines", *The Daily Chronicle*, January 19 1899.

"The Causes of War, and the Remedies", *The Clarion*, July 8 1899.

"White Men in the Tropics", *The Independent*, vol. 51, pp. 667–70, 1899.

Studies Scientific and Social [a collection of essays], 2 volumes, Macmillan & Co., London, 1900. [SSS I & II]

Man's Place in the Universe: A Study of the Results of Scientific Research in Relation to the Unity or Plurality of Worlds, Chapman & Hall Ltd, London, 1903. [MPU]

"If There Were a Socialist Government – How Should It Begin?", *The Clarion*, August 18 1905.

My Life: A Record of Events and Opinions, 2 volumes, Chapman & Hall Ltd, London, 1905. [ML I & II. Dodd, Mead & Co., New York, 1905]

"The Native Problem in South Africa and Elsewhere", *Independent Review*, vol. 11, pp. 174–82, 1906.

Is Mars Habitable? A Critical Examination of Professor Percival Lowell's Book "Mars and Its Canals," With an Alternative Explanation, Macmillan & Co., London, 1907.

"Nationalisation, Not Purchase, of Railways", *The New Age*, September 19 1908.

"Dr. A. R. Wallace and Woman Suffrage [note including letter from Wallace]", *The Times*, February 11 1909.

The World of Life: A Manifestation of Creative Power, Directive Mind and Ultimate Purpose, Chapman & Hall Ltd, London, 1910. [WL]

The Revolt of Democracy, Cassell & Co. Ltd, London, New York, Toronto and Melbourne, 1913. [RD]

Social Environment and Moral Progress, Cassell & Co. Ltd, London, New York, Toronto and Melbourne, 1913. [SEMP]

A Selection of Publications on or about Wallace

Listed below is a selection of articles and books on Wallace, most of them cited in the text. Again, for a comprehensive list, refer to Charles Smith's Wallace website, http://www.wku.edu/~smithch/second.htm. It is pleasing to be able to report that, with a number of major studies currently in the works, this list will soon be out of date.

Beddall, Barbara G., ed., *Wallace and Bates in the Tropics: An Introduction to the Theory of Natural Selection*, Macmillan, London, 1969.

Bowler, Peter J., "Alfred Russel Wallace's Concepts of Variation",

Journal of the History of Medicine and Allied Sciences, vol. 31(1), pp. 17–29, 1976.

Brackman, Arnold C., *A Delicate Arrangement: The Strange Case of Charles Darwin and Alfred Russel Wallace*, Times Books, New York, 1980.

Brooks, John L., *Just Before the Origin: Alfred Russel Wallace's Theory of Evolution*, Columbia University Press, New York, 1984.

Camerini, Jane R., "Wallace in the Field", *Osiris*, vol. 11 (2nd Series), pp. 44–65, 1996.

Daws, Gavan and Fujita, Marty, *Archipelago: The Islands of Indonesia: From the Nineteenth-century Discoveries of Alfred Russel Wallace to the Fate of Forests and Reefs in the Twenty-first Century*, University of California Press, Berkeley, 1999.

Durant, John R., "Scientific Naturalism and Social Reform in the Thought of Alfred Russel Wallace", *British Journal for the History of Science*, vol. 12(40), pp. 31–58, 1979.

Fichman, Martin, *Alfred Russel Wallace*, Twayne Publishers, Boston, 1981.

Fichman, Martin, "Science in Theistic Contexts: A Case Study of Alfred Russel Wallace on Human Evolution", *Osiris*, vol. 16 (2nd Series), pp. 227–50, 2001.

George, Wilma, *Biologist Philosopher: A Study of the Life and Writings of Alfred Russel Wallace*, Abelard-Schuman, London, 1964.

Gould, Stephen Jay, "Wallace's Fatal Flaw", *Natural History*, vol. 89(1), pp. 26–40, 1980.

Gould, Stephen Jay, "War of the Worldviews", *Natural History*, vol. 105(12), pp. 22–33, 1996.

Knapp, Sandra, *Footsteps in the Forest: Alfred Russel Wallace in the Amazon*, Natural History Museum, London, 1999.

Kottler, Malcolm Jay, "Alfred Russel Wallace, the Origin of Man, and Spiritualism", *Isis*, vol. 65, pp. 144–92, 1974.

McKinney, Lewis H., *Wallace and Natural Selection*, Yale University Press, New Haven and London, 1972.

Marchant, James, ed., *Alfred Russel Wallace: Letters and Reminiscences*, Harper & Brothers, New York, 1916. [M. Reprint, Arno Press, New York, 1975].

Milner, Richard, "Charles Darwin and Associates, Ghostbusters", *Scientific American*, vol. 275(4), pp. 96–101, 1996.

Moore, James, "Wallace's Malthusian Moment: The Common Context Revisited", in Bernard Lightman, ed., *Victorian Science in Context*, University of Chicago Press, Chicago, 1997.

Raby, Peter, *Alfred Russel Wallace*, Chatto & Windus, London, 2001.

Severin, Timothy, *The Spice Islands Voyage: The Quest for the Man Who Shared Darwin's Discovery of Evolution*, Carroll & Graf Publishers, New York, 1998.

Smith, Charles H., ed., *Alfred Russel Wallace: An Anthology of His Shorter Writings*, Oxford University Press, Oxford, 1991.

Van Oosterzee, Penny, *Where Worlds Collide: The Wallace Line*, Cornell University Press, Ithaca, NY, and London, 1997.

Williams-Ellis, Amabel, *Darwin's Moon: A Biography of Alfred Russel Wallace*, Blackie, London, 1966.

Wilson, John G., *The Forgotten Naturalist: In Search of Alfred Russel Wallace*, Australia Scholarly Publishing, Melbourne, 2000.

Index

Note: Bold titles refer to Wallace's works; bold page references refer to major extracts from those works; italic page references refer to illustrations.